THE CONSTRUCTION OF BUILDINGS

VOLUME 2

WINDOWS, DOORS, FIRES, STAIRS, FINISHES

R. BARRY
Architect

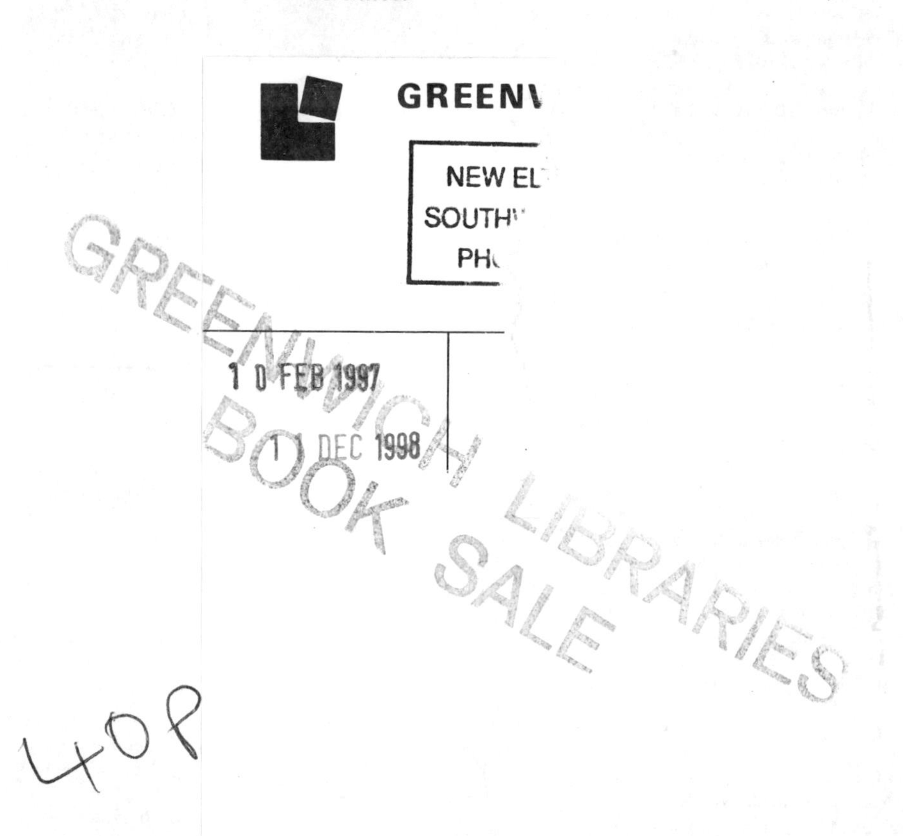

OXFORD
BLACKWELL SCIENTIFIC PUBLICATIONS
LONDON EDINBURGH BOSTON
MELBOURNE PARIS BERLIN VIENNA

Blackwell Scientific Publications
Editorial Offices:
Osney Mead, Oxford OX2 0EL
25 John Street, London WC1N 2BL
23 Ainslie Place, Edinburgh EH3 6AJ
238 Main Street, Cambridge
Massachusetts 02142, USA
54 University Street, Carlton
Victoria 3053, Australia

Other Editorial Offices:
Librairie Arnette SA
1, rue de Lille
75007 Paris
France

Blackwell Wissenschafts-Verlag GmbH
Düsseldorfer Str. 38
D-10707 Berlin
Germany

Blackwell MZV
Feldgasse 13
A-1238 Wien
Austria

First Edition published by Crosby Lockwood & Son Ltd 1960
Reprinted 1963, 1965, 1967, 1969
Second Edition (metric) published 1970
Reprinted by Granada Publishing Ltd in
Crosby Lockwood Staples 1972, 1974, 1975, 1979
Third Edition published by Granada Publishing Ltd
Technical Books Division 1982
Reprinted by Collins Professional and Technical Books 1986
Reprinted by BSP Professional Books 1988, 1989, 1991
Fourth Edition published by Blackwell Scientific Publications 1992
Reprinted 1994

Set by DP Photosetting, Aylesbury, Bucks
Printed and bound in Great Britain by
Hartnolls Ltd, Bodmin, Cornwall

DISTRIBUTORS

Marston Book Services Ltd
PO Box 87
Oxford OX2 0DT
(*Orders:* Tel: 0865 791155
Fax: 0865 791927
Telex: 837515)

USA
Blackwell Scientific Publications, Inc.
238 Main Street,
Cambridge, MA 02142
(*Orders:* Tel: 800 759-6102
617 876-7000)

Canada
Oxford University Press
70 Wynford Drive
Don Mills
Ontario M3C 1J9
(*Orders:* Tel: 416 441-2941)

Australia
Blackwell Scientific Publications Pty Ltd
54 University Street
Carlton, Victoria 3053
(*Orders:* Tel: 03 347-5552)

British Library
Cataloguing in Publication Data

A catalogue record for this book
is available from the British Library

ISBN 0–632–03289–8

Library of Congress
Cataloging in Publication Data

Barry, R. (Robin Lewis)
The construction of buildings.
Vol. 2 published by Blackwell Scientific, Oxford.
Includes indexes.
Contents: v. 1. Construction and materials—
v. 2. Windows, doors, fires, stairs, finishes
1. Building. I. Title.
TH146.B3 1980 690 81–463308
ISBN 0–632–03289–8

CONTENTS

PREFACE

Since this volume was last revised, changes in the Building Regulations have led to the Building Regulations 1991, which come into effect on 1 June 1992. Nine Approved Documents, giving practical guidance to meeting the requirements of the regulations, have so far been published.

In the past, both local and national building regulations were primarily concerned to specify standards of materials and workmanship necessary to produce reasonably sound buildings, related to structure, adequate fire resistance, reasonable weathering, hygiene and sufficient daylighting and heat insulation. The emphasis was on the form and construction of buildings.

In the most recent Building Regulations (1991) the emphasis has moved, in the main, to a concern for the safety of the occupants and users of buildings rather than the building itself. This is reflected both in the preface and the titles of Approved Documents. In the 'Use of Guidance Notes' to Approved Document A it is stated that: 'Building Regulations are made for specific purposes; health and safety, energy conservation and the welfare and convenience of disabled people'. The titles to the latest Approved Documents, such as *Fire safety*, *Stairs ramps and guards* and *Glazing – materials and protection* reflect the same change of emphasis.

In this revision, Chapters one and two have been rearranged to give greater emphasis to a description of functional requirements generally and the manner in which the assembly and fixing of windows and doors meet those requirements in particular. Details of properties and uses of aluminium and uPVC windows and doors and details of glass, double glazing, IG units, weatherstripping and sealing have been expanded.

Where relevant the practical guidance in the nine Approved Documents, that have so far been published as guidance to the Building Regulations 1991, has been included in this revision, including a new Approved Document N: *Glazing – materials and protection.* Where no new Approved Document has yet (January 1992) been published, reference is made to the documents that gave guidance to the Building Regulations 1985.

R. BARRY

ACKNOWLEDGEMENTS

Table 9 is reproduced from Building Research Establishment Digest 38 and Table 14 is reproduced from Building Research Establishment Digest 196 by permission of the Controller of HMSO: Crown copyright 1992.

Tables 6, 11 and 12 are reproduced with the permission of the Controller of Her Majesty's Stationery Office.

Extracts from British Standards are reproduced with the permission of BSI. Complete copies of the standard can be obtained by post from BSI Sales, Linford Wood, Milton Keynes, MK14 6LE.

AUTHOR'S NOTE

For linear measure all measurements are shown in either metres or millimetres. A decimal point is used to distinguish metres and millimetres, the figures to the left of the decimal point being metres and those to the right millimetres. To save needless repetition, the abbreviations 'm' and 'mm' are not used, with one exception. The exception to this system is where there are at present only metric equivalents in decimal fractions of a millimetre. Here the decimal point is used to distinguish millimetres from fractions of a millimetre, the figures to the left of the decimal point being millimetres and those to the right being fractions of a millimetre. In such cases the abbreviations 'mm' will follow the figures e.g. 302.2 mm.

R. BARRY

CHAPTER ONE

WINDOWS

A window is an opening formed in a wall or roof to admit daylight through some transparent or translucent material fixed in the opening. This primary function of a window is served by a sheet of glass fixed in a frame in the window opening. This simple type of window is termed a fixed light or dead light because no part of the window can be opened.

As the window is part of the wall or roof envelope to the building, it should serve to exclude wind and rain, and act as a barrier to excessive transfer of heat, sound and spread of fire in much the same way as the surrounding wall or roof does. The functional material of a window, glass, is efficient in admitting daylight and excluding wind and rain but is a poor barrier to the transfer of heat, sound and the spread of fire.

The traditional window is usually designed to ventilate rooms through one or more parts that open to encourage an exchange of air between inside and outside. Ventilation is not a necessary function. Ventilation can as well be provided through openings in walls and roofs that are either separate from windows or linked to them to perform the separate function of ventilation. The advantage of separating the functions of daylighting and ventilation is that windows may be made more effectively wind and weathertight and ventilation can be more accurately controlled.

FUNCTIONAL REQUIREMENTS

The primary function of a window is:

Admission of daylight

and the secondary functions are:

A view out, and
Ventilation.

Daylighting is a necessary function and a view out is generally a desirable function. Ventilation is neither a necessary nor a desirable function yet it is traditionally expected of a window.

The functional requirements of a window as a component part of a wall or roof are:

Structure – strength and stability
Resistance to weather
Durability and freedom from maintenance
Fire safety
Resistance to the passage of heat
Resistance to the passage of sound
Security.

DAYLIGHT

The prime function of a window is to admit adequate daylight for the efficient performance of daytime activities. Good sense dictates taking the maximum advantage of this free source of illumination when the modern alternative, electric light, is so extravagantly wasteful of natural fuel sources and grossly expensive.

The quantity of light admitted depends in general terms on the size of the window or windows in relation to the area of the room lit, and the depth inside the room to which useful light will penetrate depends on the height of the head of windows above floor level. Common sense and observation suggest that the quantity of daylight in rooms is proportional to the area of glass in windows relative to floor area and this is confirmed by measurement.

The intensity of daylight at a given point diminishes progressively into the depth of the room away from windows. For general activity purposes, such as in living rooms, an adequate overall level of daylight illumination is sufficient, whereas a minimum level of illumination in a particular area is necessary for such activities as drawing.

Unlike artificial lighting, daylight varies considerably in intensity both hourly and daily due to the rotation of the earth and the consequent relative position of the sun, and also due to climatic variations from clear to overcast skies. In order to make a prediction of the relative level of daylight indoors, it is necessary to make an assumption. In Britain and north-west Europe it is current practice to calculate daylight in terms of a 'daylight factor' which is the ratio of internal illumination to the illumination occurring simultaneously out of doors from an unobstructed sky, rather than using the absolute value, that is lux, commonly used for artificial lighting. In the calculation of the daylight factor it is assumed that the illumination from an unobstructed sky, in the latitude of Britain, is 5000 lux and that a daylight factor of 2% means that 2% of the 5000 lux outdoors is available as daylight illumination at a specified point inside.

The assumption of a standard overcast sky, which represents the condition of poor outdoor illumination that may occur in autumn, winter and spring in northern Europe, is taken as a minimum standard on which to make assumptions. The term 'unobstructed sky' defines the illumination available from a hemisphere of sky free of obstructions such as other buildings, trees and variations in ground level, a condition that rarely occurs in practice.

The concept of a daylight factor has the advantage that it is a comparative value of the intensity of daylight indoors at different points so that even though the intensity of daylight outdoors will vary, the relative indoor intensity will remain more or less the same. The daylight factor concept provides a better indication of the subjective impression of daylight than would be the case were an absolute value given. In the assumption of a standard overcast sky the effect of direct sunlight is excluded. The International Commission on Illumination (CIE) defines daylight factor as 'the ratio of the daylight illumination of a given point on a given plane due to the light received directly or indirectly from a sky of assumed or known luminance distribution, to the illumination on a horizontal plane due to an unobstructed hemisphere of this sky. Direct sunlight is excluded for both values of illumination'.

The intensity of illumination or luminance of the standard sky is assumed to be uniform to facilitate calculation of levels of daylight. In practice sky luminance varies, with luminance at the horizon being about one third of that at the sun's zenith.

Where artificial illumination is used to supplement daylighting it is often practice to determine a working level of illumination in values of lux and convert this value to an equivalent daylight factor by dividing the lux value by 50 to give the daylight factor. For example, a lux value of 100 is equivalent to a daylight factor of 2. The average daylight factor in side-lit rooms is roughly equal to one fifth of the percentage ratio of glass to floor area. In a room with windows on one long side with no external obstructions and a room surface reflectance of 40%, where the glass area is one fifth or 20% of the floor area, the average daylight factor will be 4 and the minimum about half that figure. Conversely, to obtain an average daylight factor of, say 6, in a room with a floor area of 12 m^2, a glass area of about $6 \times 12 \times 5/100 = 3.6\ m^2$ will be required. This broad average calculation is generally sufficient when used for general activity purposes such as in living rooms, and it is an adequate base for preliminary assumptions of window to floor area which can be adjusted later by a more accurate calculation of the light required for activities in which the lighting is critical.

A broad measure of the penetration of useful daylight into rooms is, taking an average figure of 2 as a daylight factor, the depth of penetration in line with the centre of windows as equal to the height of the window head above floor level.

The quantity and quality of daylight illumination in side-lit rooms is affected to an extent by the light reflected from floors, walls and ceilings which will augment light coming directly through windows. Plainly the effect of this reflected light will be affected by the colour and texture of the reflective surfaces. Similarly some daylight, reflected from pavings and nearby external obstructions such as buildings and trees, will to an extent add to both the direct penetration of light and internally reflected light.

In the assumption of a daylight factor, account is taken of the contribution of what is termed 'the internally reflected component' and the 'externally reflected component' of indoor daylight illumination. Obviously the extent to which both the internal and external reflected light adds to or augments the indoor

Table 1. Recommended average daylight factors

Building type	Location	Daylight factor
Dwellings	Living rooms	1.5
	Bedrooms	1
	Kitchens	2
Work places	Offices Libraries Schools Hospitals Factories	5
All buildings	Residential spaces	2
All buildings	Entrances Public areas Stairs	2

Taken from DD73:1982.

lighting will be least with low levels of overall daylight and dark, rough textured reflective surfaces, and will be most with higher levels of overall daylight and light coloured, smooth textured reflective surfaces. Average daylight factors for various activities are given in Table 1.

The shape, size and position of windows affect the distribution of daylight in rooms and the view out. Tall windows give a better penetration of light than low windows, separate windows give a less uniform distribution of light than continuous windows, and windows in adjacent walls give good penetration and reduce glare by lighting the area of wall surrounding the adjacent window, as illustrated in Fig. 1. Windows in opposite walls of narrow rooms give good penetration and reduce glare by lighting opposite walls around windows.

In the calculation of daylight factors it is usual to determine the quantity of daylight falling on a horizontal working plane 850 above floor level to correspond with the height of working surfaces such as tables, desks or benches.

It is advantageous to be able to make a reasonably accurate estimate of the area of glass in windows necessary to provide the average daylight factor (Table 1) recommended for the activity for which the room or space is designed. The averaged or average daylight factor represents the overall visual impression of the daylighting in a room or space taking into account the distribution of light in the space and the effect of reflected light.

Table 2 gives reasonably accurate net glass areas required to provide average daylight factors of 1%, 2% and 5% in side-lit rooms of 3 m height with high

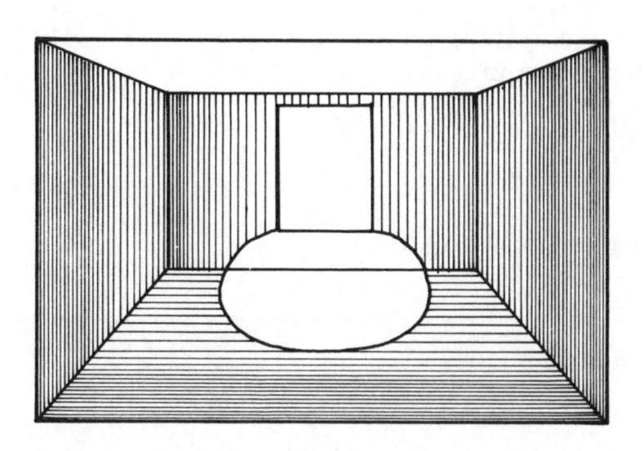

Tall narrow window

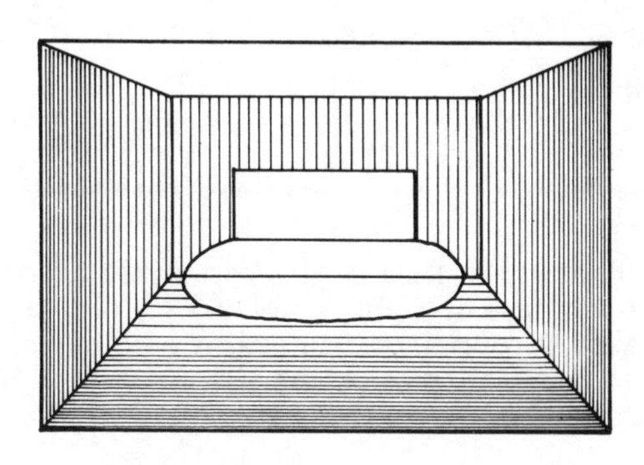

Long low window

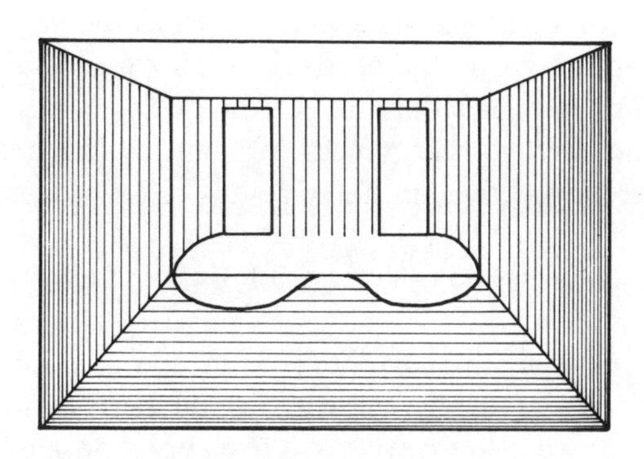

Two tall narrow windows

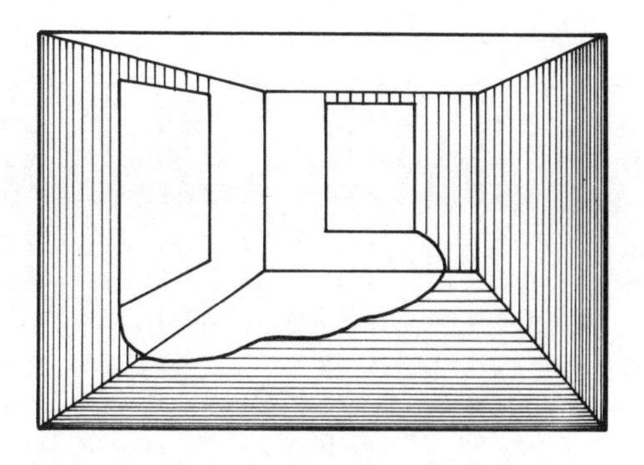

Windows in adjacent walls

Diagram illustrating effect of shape and position of window on distribution of daylight

Fig. 1

Table 2. High reflectances: the net glass areas required to provide average daylight factors of 1%, 2% and 5% in side-lit rooms, of 3 m height with ceiling, wall and floor reflectances of 70%, 50% and 30% respectively and with an external obstruction of 15°

Room width (m)‡	Room depth (m)	Average daylight factor (%)		
		1	2	5
		Glass area (m²)		
3.0	3.0	0.64	1.28	3.22
3.0	4.5	0.88	1.75	4.39
3.0	6.0	1.11	2.22	5.61 ▮
4.5	3.0	0.83	1.66	4.18
4.5	4.5	1.12	2.25	5.66
4.5	6.0	1.42	2.85	7.18
6.0	3.0	1.03	2.05	5.16
6.0	4.5	1.39	2.78	6.93
6.0	6.0	1.74	3.47	8.66
6.0	9.0	2.46	4.92	12.3 ▮
9.0	4.5	1.88	3.75	9.35
9.0	6.0	2.36	4.72	11.8
9.0	9.0	3.37	6.74	16.8 ▮

Low reflectances: the net glass areas required to provide average daylight factors of 1%, 2% and 5% in side-lit rooms, of 3 m height with ceiling, wall and floor reflectances of 50%, 30% and 10% respectively and with an external obstruction of 15°

Room width (m)‡	Room depth (m)	Average daylight factor (%)		
		1	2	5
		Glass area (m²)		
3.0	3.0	0.80	1.60	4.00
3.0	4.5	1.10	2.20	5.50 ▮
3.0	6.0	1.41	2.81	7.02 ▮
4.5	3.0	1.03	2.05	5.12
4.5	4.5	1.41	2.81	7.02
4.5	6.0	1.79	3.58	8.92 ▮
6.0	3.0	1.27	2.54	6.33
6.0	4.5	1.71	3.42	8.53
6.0	6.0	2.14	4.29	10.7 ▮
6.0	9.0	3.11	6.21	15.5 ▮
9.0	4.5	2.32	4.63	11.5
9.0	6.0	2.95	5.90	14.7
9.0	9.0	4.22	8.44	21.0 ▮

‡ Width of window wall
▮ These glass areas are for rooms lit from two sides

Taken from DD 73:1982

Table 3. Reflectances of some building materials

Material	Reflectances
Ceilings	(%)
White distemper on plain plaster or plasterboard	80
White distemper on acoustic perforated plasterboard	70
Floor and furniture	
Carpet, dark brown	10
Cement screed	45
Clay flooring tiles, red	10
Clay quarry tiles, red	15
Cork tiles, polished	20
Wood block, light oak	25
Wood block, medium oak	20
Wood block, dark oak	10
Walls	
Brick, concrete, light	40
Brick concrete, dark	20
Brick, fletton type	30
Brick, Leicestershire sandfaced	15
Brick, sand lime	40
Brick, yellow London stock	25
Concrete, smooth	30
Concrete, rough	20
Distemper, white	80
Distemper, light cream	60
Paint, glossy white	85

Taken from DD 73:1982

reflectances of 70%, 50% and 30% for ceilings, walls and floor respectively. Similarly Table 2 gives net glass areas for rooms with low reflectances of 50%, 30% and 10% for ceilings, walls and floor respectively. The average daylight factor of 5% cannot generally be reached in rooms that are lit from one side only.

The average reflectances of some typical building materials are given in Table 3.

Measurement and calculation of daylight factor

Where daylighting by itself or in combination with artificial lighting is critical for the performance of activities such as drawing at a fixed point or points in a room, it is necessary to estimate the minimum daylighting available at a point. For this purpose there are a number of aids, such as the artificial sky and the overlays for scale drawings.

An artificial sky provides luminance comparable to the standard overcast sky, through an artificially lit dome which is laid over a scale model of the building in

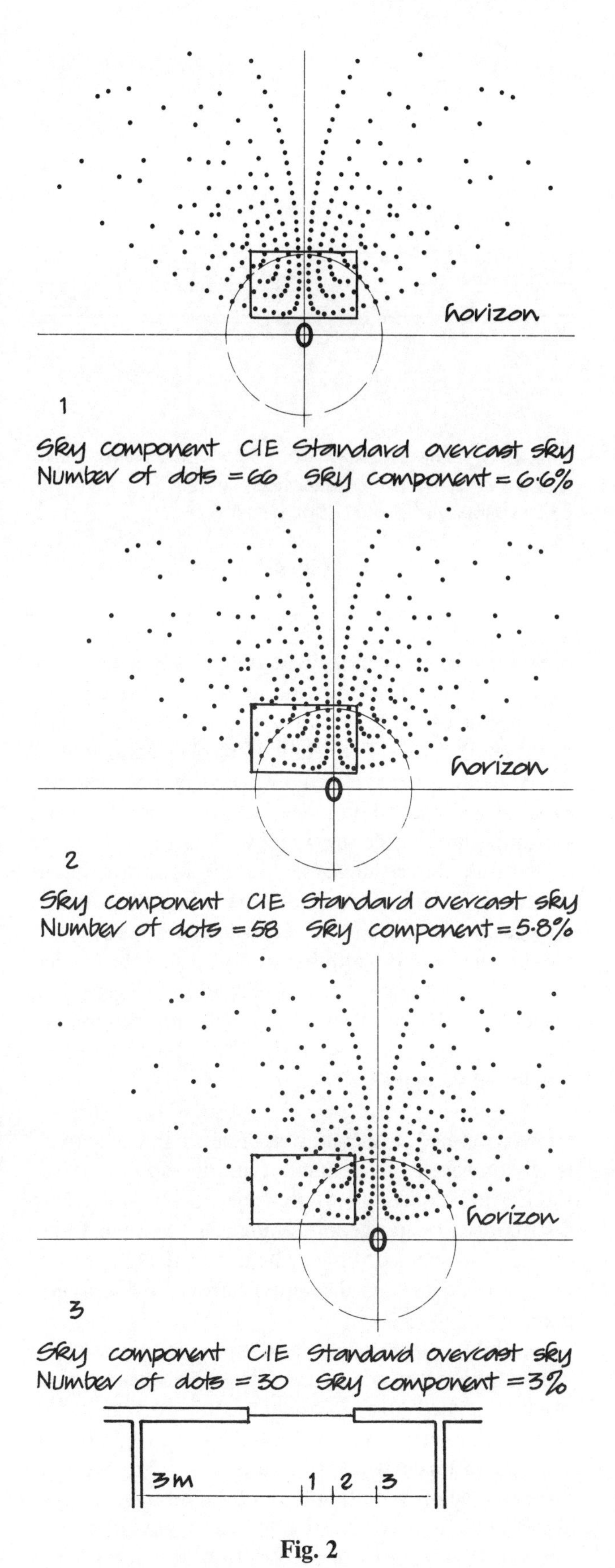

Fig. 2

which photometers are used to measure the light available. The graphical aids in the form of overlays include Waldram diagrams, BRS protractors and the dot or pepper pot diagrams of which the dot diagram is the most straightforward in use.

The pepper pot diagram is a transparent overlay on which dots are printed above a horizontal line representing the horizon. The diagram is drawn to a scale of 1:100 as an overlay to drawings to the same scale. Each dot represents 0.1% of the sky component. The overlay shown in Fig. 2 is for daylight through side-lit windows with the CIE standard overcast sky. To use the overlay, draw the outline of a window to a scale of 1:100 so that the outline represents the glass area to scale. The diagram is designed to determine the sky component of daylight on a line 3 m back from the window. Place the overlay on the scale elevation of the window with the horizontal line of the overlay on the line of the working plane, that is 850 above the floor, drawn to scale on the window elevation. To determine the sky component at a point 3 m back on the centre of the window, place the vertical line of the overlay on the centre of the window as illustrated in Fig. 2 then count the dots inside the window outline. The 66 dots inside the window outline represent a sky component of 66/10 that is 6.6% at a point 3 m back from the centre of the window on the working plane. To find the sky component on the line 3 m back from the window at other points, slide the overlay horizontally across the window outline until the vertical line of the overlay coincides with the chosen point inside the room, either inside or outside the window outline, as illustrated in Fig. 2. Count the dots inside the window outline to determine the sky component at the chosen points.

To find the sky component at points on a line other than the line 3 m back from the window drawn to a scale of 1:100, it is necessary to adjust the scale of the window outline. If the scale of the window is doubled, it will represent the sky component at points $1\frac{1}{2}$ m back from the window, and if the scale is halved, 6 m back from the window, as illustrated in Fig. 3 for points on the centre of the window. In adjusting the scale of the window outline it is also necessary to adjust the scale height of the working plane above the floor by doubling or halving the scale as shown in Fig. 3.

The particular use of this diagram is to test the sky component of daylight inside rooms at an early stage in the design of buildings. By the use of window outlines drawn freehand to scale on graph paper, with the overlay, a comparative assessment of the effect of window size and position on the sky component of

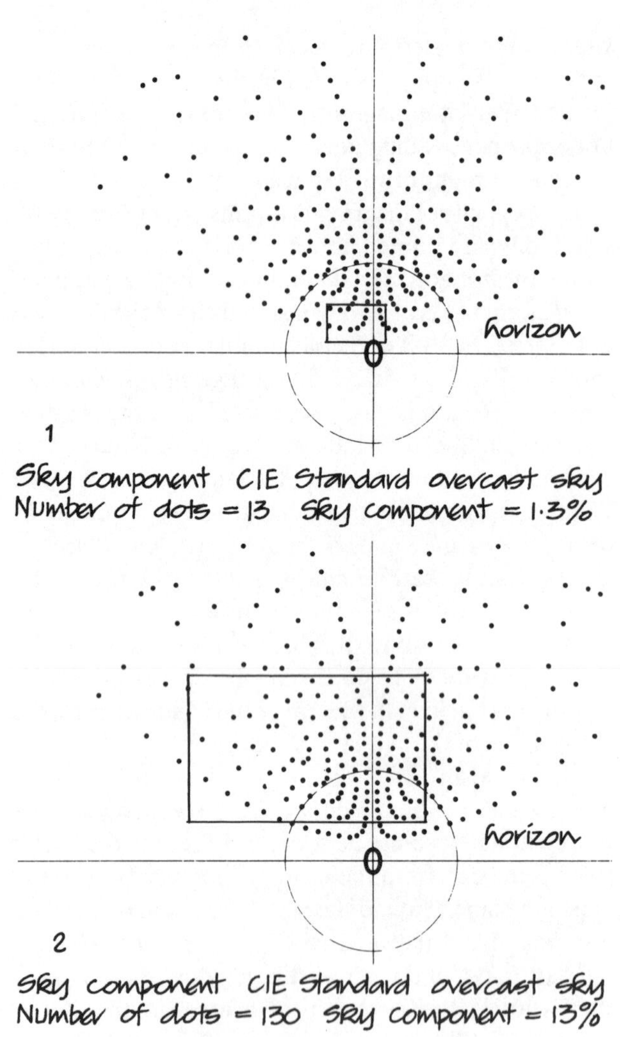

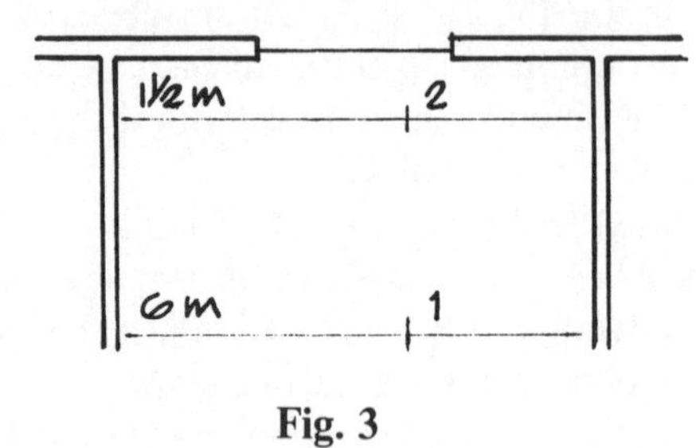

Fig. 3

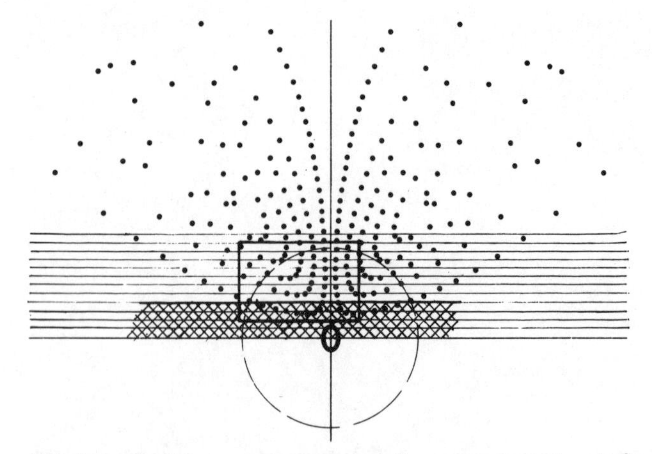

Fig. 4

daylight inside rooms can quickly be made. This will provide a reasonably accurate assessment of comparative daylight levels in rooms to be used for many activities where the exact level of daylight is not critical.

Where there are obstructions outside windows, such as adjacent buildings, which obscure some of the daylight, the overlay can be used to determine both the loss of light due to the obstruction and the externally reflected component of light due to reflection of light off the obstruction and into the room through the window. A simple example of this is where a long low building will obstruct daylight at a point 3 m inside the room on the centre of the window at the working plane. The outline of the long obstruction is shown on Fig. 4 by the shaded area. The height of the obstruction above the horizon is represented by the height to distance ratio of the obstruction relative to the point on the working plane inside the window. This ratio is 0.1 for each 3 mm above the horizon on the scale drawing of the window. The number of dots inside the window outline above the shaded obstruction gives the sky component, and the number of dots inside the shaded area, the externally reflected component. These dots represent 0.01% of the externally reflected component.

Quality of daylight

Glare is defined as 'a condition of vision in which there is discomfort or a reduction in the ability to see significant objects or both, due to an unsuitable distribution or range of luminance, or to extreme contrasts in space or time'. The two distinct aspects of glare are defined as disability glare and discomfort glare.

Disability glare, which is defined as 'glare which impairs the vision of objects without necessarily causing discomfort', is caused when a view of bright sky obscures objects close to the source of glare. An example of this is where a lecturer is standing with his back to a window so that he is obscured by the bright sky behind him. Disability glare can be avoided by a sensible arrangement of the position of windows and people, whose vision of objects might otherwise be obscured.

Discomfort glare, defined as 'glare which causes discomfort without necessarily impairing the vision of objects', is created by large areas of very bright sky viewed from inside a building which causes distraction, dazzle and even pain. With vertical windows discomfort glare is caused, in the main, by the contrast between visible sky and the room lighting and this contrast can be usefully reduced by splaying window reveals and painting them a light colour to provide a graded contrast between the bright sky and the darker interior. This 'contrast grading' effect can be used with many window shapes and sizes. With very large windows such as the continuous horizontal strip windows which face southwards, discomfort glare is difficult to avoid owing to the large unbroken area of glazing and here some form of shading device will be required.

The degree of glare can be determined numerically and stated in the form of a 'glare index' from a formula suggested by The Building Research Station.

Form and texture For the best visual enjoyment of solid objects the direction of light is important in relation to modelling by shadows and for appreciation of texture the quantity and area of the light source has most effect. These known subjective effects are difficult to quantify. It is generally accepted that light from the side is more agreeable than light from above and that side lighting from tall vertical windows provides better modelling of solid objects than large wide side windows. A single bright source of light emphasises texture and gives hard shadows whereas a light from a large diffuse source such as a window gives softer shadows and texture but appreciable modelling of form, and a very large diffuse source, such as roof lighting or overall ceiling illumination, may cause all but the most coarse texture to disappear and give poor modelling.

View out

As well as admitting daylight it is generally accepted that windows perform the useful function of providing a view out of buildings as a link with the outside and to provide the variations of interest that stimulate and break the monotony of repetitive tasks. Studies have been made to deduce possible optimum sizes and spacing of windows to provide a view out. These studies have been inconclusive in detail but have established that the majority of people in sedentary occupations, such as office workers, derive benefit from a view out.

SUNLIGHT

In the late nineteenth century, concern for what were considered to be the poor living conditions in urban areas of northern European countries turned to the effects of sunlight on health. Early research sought to relate mortality and disease to the availability of sunlight in rooms and courtyards of, for example, back-to-back dwellings. It is now plain that sunlight is not essential for hygiene, biological or therapeutic purposes. Later research seeking a norm of preference for sunlight in buildings has been inconclusive in determining a chosen minimum amount of sunlight because preferences varied so widely. From all the surveys that have been undertaken it is apparent that the majority preference is for a satisfying view, some sunlight, particularly in living rooms, and visual privacy.

In this country, where the norm seems to be overcast dull skies, the cheerful aspect of sunlight is cherished. The fashion in recent years for large windows, sometimes called picture windows, reflects the wish of a mainly indoor people to enjoy sunlight and a view. With the recent very steep increase in the cost of fuels there has been a move towards smaller windows to reduce heat losses and solar heat gain that can cause discomfort in summer. The pendulum of fashion that has swung from maximum glass area towards minimum glass area has yet to settle towards a sensible mean between the two.

The fashion for large windows and large areas of glass ('curtain walls' in Volume 4), prompted by the comparatively low cost of glass, has changed as glass is no longer a comparatively cheap building envelope material and its disadvantages as a thermal and sound insulator are now more widely known. Nonetheless the subjective preference for sunlight and a view out, and the economic advantage of freely available daylight and controlled solar heat gain, prompt the optimum use of glazing compatible with reasonable thermal and sound insulation.

The current Code of Practice CP3 recognizes the preference for sunlight in rooms by setting out recommended minimum periods of sunlight penetration. The BSI Draft for Development (DD 67:1980) includes the criteria for the insolation of dwellings and will lead to the publication of a British Standard giving recommendations for the orientation of rooms and therefore the planning and siting of buildings. Standards do not include recommendations for the size and shape of windows. The criteria for insolation (exposure to sun's rays) suggest minimum possible or probable standards for sunlight on buildings as a guide in the design and layout of dwellings to gain the advantage of both sunlight and solar heat.

Sunlight causes most coloured materials to fade. It is the ultra-violet radiation in sunlight that has the most pronounced effect on coloured materials by causing the chemical breakdown of the colour in such materials as textiles, paints and plastics by oxidative bleaching, that is fading. The bleaching effect is more rapid and more noticeable with bright colours. The lining of colour-sensitive curtains on the window side with a neutral coloured material and the use of window blinds are necessary precautions to prolong the life of colour-sensitive materials.

Calculation of sunlight penetration Because of the rotation of the earth round the sun and the simultaneous rotation of the earth on its axis inclined at 22.5° to the plane of orbit, the apparent movement of the sun around the earth varies throughout the solar year and penetration of sunlight through windows varies in intensity and depth. To plot the penetration of sunlight throughout the year would be a lengthy and tedious task. For the majority of buildings in the temperate climate of Britain such an exercise is unnecessary except where it is anticipated that bright sunlight could cause discomfort or danger in performing tasks in static positions inside rooms or buildings or where solar heat gain might cause considerable discomfort or uneconomic use of internal heat.

There are geometric sunpath diagrams that may be used to check whether the face of a building will receive sunlight and when, the depth of penetration and the resultant patch of sunlight on room surfaces and the shading by obstructions at various times of the day throughout the year. An example of the use of 'gnomonic' projections to deduce sunlight patterns on room surfaces throughout the day is shown in Fig. 5. These sunpath diagrams may also, with suitable overlays, be used to predict the intensities of direct and diffuse solar radiation and the consequent solar heat gain. Recently a computer program has become available that will predict energy consumption for heat loss and heat input calculations and will make allowances for the variable of solar heat gain through windows so that modifications in both window sizes and the heat input from heating plant can be adjusted at the design stage. This facility has taken over the tedium of calculation by sunpath diagrams.

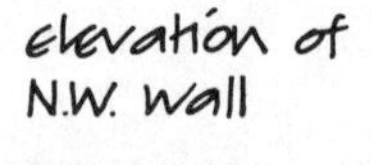

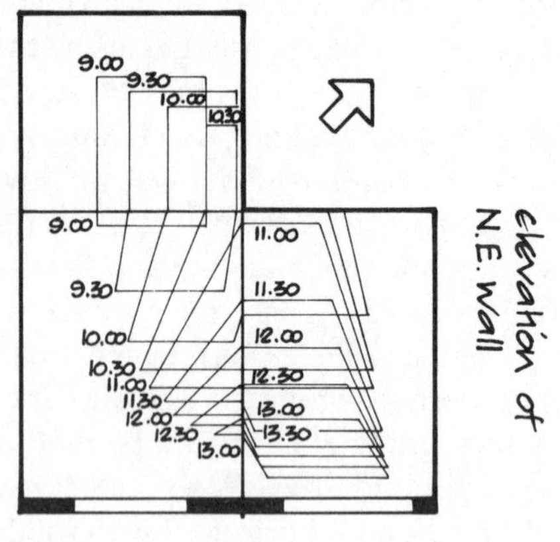

Patterns of sunlight on floor and walls at half hourly intervals on 15th January

Fig. 5

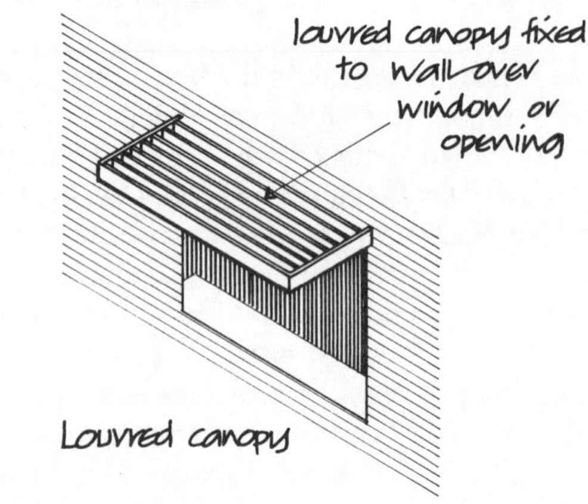

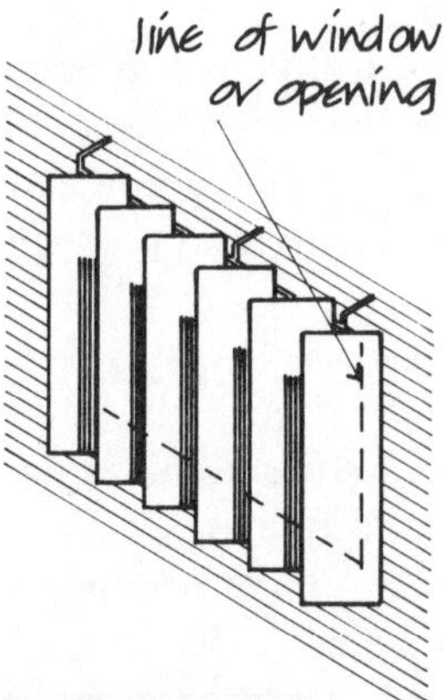

Fixed sun controls

Fig. 6

Sun controls and shading devices The traditional temperate climate means of controlling the penetration of sunlight to rooms are the slatted wooden louvre shutters common to the French window, and awnings and blinds that can be open or closed. These controls are adjustable between winter and summer conditions, graduated from no shade and the maximum penetration of daylight in winter through some shade and some daylight to full shading in high summer.

Fixed projections above windows, such as canopies and balconies, are also used as sun controls in temperate climates to provide shade from summer sun while allowing penetration of sun at other times of the year for the advantage of sunlight and solar heat gain, as illustrated in Fig. 7.

In tropical and semi-tropical climates fixed sun controls or shading devices in the form of canopies, screens or louvres are used, as illustrated in Fig. 6.

Sun controls serve to exclude sunlight to reduce glare or solar heat gain or both. To control and reduce solar heat gain sun controls should be fixed outside windows where they absorb solar heat which is then dissipated to the outside, whereas where sun controls are fitted internally e.g. blinds, the solar heat they absorb is dissipated inside the room.

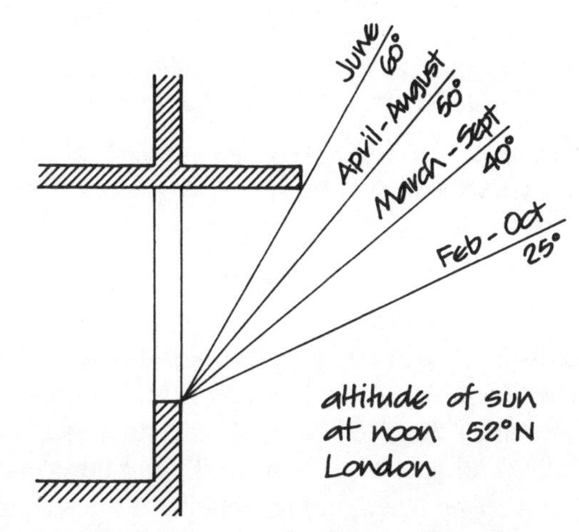

Fixed projection over window giving full shade in June and some sunlight for remainder of the year

Fig. 7

VENTILATION

Up to the middle of the twentieth century the principal means of heating was by solid fuel burning fires and stoves. The considerable intake of air for combusion of wood, coal or coke in fires and stoves at the same time provided more than adequate changes of air for the ventilation of rooms, to the extent that cold draughts of air drawn in during winter months through cracks around opening windows and doors caused discomfort. At the time the concern was to control draughts of incoming cold air rather than considerations of ventilation. With the introduction of oil and gas fired central heating boilers, it was practical to heat the whole of buildings from one central boiler that drew air for combustion directly from outside and so reduced draughts of cold air from outside. The rapidly increasing use of oil and gas from the middle of the twentieth century prompted concern for the need to conserve the limited sources of energy. Initially regulations required minimum standards of insulation in the roofs and walls of new buildings.

The current trend towards conservation of energy, by more efficient use of insulation against excessive transfer of heat, has led to the installation of double glazing to windows in both new and old buildings and the fitting of effective weather-stripping around the opening parts of windows and doors to reduce draughts of cold air entering the building. Open fires are uncommon in modern buildings and many open fireplaces in older buildings have been sealed so blocking flues that provided some ventilation. This means that there is less provision for permanent changes of air. The air in rooms may become 'stuffy' and uncomfortable and at worst unhealthy.

So that there is some provision for natural ventilation the Building Regulations 1991 now require means of ventilation to habitable rooms, kitchens, bathrooms and sanitary accommodation to provide air change by natural or mechanical ventilation and also to reduce condensation in rooms where warm, moisture vapour laden air may condense to water. The provisions are for opening windows and vents and some mechanical ventilation to kitchens, bathrooms and sanitary accommodation.

For the comfort and well-being of people it is necessary to ventilate rooms by allowing a natural change of air between inside and outside or to cause a change by mechanical means. The necessary rate of change will depend on the activities and numbers of those in the room. The rate of change of air may be given as air changes per hour, as for example one per hour for living and up to 4 for work places, or as litres per second as a more exact requirement where mechanical ventilating is used, because it gives a clear indication of the size of inlets, extracts, ducts and pressures required.

The size of a ventilating opening, by itself, gives no exact indication of the likely air change as the ventilating effect of an opening depends on air pressure difference between inside and outside and the size of the opening or openings through which air will be evacuated to cause air flow. The actual ventilating effect of a window, by itself, is unpredictable as it will, when open, in all likelihood act to intake and extract air at the same time.

The rate of exchange of air will depend on variations between inside and outside pressure and heat, and the size and position of other openings in the room such as doors and open fireplaces that may play a part in air exchange. An open window, by itself, may well not thoroughly ventilate a room. For thorough ventilation, that is complete air change, circulation of air is necessary between the window and another or other openings distant from the window, otherwise pockets of stagnant air may be undisturbed in those parts of the room distant from the window.

Ventilation air changes are necessary to minimise condensation which is caused when warm air-borne moisture vapour precipitates in droplets on cold surfaces such as glass and metal. By ventilating, the warm moist air is exchanged with drier air that is less likely to cause condensation.

The probable ventilating action of the various types of window in comparatively still air conditions due to the exchange of warmed inside and cooler outside air is illustrated in Fig. 8.

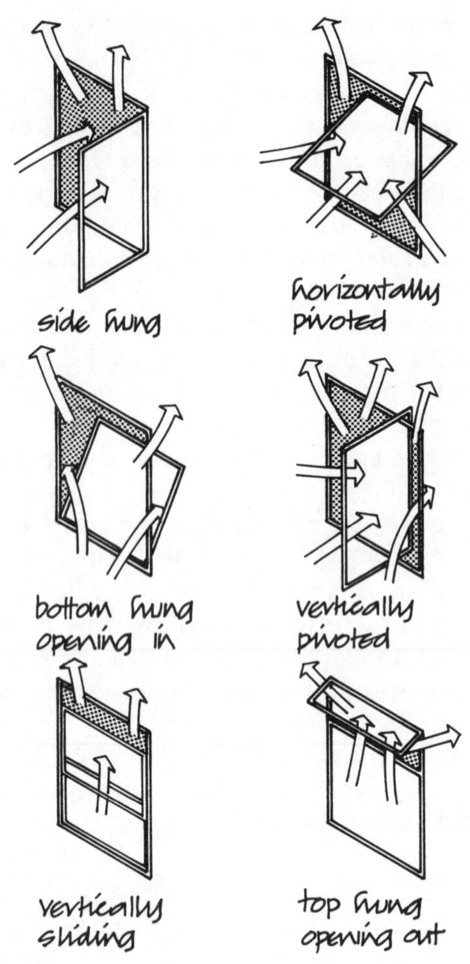

Fig. 8

The traditional method of ventilating is through opening lights in windows. The advantage of opening lights is that they can be opened or closed to suit the individual choice of the occupant of rooms regardless of notional optimum rates of air change for comfort and well being. The facility of 'flinging wide the casement to fresh air' has long been cherished and is unlikely to be abandoned in the foreseeable future. The disadvantages of opening lights are that they are difficult to open just sufficient for ventilation without letting in cold draughts or gusts of wind; the necessary clearance gaps around opening lights may allow an excess of air leakage and rain leakage; the necessary framing around them reduces the area available for glass; and they present a high security risk.

For control of ventilation the vertically sliding window is the most efficient as it can be operated to provide either small gap ventilation between meeting rails and sashes and frame, or opened to nearly half its total area, and the degree of opening can be closely controlled between these extremes. Side-hung casements are less efficient as they are difficult to open to provide closely controlled gap ventilation around the three open edges of the sash and for this reason top hung ventlights are often used. Top-hung lights are reasonably efficient but less readily controlled than the sliding sash in that there is the likelihood of both the extraction of air from below and the intake of air from the sides. Bottom-hung windows will operate to encourage the intake of air over them and extraction from the sides. Pivot windows are generally less efficient in the control of ventilation as they are difficult to open sufficiently to prevent variable gusts of wind and cold draughts being directed in at low level. In addition large pivot windows, when open, may be distorted by heavy gusts of wind and may then be difficult to close tight unless there is a mechanism to lock the sash shut at several points. Horizontally pivoted windows should be capable of being locked shut both top and bottom else the top, opening in part of the sash, may be forced in by high winds and allow considerable air seepage.

Apart from the wish to fling windows wide open there is every reason to dispense with opening lights and replace them with ventilators designed to control air movement only. These ventilators can be included in windows either in place of part of the glass or as part of the window head or cill construction, or they may be fixed separate from the windows. For ventilation alone these ventilators need only small apertures that can be opened and closed by means of simple 'hit and miss' controls or hinged or pivoted flaps operated by cord and pulley or winding gear (Fig. 9).

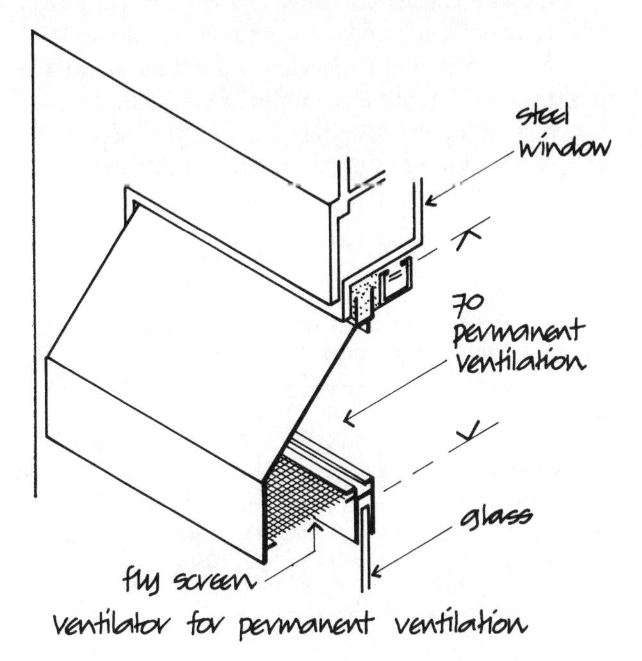

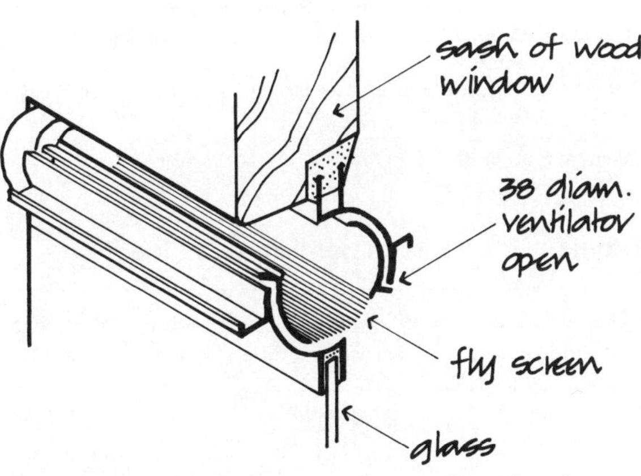

Ventilators fixed in glazing rebate of sash or fixed light.

Fig. 9

Approved Document F gives practical guidance to meeting the requirements of the Building Regulations 1991 for the provision of means of ventilation for dwellings. The requirements are satisfied for habitable rooms, such as living rooms and bedrooms when there are:

(a) For rapid ventilation – one or more ventilation openings, such as windows, with a total area of at least $1/20$ of the floor area of the room, with some part of the ventilating opening at least 1.75 m above the floor.
(b) For background ventilation – a ventilation opening or openings having a total area of not less than 4000 mm, which is controllable, secure and located so as to avoid undue draughts, such as a trickle ventilator.

For kitchens the requirements are satisfied when there is both:

(a) Mechanical extract ventilation for rapid ventilation, rated as capable of extracting at a rate of not less than 60 litres per second (or incorporated within a cooker hood and capable of extracting at 30 litres a second) which may be operated intermittently for instance during cooking, and
(b) Background ventilation, either by a controllable and secure ventilation opening or openings having a total area of not less than 4000 mm^2, located so as to avoid draughts, such as a trickle ventilator or by the mechanical ventilation being in addition capable of operating continuously at nominally one air change per hour.

For bathrooms the requirements are satisfied by the provision of mechanical extract ventilation capable of extracting at a rate of not less than 15 litres a second, which may be operated intermittently.

For sanitary accommodation the requirements are satisfied by either:

(a) Provision for rapid ventilation by one or more ventilation openings with a total area of at least $1/20$ of the floor area of the room and with some part of the ventilation opening at least 1.75 m above the floor level, or
(b) Mechanical extract ventilation, capable of extracting air at a rate of not less than 3 air changes per hour, which may be operated intermittently with 15 minutes overrun.

STRENGTH AND STABILITY

A window should be strong enough when closed to resist the likely pressures and suctions due to wind, and when open be strong and stiff enough to resist the effect of gale force winds on opening lights. A window should also have sufficient strength and stiffness against pressures and knocks due to normal use and appear to be safe, particularly to occupants in high buildings. A window should be securely fixed in the wall opening for security, weathertightness and the strength and stiffness given by fixings.

Wind loading The direction and strength of wind fluctuates to the extent that sophisticated electronic equipment is necessary to measure the changes in pressure that occur. To determine the wind pressures that a window is likely to suffer it is convenient to define these as maximum gust speed, averaged over 3 second periods, which are likely to be exceeded on average only once in 50 years. These gust speeds have been measured by the Meteorological Office and plotted as basic wind speeds on a map of the United Kingdom (Fig. 10).

To determine probable wind loads on buildings the method given in BS 6262 can be used for buildings that are of simple rectangular shape and up to 10 m high from eaves to ground level. The basic wind speed is determined from the map of the United Kingdom (Fig. 10). The basic wind speed is then multiplied by a correction factor that takes account of the shelter afforded by obstructions and ground roughness as set out in Table 4, to arrive at a design wind speed. The probable maximum wind loading is then obtained from Table 5 by reference to the design wind speed. The wind loading is used to select the test pressure class of window construction necessary and graphs are used to select the required thickness of glass.

Windows are tested in a laboratory to determine test pressure classes; a sample of manufactured windows complete with opening lights and glass is mounted in a frame to represent the surrounding walls. The criterion of success in the pressure test is that after the test the window should show no permanent deformation or other damage and there should be no failure of fastenings.

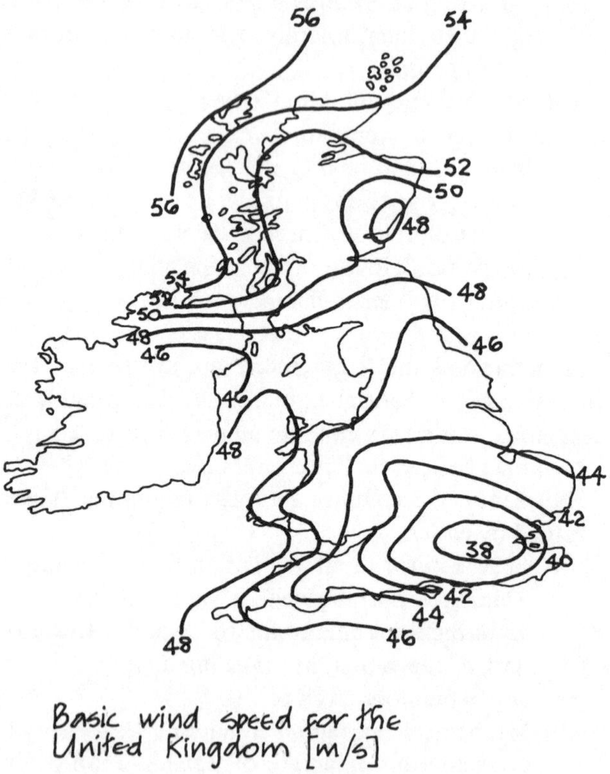

Basic wind speed for the United Kingdom [m/s]

Fig. 10

Table 4. Correction factors for ground roughness and height above ground

Height above ground (m)	Category 1	Category 2	Category 3	Category 4
3 or less	0.83	0.72	0.64	0.56
5	0.88	0.79	0.70	0.60
10	1.00	0.93	0.78	0.67

Category 1: Open country; with no obstructions. All coastal areas.
Category 2: Open country; with scattered wind breaks.
Category 3: Country; with many wind breaks; e.g. small towns; city outskirts.
Category 4: Surfaces with large and frequent obstructions; e.g. city centres.

Taken from BS 6262

Table 5. Probable maximum wind loading

Design wind speed (m/s)	Wind loading (N/m²)	Design wind speed (m/s)	Wind loading (N/m²)
28	670	42	1510
30	770	44	1660
32	880	46	1820
34	990	48	1980
36	1110	50	2150
38	1240	52	2320
40	1370		

Taken from BS 6262

RESISTANCE TO WEATHER

Air permeability (airtightness)

To conserve heat and avoid cold draughts it is good practice to design windows so that there is little unnecessary leakage of air. Air movement through closed windows may occur between the window frame and the surrounding wall, through cracks between glass and the framing, through glazing joints, and more particularly through clearance gaps between opening lights and the window frame. Leakage of air around window frames, around glass and through glazing joints can be avoided by care in design, construction and maintenance. The necessary clearance gaps around opening lights can be made reasonably airtight by care in design and the use of weatherstripping.

For comfort in living and working conditions in buildings some regular change of air is necessary. The necessary ventilation should be provided through controlled ventilators, through opening lights, or by mechanical ventilation. It is not satisfactory to rely on leakage of air through windows for ventilation as this leakage cannot be controlled, and it may be excessive for ventilation and conservation of heat or too little for ventilation.

While air leakage through windows will contribute to wastage of heat by an excess of cold air entering, other parts of the building envelope may add considerably to heat loss by leakage of air through construction cracks. An example of this is where a weep hole in the external brick leaf of a cavity wall faces construction gaps around timber joist ends built into the inner skin, so that in high wind measurable volumes of cold air blow into the timber floor. The need for and use of weep holes in cavity walls is questionable particularly as they will allow cold air to enter the cavity and so reduce the insulating properties of this construction. In many traditionally constructed buildings some one third or more of all air leakage is through construction gaps and cracks. Close attention should therefore be paid to the solid filling or sealing of all potential construction gaps and cracks as well as controlling leakage through windows.

The flow of air through windows is caused by changes in pressure and suction caused by wind that may cause draughts of inward flowing cold air and loss of heat by excessive inflow of cold and outflow of warmed air. It is to control this air movement that systems of check rebates and weatherstripping are used in windows.

The performance of windows with regard to airtightness is based on predicted internal and external pressure coefficients which depend on the height and plan of the building. These are related to the design wind pressure which is determined from the exposure of the window and basic wind speed from the map in Fig. 10. From these, test pressure classes are established for use in the tests for air permeability and watertightness to set performance grades.

Watertightness Penetration of rain through cracks around opening lights, frames or glass occurs when rain is driven on to vertical windows by wind, so that the more the window is exposed to driving rain the greater the likelihood of rain penetration.

Because of the smooth, impermeable surface of glass, driven rain will be driven down, across and up the surface of glass thus making seals around glass and clearance gaps around opening lights vulnerable to rain penetration.

The tests for watertightness of windows are based on predictions similar to those used for air infiltration in determining design wind speed, exposure grades and test pressure classes to set performance standards.

On sites where there is high exposure to wind-driven rain it may be reasonable to adopt a higher performance for watertightness than that used for strength and stability, to ensure watertightness and to avoid the need for thicker mullions, transomes and glass.

To minimise the penetration of driven rain through vertical windows, it is advantageous to:

(a) Set the face of the window back from the wall face so that the projecting head and jamb will to some extent give protection by dispersing rain.
(b) Ensure that external horizontal surfaces below openings are as few and as narrow as practicable to avoid water being driven into the gaps.
(c) Ensure that there are no open gaps around opening lights by the use of lapped and rebated joints and that where there are narrow joints that may act as capillary paths there are capillary grooves.
(d) Restrict air penetration by means of weatherstripping on the room side of the window so that the pressure inside the joint is the same as that outside; a pressure difference would drive water into the joint.
(e) Ensure that any water entering the joints is drained to the outside of the window by open drainage channels that run to the outside.

In modern window design weatherstripping is used on the room side of the gaps around opening lights to exclude wind and reduce air filtration, and rebates and drain channels are used on the outside to exclude rain.

DURABILITY AND FREEDOM FROM MAINTENANCE

The durability of the traditional material for windows – wood – has been established over centuries. The majority of wood windows are of softwood that suffers moisture movement with change of moisture content and may rot where water enters open joints, if it is not adequately protected by paint or other protective coating. A wood window strongly framed from sound, well-seasoned wood protected by a sufficiently elastic paint coating, that is adequately maintained, may have a useful life comparable to that of most buildings. The disadvantage of softwood windows is that they need comparatively frequent maintenance expenditure at intervals of five to seven years. It is the maintenance costs of wood windows that has led, over recent years, to the large market in 'replacement windows' of uPVC and aluminium.

Steel windows have acquired a bad name due to the progressive, corrosive rusting that occurred with the early use of mild steel sections which were not protected with a galvanised zinc coating. Steel windows have been unable to regain favour in competition with uPVC and aluminium windows, and because a galvanised coating does not give total protection against corrosion, these windows need comparatively frequent painting.

On exposure to air, aluminium forms an oxide that generally protects the aluminium below it from further corrosion. The oxide coating that forms on aluminium is coarse textured, dull and silver-grey in colour which readily collects dirt, is not easily cleaned and has an unattractive appearance. For these reasons aluminium is usually coated by anodising, polyester powder, organic or acrylic coatings, to inhibit corrosion and for appearance sake. Anodised finishes may fail after some years, whereas organic powder coating and acrylic coatings survive for many years and require cleaning by washing with water from time to time to maintain appearance. The powder and acrylic coatings are applied in a full range of colours. White is preferred as it does not suffer colour bleaching as do the stronger colours.

Windows made from PVC sections have been in use for more than 30 years. The material has maintained its original characteristics over this period in various climatic conditions and there is reason to suppose uPVC windows have a useful life similar to that of most buildings. Strongly coloured uPVC will, after some years, bleach due to the effect of ultra violet light. The colour loss is irregular and unsightly and overpainting of uPVC is not generally successful. The use of white or off-white is recommended. The smooth surface of this material will, after some time, collect a layer of grime that can be easily removed by washing with water. Other than occasional washing these windows need no maintenance.

A layer of grime will collect on the surface of glass over the course of a month or two, to the extent that it is unsightly and reduces light transmission. To maintain its lustrous, fire-glazed finish, glass needs cleaning at intervals of one to two months by washing with water and polishing dry with a linen scrim cloth.

An extremely thin protective coating of copolymer which can be sprayed over the surface of glass, appreciably reduces the build-up of a dirt film and facilitates cleaning. This sprayed on coating can only be applied in factory conditions to glass cut ready for glazing.

FIRE SAFETY

The requirements from Part B of Schedule 1 to the Building Regulations 1991 are concerned to:

(a) Provide adequate means of escape.
(b) Limit internal fire spread (linings).
(c) Limit internal fire spread (structure).
(d) Limit external fire spread.
(e) Provide access and facilities for the fire service.

The current advisory document giving practical guidance to meeting the requirements of the Building Regulations 1991 is Approved Document B. Entitled *Fire Safety* it is concerned with the escape of people from buildings after the outbreak of fire rather than the protection of the building and its contents.

Means of escape The main dangers to people in buildings, in the early stages of fire, are the smoke and noxious gases produced which cause most of the casualties and may also obscure the way to escape routes and exits. The regulations are concerned to:

(a) Provide a sufficient number and capacity of escape routes to a place of safety.
(b) Protect escape routes from the effects of fire by enclosure where necessary.
(c) Ensure that escape routes are adequately lit and exits suitably indicated.

The regulations accept that it is, in general, possible for the occupants of one and two storey houses to escape without outside help. The provisions in Approved Document B are, therefore, limited to ensuring that each habitable room opens directly to a hall or stair or that the rooms have a window or door through which escape can be made and that means of early warning, by automatic smoke alarms, be provided.

For flats and maisonettes of more than two storeys, and all other buildings, means of escape are required by escape routes and escape stairs so that there are alternative means of escape to a place of safety depending on the size, height, use and occupancy of the building.

Internal fire spread (linings) Fire may spread within a building over the surface of materials covering walls and ceilings. The Regulations prohibit the use of lining materials that encourage spread of fire across their surfaces when subject to intense radiant heat and those which give off appreciable heat when burning.

Internal fire spread (structure) A building shall be designed and constructed so that:

(a) Its stability will be maintained for a reasonable period (fire resistance).
(b) Fire resisting compartments are formed to limit the spread of fire (compartments).
(c) Concealed spaces have barriers to the spread of smoke and flames (concealed spaces).

Fire resistance As a measure of ability to withstand the effects of fire the elements of a structure are given notional fire resistance times, in minutes, based on tests. Elements are tested for the ability to withstand the effects of fire in relation to:

(a) Resistance to collapse (loadbearing capacity) which applies to loadbearing elements.
(b) Resistance to fire penetration (integrity) which applies to fire separating elements.
(c) Resistance to the transfer of excessive heat (insulation) which applies to fire separating elements.

The notional fire resistance times, which depend on the size, height and use of the building, are chosen as being sufficient for the escape of occupants in the event of fire.

Compartments To prevent rapid fire spread, which could trap occupants, and to reduce the chance of fires growing large, it is necessary to sub-divide buildings into compartments separated by walls and/or floors of fire-resisting construction. The degree of sub-division into compartments depends on:

(a) The use and fire load (contents) of the building.
(b) The height to the floor of the top storey as a measure of ease of escape and the ability of fire services to be effective.
(c) The availability of a sprinkler system which can slow the rate of growth of fire.

The necessary compartment walls and/or floors are of solid construction sufficient to resist the penetration of fire for the stated notional period of time in minutes.

Concealed spaces Smoke and flame may spread through concealed spaces such as voids above suspended ceilings, roof spaces and enclosed ducts and wall cavities in the construction of a building. To restrict the unseen spread of smoke and flame through such spaces cavity barriers and stops should be fixed as a tight fitting barrier to the spread of smoke and flame.

External fire spread To limit the spread of fire between buildings, limits to the area of 'unprotected areas' in walls and finishes to roofs, close to boundaries, are imposed by the Building Regulations. The term 'unprotected area' is used to include those parts of external walls that may contribute to the spread of fire between buildings. Windows are unprotected areas, as glass offers negligible resistance to the spread of fire. In Approved Document B rules are set out that give practical guidance to meeting the requirements of the Building Regulations 1991 in regard to minimum distances of walls from boundaries and maximum unprotected areas.

Access and facilities for the Fire Services The Building Regulations 1991 require that buildings shall be designed and constructed so that facilities to assist fire fighters shall be provided within the site of the building for access for vehicles and in the building for access for fire fighting personnel and also the provision of fire mains.

RESISTANCE TO THE PASSAGE OF HEAT

A window, which is a component part of a wall or roof, will affect thermal comfort in two ways, firstly by transmission (passage) of heat and secondly through the penetration of radiant heat from the sun, that causes 'solar heat gain'. Glass, which forms the major part of a window, offers poor resistance to the passage of heat and readily allows penetration of solar radiation.

The transfer of heat through a window is a complex of conduction, convection and radiation. Conduction is the direct transmission of heat through a material, convection the transmission of heat in gases by circulation of the gases, and radiation the transfer of heat from one body of radiant energy through space to another. Because of the variable complex of these modes of transfer it is convenient to adopt a standard average thermal transmittance coefficient (*U*) as a comparative practical measure of heat loss through materials in steady state conditions. This comparative standard measure of heat transfer, known as the *U* value, is the heat in watts that will be transferred through 1 m^2 of a construction where there is a difference of 1°C between the temperature of the air on opposite sides. In using this unit of measure of heat transfer, assumptions are made about the moisture content of materials, the rate of heat transfer to surfaces by radiation and convection, the rates of air flow in ventilated spaces, and heat bridge effects.

Glass has low insulation and high transmittance value. The *U* value of a single sheet of 6 thick glass (single glazing) is 5.4 W/m^2K and that of a double glazed unit with two 6 thick sheets of glass spaced 10 apart is 3.1 W/m^2K, as compared to that of an insulated cavity wall of 0.45 W/m^2K. Because glass has relatively poor resistance to the passage of heat as compared to that of an insulated wall, it is advantageous to limit the area of glass in buildings for the conservation of energy. This is the assumption in the Building Regulations 1991. In Approved Document L giving practical guidance to meeting the requirements of the Building Regulations 1991, maximum *U* values for elements of construction are given and maximum areas of windows and rooflights are stated as a percentage of perimeter wall and roof areas, as set out in Table 6.

In calculating the area of perimeter wall, measured between internal finished surfaces, all openings in the wall should be included. An external door with 2 m square of glass or more should be treated as a window.

Table 6. Maximum U values and glazed areas

Maximum U values (W/m^2K)

	Dwellings	**All other buildings**
Exposed walls Exposed floors Ground floors	0.45	0.45
Roofs	0.25	0.45
Semi-exposed walls and floors	0.6	0.6

Maximum single glazed areas of windows and rooflights

Building type	**Windows**	**Rooflights**
Dwellings	windows and rooflights together. 15% of total floor area	
Other residential (including hotels and institutional)	25% exposed wall area	20% of roof area
Places of assembly, offices and shops	35% exposed wall area	20% of roof area
Industrial and storage	15% exposed wall area	20% of roof area

Notes

(a) In any building, the maximum glazed area may be doubled where double glazing is used, and trebled where double glazing is, in addition, coated with a low emissivity coating (emissivity not greater than 0.2).

(b) Triple glazing may be considered equivalent to double glazing with a low emissivity coating.

Taken from Approved Document L
The Building Regulations 1991, HMSO

An opening in a wall, other than a window and a meter cupboard recess, may be treated as wall area and lintels, jambs and sills to windows may be included in the area of the window. The inclusion of lintels, jambs and sills in the area of the window is to make allowance for solid construction where cavity or other insulation does not extend up to the back of the window frame. Any part of a roof with a pitch (slope) of 70° or more is treated as a wall.

Instead of limiting the areas of glass relative to areas of wall and roof, Approved Document L provides alternative methods of determining areas of glass by 'calculated trade-off' or 'calculated energy use', which does not apply to dwellings. The purpose of these calculations is to allow for the use of larger areas of glass than those set out in Table 6 where, for example, double glazing is used or improved wall or roof insulation is employed to the extent that the overall transfer of heat or use of energy will be no greater than it would be if the assumptions in Table 6 were accepted.

In setting limits to areas of glass in windows, doors and rooflights no allowance is made for what, in layman's terms, are called draughts; that is draughts of cold air that may enter through cracks around opening windows and doors and construction gaps or cracks. There may be considerable heat loss to cold air entering through clearance gaps around unsealed or poorly sealed, opening windows and doors, and equal loss to air through construction gaps where pipes penetrate walls, joist ends are built into walls, weep holes are created in cavity walls, and other construction gaps.

The move in recent years to fitting replacement windows and glazed doors to existing buildings and windows and glazed doors to new buildings with effective draught seals or weatherstripping, which has appreciably reduced draughts of cold air, has tended to substantially reduce air changes in buildings and prompted the regulations requiring ventilation.

Solar heat gain The term 'radiation' describes the transfer of heat from one body through space to another. When the radiant energy from the sun passing through a window reaches, for example, a floor, part of the radiant energy is reflected and part absorbed and converted to heat. The radiant energy reflected from the floor will in part be absorbed by a wall and converted into heat and partly reflected. The heat absorbed by the floor and wall will in turn radiate energy that will be absorbed and converted to heat. This process of radiation, reflection, absorption, conversion to heat and radiation will produce rapidly diminishing generation of heat. The heat generated by radiation will be dissipated by conduction in solid materials such as walls, and by convection in air.

The wavelength of radiant energy depends on the temperature of the radiating body: the higher the temperature the shorter the wavelength. Part of the radiation of energy from the extremely high temperature of the sun is short wave which will pass through clear glass with little absorption, whereas the comparatively low temperature and long wavelength of an electric fire and a floor or wall will mostly be absorbed by glass.

Where the balance of gain of heat from radiation is greater than that dissipated by conduction and convection, there will be a gradual build-up of heat that can cause discomfort in rooms due to solar heat gain.

Plainly the degree of solar heat gain is affected by the size and orientation of windows. Large windows facing south in the northern hemisphere will be more affected than those facing east or west. The time of year will also have some effect between the more intense summer radiation which will not penetrate deeply into rooms at midday to the less intense but more deeply penetrating radiation of spring and autumn.

In the temperate climate of northern Europe discomfort from solar heat has not, until recently, been a concern. Sunlight is welcome as a relief from preponderant, dull overcast days. In middle and southern Europe systems of shutters and blinds are used to provide shade from the more intense radiation of summer sun.

Discomfort from solar heat gain has mainly been a consequence of the fashion to use large areas of glass as a sealed walling material for offices and other non-domestic buildings, where the build-up of heat can make working conditions uncomfortable. The transmission of solar radiation can be effectively reduced by the use of body tinted, surface modified or surface coated glass to control solar heat gain.

RESISTANCE TO THE PASSAGE OF SOUND

Sound is the sensation produced through the ear by vibrations caused by air pressure changes superimposed on the comparatively steady atmospheric pressure. The rate or frequency of the air pressure changes determines the pitch as high pitch to low pitch sounds. The audible frequencies of sound are from about 20 Hz to 15,000 or 20,000 Hz, the abbreviation Hz representing the unit Hertz where one Hertz is numerically equal to one cycle per second. The sound pressure required for audibility is generally greater at very low frequencies than at high frequencies.

Because of the variation in the measured sound pressure and that perceived by the ear over the range of audible frequencies, a simple linear scale will not suffice for the measurement of sound. The measurement that is used is based on a logarithmic scale that is adjusted to correspond to the ears' response to sound pressure.

The unit of measurement used for ascribing values to sound levels is the decibel (dB). Table 7 gives sound pressure levels in decibels for some typical sounds. Because the sensation of sounds at different frequencies, although having the same pressure or energy, generally appears to have different loudness, a sound of

Table 7. Sound pressure levels for some typical sounds

Sound	Sound pressure level (dB)
Threshold of hearing	0
Leaves rustling in the wind	10
Whisper or ticking of a watch	30
Inside average house, quiet street	50
A large shop or busy street	70
An underground train	90
A pop group at 1.25 m	110
Threshold of pain	120
A jet engine at 30 m	130

Subjectively, a sound of 100 dB is not twice as loud as one of 50 dB, it is very much louder.
An approximate relationship between change in sound pressure levels and change in apparent loudness is:

Change in SPL (dB)	Apparent change in loudness
3	Just perceptible
5	Clearly noticeable
10	Twice or half as loud
20	Much louder or much quieter

100 dB is not twice as loud as one of 50 dB, it is very much louder. The scale of measurement used to correlate to the subjective judgement of loudness, which is particularly suitable for traffice noise, is the A weighting with levels of sound stated in db(A) units.

The word loud is commonly taken to indicate the degree of strongly or clearly audible sound, and the word noise as distracting sound.

To provide a measure of generally accepted tolerable levels of audible sound, which will not distract attention or be grossly intrusive, tolerance noise levels are set out in Table 8.

Table 8. Tolerance noise levels

	dB(A)
Large rooms for speech such as lecture theatre, conference rooms etc.	30
Bedrooms in urban areas	35
Living rooms in country areas	40
Living rooms in suburban areas	45
Living rooms in busy urban areas	50
School classrooms	45
Private offices	45–50
General offices	55–60

Taken from BS CP153 Part 3:1972

Sound is produced when a body vibrates, causes pressure changes in the air around it and these pressure changes are translated through the ear into the sensation of sound. Sound is transmitted to the ear directly by vibrations in air pressure – airborne sound, or partly by vibrations through a solid body that in turn causes vibrations of air that are heard as sound – impact sound. The distinction between airborne and impact sound is made to differentiate the paths along which sound travels, so that construction may be designed to interrupt the sound path and so reduce sound levels. Airborne sound is, for example, noise transmitted by air from traffic through an open window into a room, and impact sound from a door slamming shut that causes vibrations in a rigid structure that may be heard some distance from the source.

The sensation of sound is affected by the general background level of noise to the extent that loud noise may be inaudible inside a busy machine shop where comparatively low levels of sound may be disturbing inside a quiet reading room.

For the majority of people, who live and work in built-up areas, the principal sources of noise are external traffic – airborne sound – and internal noise from neighbouring radios, televisions and impact of doors and footsteps on hard surfaces – impact and airborne sounds.

The noise levels from traffic at different locations have been predicted and tabulated at points up to 300 m from the source of sound. Two indices are used, the L_{10} and the L_{90}, and these levels in dB(A) are only exceeded for 10% and 90% of the time respectively. Table 9, taken from BRE Digest 38, 1963, gives a range of noise levels at locations where traffic noise predominates.

Windows and doors are a prime source for the entry of airborne sound both through glass, which affords little insulation against sound, and by clearance gaps around opening parts of windows and doors. Appreciable reduction of intrusive airborne sound can be effected by weather-stripping around the opening parts of windows and doors.

The transmission of sound through materials depends mainly on their mass; the more dense and heavier the material the more effective it is in reducing sound. The thin material of a single sheet of glass provides poor insulation against airborne sound.

A small increase in insulation or sound reduction of glass can be effected by the use of thicker glass, where an average reduction of 5 dB is obtained by doubling the thickness of glass. There is no appreciable sound

Table 9. Range of noise levels at locations in which traffic noise predominates

Location	Noise dB(A)			
	Day		Night	
	L10	L90	L10	L10
Arterial roads with many heavy vehicles (kerbside)	80	68	68	50
Major roads with heavy traffic Side roads within 20m of major road	75	63	61	48
Main residential roads Side roads within 50m of heavy traffic routes Courtyards screened from heavy traffic	70	60	54	44
Residential roads, local traffic	65	57	52	44
Minor roads, gardens, traffic 100m distant	60	52	48	43
Parks, gardens well away from traffic noise	55	50	46	41
Places of few local noises	50	47	43	40

Taken from *Building Research Establishment Digest* **38** (1963)

reduction by using the sealed double glazed units that are effective in heat insulation as the small cavity is of no advantage, so that sealed double glazing is no more effective than the combined thickness of the two sheets of glass. For appreciable reduction in sound transmission double windows are used where two separate sheets of glass are spaced from 100 to 300 apart. An average reduction of 39 dB with 100 space and 43 dB with 200 space can be obtained with 4 glass. This width of air space is more than the usual window section can accommodate and it is necessary to use some form of double window. With double windows it is necessary to provide for cleaning both sheets of glass both sides and to ventilate the cavity to avoid condensation.

SECURITY

Windows and doors are the principal route for illegal entry to buildings. Of the recorded cases of illegal entry – burglary – about 30% involve entry through unlocked doors and windows. Of the remaining 70% some 20% involve breaking glass to gain entry by opening catches, and the remaining 80% by forcing frames or locks. As speed is of the essence in successful burglary, well-lit and exposed windows and doors are less likely to be attacked than out of sight rear windows and doors.

Locks, bolts and catches to windows and doors are forced open by inserting a tool in the clearance gap between the opening parts of windows and doors and the frame so that the lock, bolt or catch is disengaged from the frame. Plainly flimsy frame, sash and door material can more readily be prised apart than solid material, and lightweight single locks that shoot shut a small distance into frames are more readily prised open than heavy locks that shoot shut some distance into frames. Similarly flimsy or ill-fitted hinges can be prised loose from frames. Window and door frames insecurely fixed can be prised away from the surrounding wall.

Of the materials used for windows, uPVC can more easily be deformed than more rigid wood, steel or aluminium sections, to prise locks open particularly where lock and bolt fittings are not secured to the steel or aluminium reinforcement in uPVC sections.

Security against locks, bolts, catches and hinges being forced open depends on reasonably rigid frame and opening window sections, and strong lock, bolt, catch and hinge material being securely fixed. Plainly where more than one substantial lock or bolt is used with sound frame and window material the better the security.

Even though glass is comparatively easy to smash or cut, breaking glass is the least favoured method of illegal entry, principally because the distinctive sound of breaking glass may alert householders. Small panes of glass in putty glazing are more difficult to break than large panes and the jagged edges of glass left in the putty are themselves a hazard to entry. The majority of uPVC and aluminium windows are glazed with beads, often fixed externally. It is fairly easy to remove these beads that are either screwed in place or are of the 'pop-in' type where the beads fit to projections in the sash or frame and are held by friction. Where there is ease of access beads should either be of the shuffle type which require considerable force to remove from outside or they should be fixed internally. Once beads have been removed it is usually easy to take out the glass. To make it more difficult the glass can be secured with double sided tape or glass retaining clips. The purpose of breaking glass is to open catches to windows from outside for ease of entry. It is only after the glass has been broken that the burglar may find that the catches are locked shut.

Wired glass, which can easily be broken, will make it more difficult to make a clear opening because much of the broken glass will remain attached to the wire and so impede access. Toughened glass, which is considerably more difficult to break than ordinary glass, may deter all but the most determined burglar. Laminated glass is the best protection against burglary as the glass, which is not easily broken, will not shatter but break to small fragments which have to be removed for access. Double glazing is only more secure than single to the extent that there are two sheets of glass to break.

All security measures involve extra cost in better quality frames, sashes, locks, bolts, hinges and glass. It is wise, therefore, to employ security measures on those windows and doors most vulnerable to attack. From recorded cases it is clear that 62% of burglaries occur at the rear of buildings where there is ease of access to the 14–17 year old age group of preponderant opportunist burglars, and where access is out of sight.

A disadvantage of security against illegal entry from outside is that means of escape to the outside is made that much more difficult in case of need. The balance of advantage is to provide reasonable security to those windows and doors most vulnerable to burglary, with some allowance for ease of escape where burglary is least likely.

The practical guidance in Approval Document N to the Building Regulations 1991 recommends the use of safety glass to windows and glazed panels up to a level of 800 above finished floor level. It is at this low level above floor that children are particularly vulnerable.

MATERIALS USED FOR WINDOWS

Up to the latter part of the seventeenth century the wood casement was the traditional window. The vertically sliding, double hung, sash window which was developed and used from the late seventeenth century onwards was generally the form of window used in England for all but the smaller domestic scale windows.

The advantage of wood as a material for framing windows was that the material was readily available and could be cut, shaped and framed with traditional hand tools. The cost of labour and materials in regular painting of these windows was accepted as a necessity to maintain the material and enhance appearance.

Following the industrial revolution it became practical and economic to produce mild steel sections which were developed by Crittal's in the early 1880s as hot rolled, steel section window frames and sashes. The comparatively small sections used for these windows were adopted more for aesthetic than practical reasons. The slim section steel window became the modern fashion of the late nineteenth and early twentieth centuries. These early steel windows were protected by paint which was not successful in preventing progressive, corrosive rusting and the steel window lost favour until the 1940s when steel windows, protected with hot-dip galvanising, were introduced. The galvanised zinc coating greatly reduced the onset of rusting. The limited sections that are practical with the hot rolled method of forming steel, limited the types of window that could be made, made it difficult to accommodate double glazing and effective draught seals and contributed to the loss of favour of these windows.

In the late nineteenth and early twentieth centuries bronze windows were used for large monumental scale buildings such as banks and civic buildings. These very expensive windows of strong, slender section metal which does not rust and maintains its attractive colour, were the fashion for many large buildings at the time.

Aluminium section windows were first used in this country in the early 1930s as a substitute for steel section windows. This metal, though not having the same advantageous weight to strength ratio as steel, does not suffer progressive, corrosive rusting and can be formed in a wide variety of sections suitable for the whole range of windows with single or double glazing and draught seals. On exposure to air aluminium forms an oxide that does not progressively corrode but which has a coarse textured and unattractive appearance. To inhibit the formation of the oxide coating, aluminium windows are protected with an anodised, powder or organic liquid coating that can be coloured to improve appearance.

In recent years plastics have been used for window sections. First used in Germany in the middle of the twentieth century, plastic windows are now extensively used both for new buildings and largely as 'replacement windows'. The advantage of the plastic window, fabricated from extruded uPVC (unplasticised, poly vinyl chloride) hollow box sections, is that it requires no maintenance such as painting throughout its useful life and can be formed to accommodate effective draught seals and double glazing units. Because the strength and rigidity of this material are somewhat suspect, it is generally necessary to include metal reinforcement inside the hollow plastic sections.

Wood The traditional material used for windows is wood, which is easy to work by hand or machine, can readily be shaped for rebates, drips, grooves and mouldings, has a favourable strength to weight ratio, and thermal properties (see also Volume 1, Timber) such that the window members do not act as a cold bridge to heat transfer.

The disadvantages of wood are the considerable moisture movement that occurs across the grain with moderate moisture changes, and liability to rot. The dimensional changes can cause joints to open to admit water which increases the moisture content that can lead to rot. It is of prime importance, therefore, that the moisture content of timber at the time of assembly be 17% or less, that the timber be treated with a preservative, and that the assembled window has a protective coating such as paint which is regularly maintained. It is important to maintain a sound paint film over the end grain of wood as it is more vulnerable than the long grain, in particular the end grain on the end of the stiles at the top of casements as it is much exposed to rain.

The majority of wood frames are cut from softwood timbers such as Baltic redwood (red and yellow deal), red pine and fir. Ideally sapwood should be excluded from timber for joinery as it is more liable to decay than heartwood (see Volume 1). In practice it is not economically possible to exclude sapwood. There is, therefore, good reason for preservative treatment of softwood to minimise the likelihood of rot. Preservative-treated softwood should none the less be protected with paint.

It is the need for regular and costly painting that is the particular disadvantage of softwood windows.

Steel Steel section windows have been in use since the latter half of the nineteenth century. The progressive corrosion, rusting, of these early windows brought them into disrepute up to the introduction of the hot-dip galvanising protective coating first used in the late 1940s. This zinc coating with regular painting may protect these windows during the useful life of most buildings.

The advantage of steel for windows is the slender sections for both frame and opening lights that are possible due to the inherent strength and rigidity of the material. The disadvantages are high thermal conductivity that makes the window framing act as a cold bridge to the transfer of heat, the very necessary regular painting required to protect the steel from rusting, and the fact that narrow sections do not readily accommodate double glazing.

The majority of steel sections for windows are made from hot-rolled steel bars which is an expensive process from which only a limited range of sections can be produced economically. In Europe, pressed or rolled sheet steel sections and cold deformed tube sections have been used to produce a greater variety of sections for window manufacture. Rolled steel section windows are much less used today than they were.

Aluminium Aluminium windows were first used in this country in the early 1930s and have been in use since then. These windows are made from aluminium alloy to BS 4873:1986, that is extruded in channel and box sections with flanges and grooves for rebates and weatherstripping. These thinwalled channel and box sections give the material adequate strength and stiffness for use as window sections. The material can be readily welded and has good resistance to corrosion.

The aluminium alloy used is resistant to corrosion that might cause loss of strength, yet the surface of the material will fairly rapidly lose lustre owing to white corrosion products and some pitting caused particularly in marine and industrially polluted atmospheres. This corrosive effect may be inhibited by anodising or liquid organic or powder coating. To maintain the initial lustre of the surface of these windows it is necessary to wash them at regular intervals.

Aluminium windows are generally more expensive than comparable wood or steel windows. The advantages of aluminium windows are the variety of sections available for the production of a wide range of window types, and the freedom from destructive corrosion. The disadvantage is the high thermal conductivity of the material which acts as a cold bridge to heat transfer. To prevent aluminium section windows acting as a thermal bridge, they are constructed as two sections mechanically linked by a plastic bridge that acts as a thermal break. As an alternative the inner face of the aluminium is covered with a plastic, clip-on facing.

Stainless steel This expensive corrosion-resistant steel product is made from an alloy of steel with chromium, nickel and molybdenum in the proportions of 18, 10 and 3 as a percentage of the whole to steel. This costly material is used in windows as a thin surface coating to other materials such as wood and aluminium for its appearance and freedom from corrosion. To keep its initial lustre the stainless steel finish requires regular washing.

Bronze Manganese brass is the material commonly used for bronze windows. The material is rolled or extruded to form window sections. This very expensive material is little used today. Its advantages are freedom from corrosion, high strength to weight ratio, and the attractive colour and texture of the material.

Plastics, uPVC The word plastics is used in a general sense to embrace a wide range of semi-synthetic and synthetic materials that soften and become plastic at comparatively low temperatures so that they can be shaped by extrusion or pressure moulding or both.

In the middle of the nineteenth century semi-synthetic plastics such as vulcanite or ebonite were produced from rubber and processed by the addition of sulphur to make tyres and imitation jewellery. Later in the century casein, which is made from milk curds treated with formaldehyde, was used to make ornamental articles. Celluloid, made from nitric acid, sulphuric acid and cellulose, was formed by heating, moulding and carving in the production of a wide range of decorative objects such as hand mirrors, combs and knife handles as a substitute for ivory and also for photographic film.

In the early years of the twentieth century the first synthetic plastics were produced in the form of a synthetic resin, Bakelite. Subsequent developments led to the synthesis and use of a range of synthetic plastics called polymers, which is the name of the range of plastics in common use today for building and a wide range of domestic products.

The polymer, polyvinyl-chloride (PVC), was first extensively used in forming window sections in Germany during the middle of the twentieth century. The polymer in the form of unplasticised (rigid) polyvinyl-chloride (uPVC) is softened by heating, extruded through a die and pressure formed to produce hollow box sections for window frames and sashes.

High impact modified uPVC More recently, modifiers such as acrylic have been added to the constituent materials of uPVC to improve the impact resistance of the material which is, by itself, fairly readily subject to damage by slight knocks or abrasions. The addition of modifiers affects the speed at which the heated material is extruded, otherwise the finished product is liable to surface ripples and variations in thickness if the speed of extrusion is too rapid.

The particular advantage of this material is that it is maintenance free and will maintain its smooth textured surface for the useful life of the material with occasional washing to remove grime. As the material is formed by extrusion it is practical to form a variety of rebates and grooves to accommodate draught seals. The basic colour of the material is off-white which is colourfast on exposure to ultra-violet light for the useful life of the material. A range of coloured plastics can be produced either with the colour integral to the whole of the material or as a surface finish. Dark colours are more susceptible to bleaching and loss of colour in ultra-violet light from sun than light ones.

Because uPVC has less strength and rigidity than metal sections, it is formed in comparatively bulky, hollow box sections that are not well suited for use in small windows such as casements. The comparatively large coefficient of expansion and contraction of the material with change of temperature and its poor rigidity require the use of reinforcing metal sections fitted into the hollow core of the sections to strengthen it and to an extent restrain expansion and contraction. The uPVC sections are screwed to the galvanised steel or aluminium reinforcement to fix the reinforcement in position, restrain deformation due to temperature movement and serve as secure fixing for hardware such as hinges, stays and bolts.

Some manufacturers use reinforcement only for frame sections over 1500 in length and casement or sash sections over 900 in length. For the advantage of a secure fixing for hardware and fixing bolts it is wise to use reinforcement for all uPVC sections.

WINDOW TYPES

Fixed lights The term fixed light or dead light is used to describe the whole or part of a window in which glass is fixed so that no part of the glazing can be opened. Typically fixed lights are one sheet of glass, several sheets of glass in glazing bars, or lead or copper lights glazed (fixed) directly to the window frame.

Opening light An opening light is the whole or part of a window that can be opened by being hinged or pivoted to the frame or which can slide open inside the frame.

Windows with opening lights are classified according to the manner in which the opening lights are arranged to open inside the frame, as illustrated in Fig. 11:

Hinged	Side hung
	Top hung
	Bottom hung
Pivoted	Horizontally pivoted
	Vertically pivoted
Sliding	Vertically sliding
	Horizontally sliding
Composite action	Side hung projected
	Top hung projected
	Bottom hung projected
	Sliding folding.

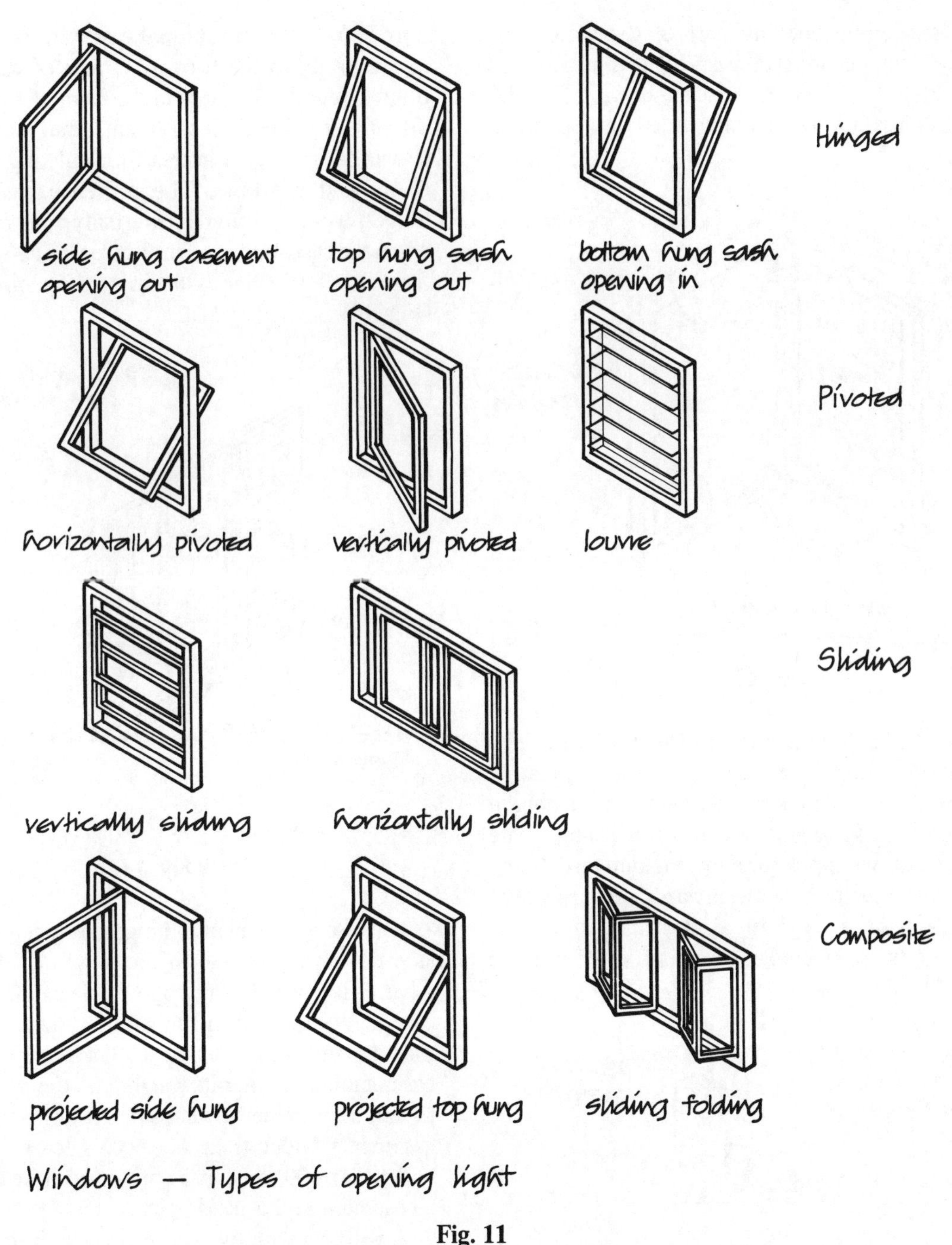

Fig. 11

HINGED OPENING LIGHTS

Side hung (casement) window

The traditional wood casement or cottage window comprised one or more comparatively small opening casements, generally with glazing bars to suit the comparatively small panes of glass that were available before the production of drawn sheet and float glass. These small windows provided sufficient daylight for indoor activities and the least heat loss through glass and draughts of cold air from cracks around casements, for the indoor comfort of the majority of the population who, before the industrial revolution, spent the major part of their lives in the open.

A casement consists of a square or rectangular window frame of wood with the opening light or casement hinged at one side to the frame to open in or out. The side hung opening part of the window is termed the casement and it consists of glass surrounded and supported by a wooden frame as shown in Fig. 12, which is an illustration of a simple one-light casement, opening out.

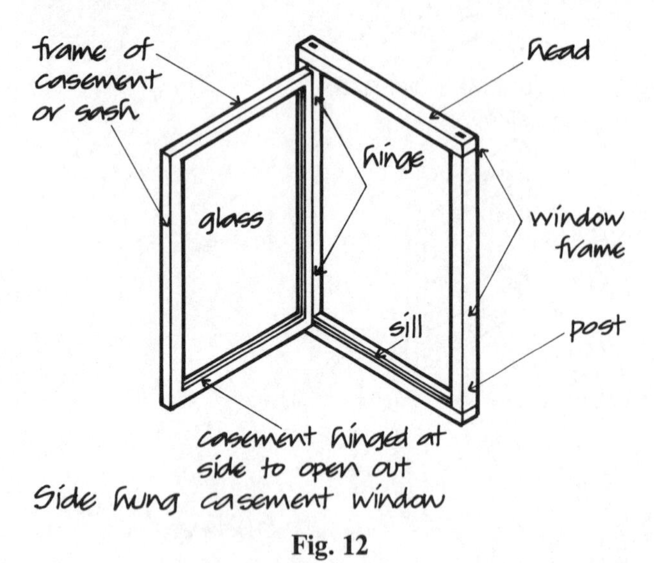

Fig. 12

The traditional English casement is hinged to open out. An outward-opening casement can more readily be made to exclude wind and rain than one opening in as the casement is forced into the outward-facing rebate in the frame by wind pressure and the outward facing rebate is more effective than the inward-facing rebate of the inward-opening casement, as illustrated in Fig. 13. Also an outward-opening casement will not be obstructed by curtains.

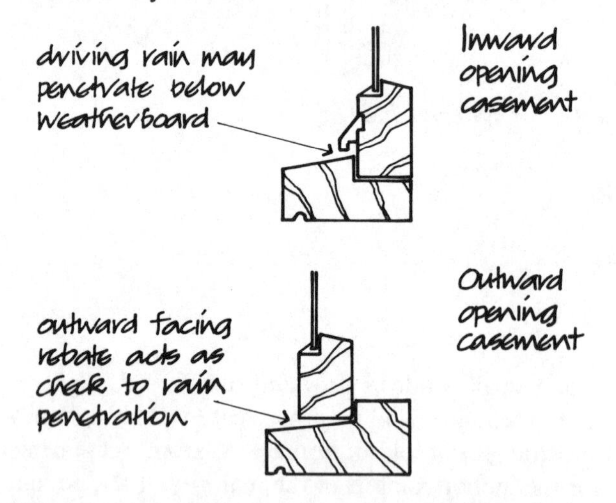

Fig. 13

In Europe the traditional casement is hung to open in, generally in the form of a pair of casements that often extend to the floor in the form of a pair of glazed doors, termed French casements, that may either serve as windows or give access to a balcony and serve as doors and windows. These French casements have hinged wooden louvres externally that can be closed when the casement is open to give ventilation and shade from the sun, hence the inward-opening casements (Fig. 14).

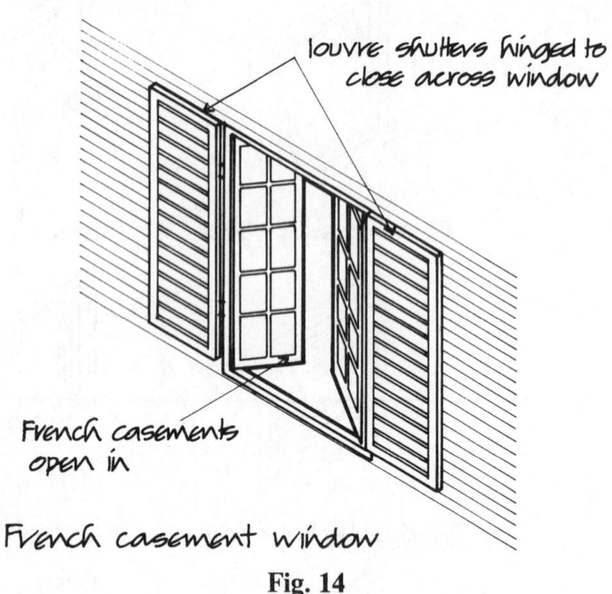

Fig. 14

Because a casement is hinged on one side, its other side tends to sink due to the weight of the casement when it is open. If any appreciable sinking occurs the casement will bind in the window frame and in time may be impossible to open. Obviously the wider a casement the greater its weight and the more likely it is to sink. It is generally considered unwise to construct casements wider than say 600. A casement window wider than 600 will consist of two or more casements or a casement and a dead light.

A window with two casements can be designed with the casements hinged so that when closed they meet in the middle of the window as illustrated in Fig. 15. The disadvantage of this arrangement is that due to expansion or sinking or both, the casements may in time jam together and be difficult to open and close. It is usually considered better to construct the window frame with vertical wood members, called mullions, to which each casement closes, as illustrated in Fig. 16.

Because a casement does not provide close control of ventilation it is common to provide small opening lights, called ventlights, which are usually hinged at the

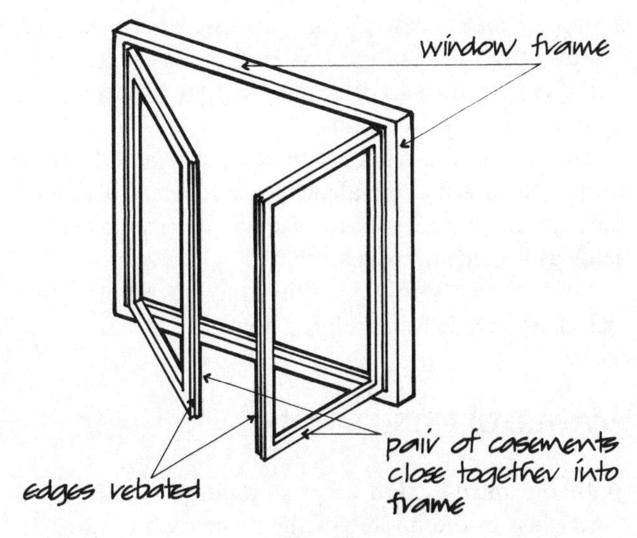

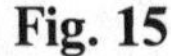
Fig. 15

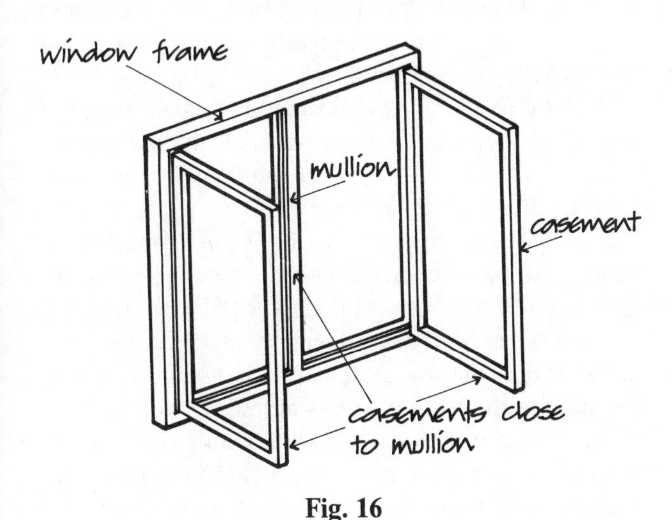

Fig. 16

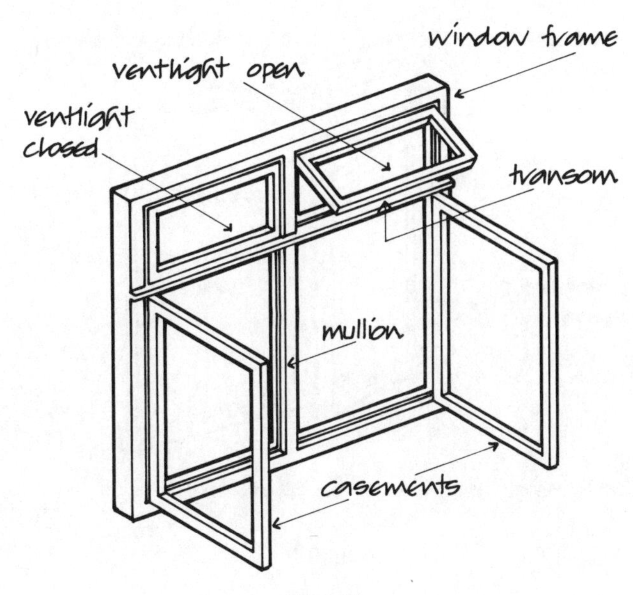

Fig. 17

top to open out. So that the ventlights can be opened independently of the casements the window frame is made with a horizontal member, called a transom, to which casements and ventlights close as illustrated in Fig. 17. Casement windows with ventlights are usually designed so that the transom is above the average eye level of people using the room, for obvious reasons.

The disadvantages of a casement window are that the casements, ventlights, mullions and transoms reduce the possible unobstructed area of glass and therefore daylight through a window of any size and many clearance gaps around opening casements and ventlights and frame members emphasise the problem of making the window weathertight. Of recent years it has been fashionable to use windows with as large an unobstructed area of glass as possible and the casement window, with its mullions and transoms and comparatively small casements, has lost favour. The manufacturers of standard casement windows now make a range of windows which combine a large dead light with a casement alongside it and a ventlight above. An outward-opening casement is difficult to clean from inside and is not suited to tall buildings where there is no outside access. The many corners of glass to the comparatively small casements and ventlights make window cleaning laborious such that corners of glass are not cleaned and become grimy, further restricting the area of clear glass available for daylight penetration.

Dead lights Casement windows may be constructed so that only a part of the window can be opened. Standard wood and metal windows are made with a casement and ventlight, the rest of the window being of glass fixed in the frame as illustrated in Fig. 18. That part of the window which cannot be opened is termed a dead light or fixed light.

Rolled steel section windows which were generally made as casements with ventlights and dead lights in much the same form as wood casements, had the advantage of maximum area of glazing due to the small section of the frame and casement framing members.

Aluminium was initially used as a substitute for rolled steel in similar sections for use in casement windows, the small section aluminium having the advantage that it did not progressively corrode.

Because of their considerable bulk, hollow uPVC window sections are not best suited for casement windows.

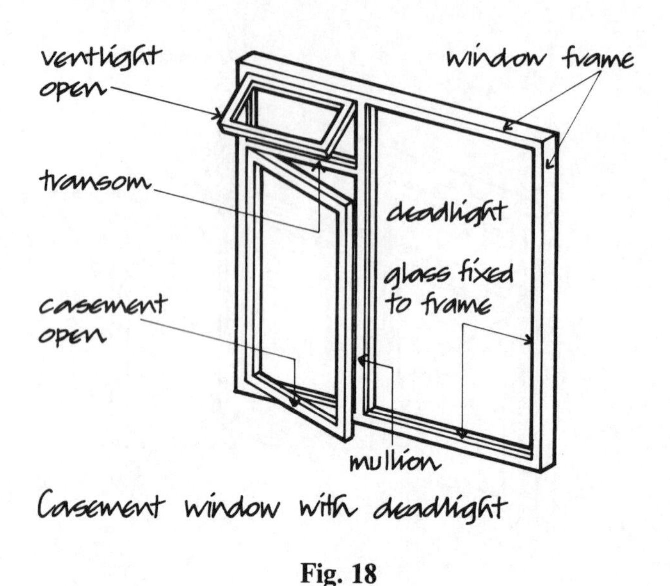

Fig. 18

Top hung and bottom hung windows

These opening lights are principally used for ventilation, the ventilation being controlled by the degree to which the light is opened. Top hung lights open out and bottom hung open in so that the slope of the sash and its glass directs rain outside the building. Usual practice is to position top hung lights at high level, as in the casement window, to encourage warmed air from inside to escape at the sides of the open sash and cold replacement air to enter below the sash as illustrated in Fig. 8. Top hung outward-opening lights are also fixed at high level so that their projection outside is at high level. Bottom hung opening-in lights are generally fixed at low level so that cold air can enter above the open light and some warmed air from inside can escape at the sides of the sash. Bottom hung opening-in lights are sometimes described as hoppers.

Top hung and bottom hung lights are often used in schools, places of assembly and factories either opened by hand or by winding gear to control circulation of air between inside and outside. Because they are top or bottom hung these lights must have a positive opening and stay mechanism, otherwise they bang shut or fully open and would be subject to wind pressure. There is therefore a limit to their opening. Being top or bottom hung the opening lights are not so subject to distortion due to their own weight as is a side hung casement, and comparatively large lights with small section frames are practical. A modification of these lights in the form of projected top and bottom hung lights has gained favour recently where the lights have a composite opening action of both sliding and pivoting to open. The disadvantage of these lights is that they may be left open and therefore be a security risk, and the bottom hung light may obstruct curtains.

These lights are made in the same way as side hung casements in wood, metal and plastic and the details of the framing of the lights and the window frame are the same as for side hung casements.

While the bottom hung lights may be cleaned both sides from inside the top hung lights cannot.

PIVOTED OPENING LIGHTS

With the introduction of continuously drawn, clear sheet glass in the middle of the nineteenth century, it became possible to use large single sheets of glass in windows. This facility was at first used to dispense with the many glazing bars previously necessary in casement and vertically sliding windows.

With the availability and demand for the comfort of central or space heating from the middle of the twentieth century, came the demand for larger unobstructed areas of glass. The width of a casement is limited by the strength of the framing in supporting its weight. The advantage of a pivoted opening light is that the weight of the frame and glass is balanced over the pivots that are fixed centrally up the height or over the width of the window, so that framing sections can be the same as those for a casement half the width.

The sashes may be either horizontally or vertically pivoted to open. Horizontally-pivoted sashes are usually pivoted at the centre of the height of the window, as illustrated in Fig. 19, to balance the weight of the sash

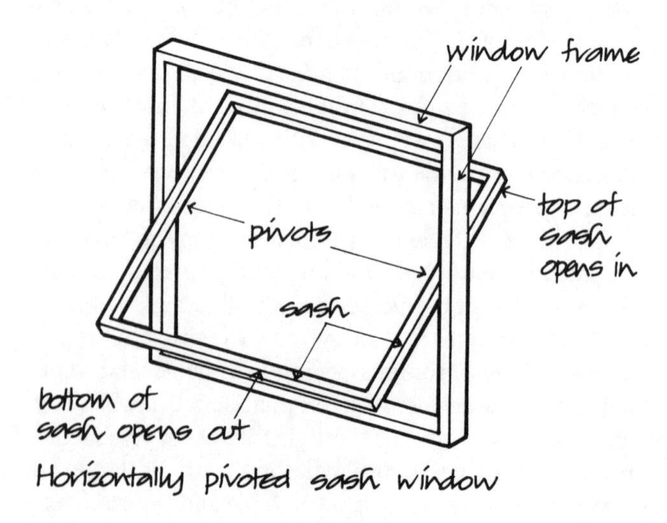

Fig. 19

over the pivots, and vertically-pivoted to open in by one third of their width to provide least obstruction inside as illustrated in Fig. 20. Because the weight of the sash is balanced over the pivots a large sash with small section framing is possible and cleaning the glass on both sides of the window is possible from inside the building. As part of pivoted sashes opens in, it obstructs the movement of curtains. Close control of ventilation with these windows is not possible as they have to open both top and bottom or both sides and they may act like a sail and catch and direct gusts of wind into the building.

Because of the pivot action of these windows the rebate between the sash and the frame has to be reversed around the pivot from inward opening to outward opening and a clearance for the opening action and the pivot has to be provided. This makes it difficult to ensure a weathertight seal around the pivot where the rebate and any weather-stripping has to be discontinued if the sash is to open. For this reason pivoted sashes are not recommended in positions of severe exposure.

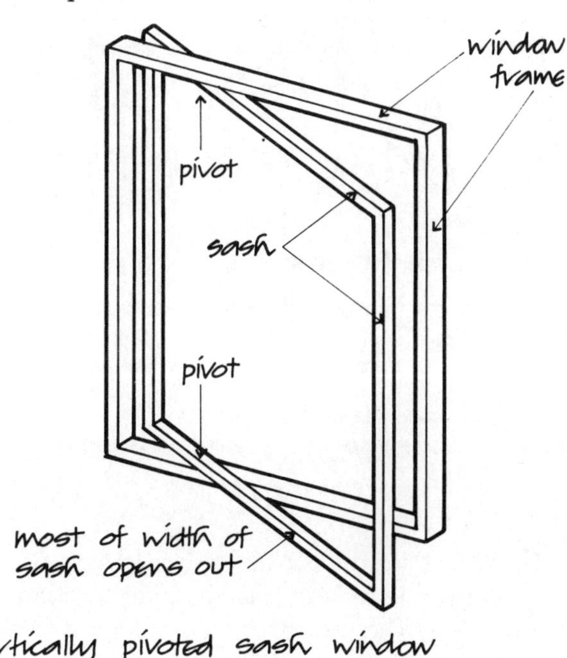

Fig. 20

SLIDING WINDOWS

The word casement is properly used to describe the framing material and glass of a side hung window. The frame material for other opening lights is termed a sash in the same sense that a sash in clothing is used to surround and support.

Vertically sliding sash window

During the seventeenth century the large casement window with two long, inward opening sashes, generally extending down to the floor, was developed in France. This French casement or French window was accepted and has remained the principal form of window on the continent of Europe.

At the same time the vertically sliding window, commonly known as a 'double hung sash window', was developed in England and became the common, singularly English window for all but small domestic or cottage windows. The earlier forms of this window operated by supporting the vertically sliding sashes in position by pegs fitted to holes in the side of the frame or by spring cams. The later method of hanging the sliding sashes was by means of ropes or chains over pulleys in the frame, connected to counter weights concealed inside the box frame of the window, as illustrated in Fig. 21.

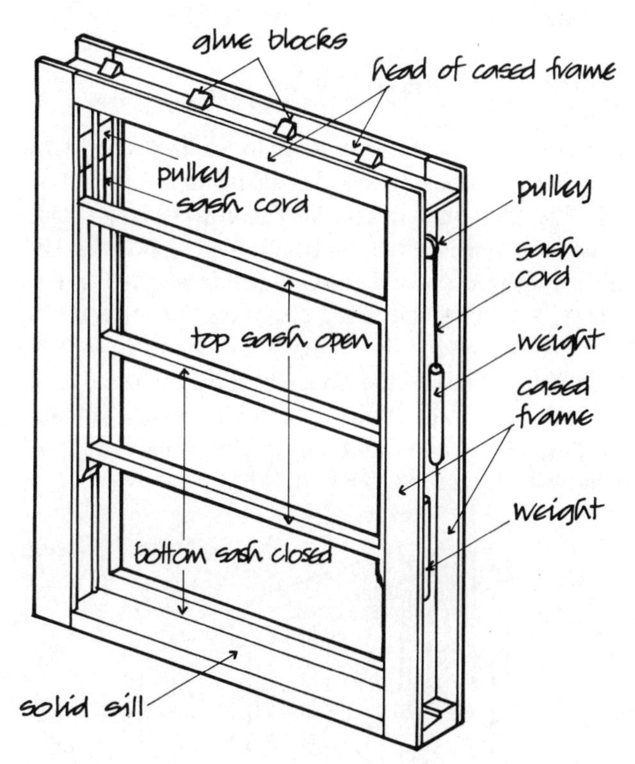

Fig. 21

The advantage of the vertically sliding sash is that as the weight of the sashes is hung vertically on ropes or chains, the sashes do not tend to distort and in consequence large sashes can be framed from small sections and large unobstructed areas of glass are possible. By setting the bulky box frame of these windows behind a rebate in the surrounding wall, the external appearance of the window is of a large area of glass framed in slim members.

Because of the sliding action, the sashes neither project into or out of the building and close control of ventilation is possible between a lower limit of a slight raising of a sash to allow some ventilation between the meeting rails to an upper limit of opening nearly half the window area. The sliding action facilitates the use of draught seals between sashes and frame.

The disadvantage of this window is that it is not easy to clean glass on both sides from inside the building. This difficulty has been overcome in recent window design in which it is possible to swing the sashes inwards for cleaning.

In time the traditional sash cords will fray and break and it is comparatively laborious to fit new ones. Sashes suspended in spring balances avoid this.

Horizontally sliding sash window

A traditional form of horizontally sliding wood window is that known as a Yorkshire light or cottage window, illustrated in Fig. 22. This crude form of small window comprised two timber-framed sashes that slid horizontally on wood runners inside a solid timber frame. As there had to be clearance for moving the sashes it was impossible to make this window weathertight and because of the tendency of the sashes to rack, i.e. move out of the vertical, they were liable to jam and be difficult to open and close. This simple form of window is little used today.

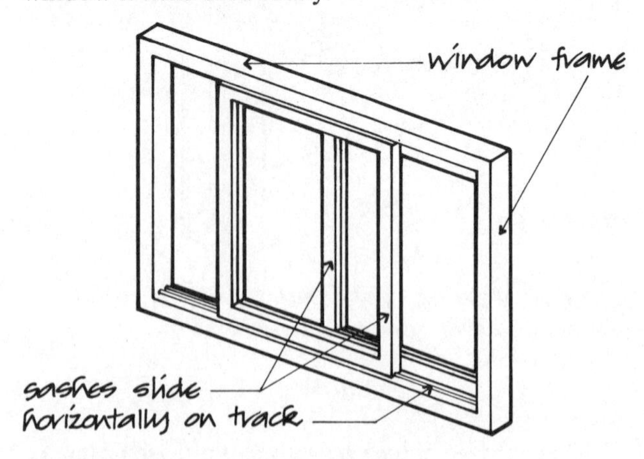

Fig. 22

The advantage of this type of window is that there are no internal or external projections from opening sashes and it can be opened to give close control of ventilation. It is difficult to clean the glass both sides from inside and the clearance required for movement of the sashes makes it difficult to weatherseal for conditions of severe exposure.

COMPOSITE ACTION WINDOWS

Projected windows

These composite action windows are designed to act like side-, top- or bottom-hung windows for normal ventilation purposes by opening on pivots which can be unlocked so that the pivots then slide in grooves in the frame and open on hinged side stays to facilitate cleaning, as illustrated in Fig. 23. Of the three methods of opening, the top-hung projected window has been the most popular.

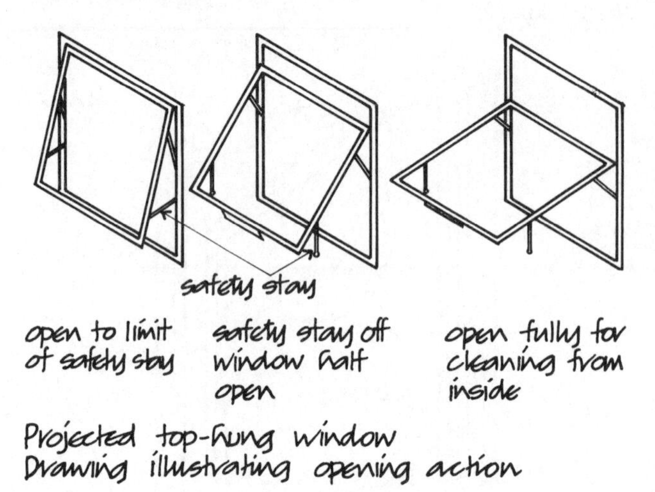

Fig. 23

To clean this window the sash is projected to a horizontal or near horizontal position to clean the outside glass. The person cleaning will need to bear some of his weight on the open horizontal sash to reach the extreme outside edge and because of this there have been some serious accidents due to the supporting pivots coming out of the grooves.

The projected top-hung window can be projected down from the top of the window to allow both top and bottom ventilation.

The advantage of cleaning both sides of these windows from inside, by means of a long-handled squeegee if the sash is of any considerable size, should be weighed against the likelihood of the complicated mechanism becoming fouled and not operating properly.

A variation of the top-hung projected sash is the projected awning window in which three or more sashes open as if top-hung and can be projected for ease of cleaning from inside. The three sashes are ganged to open and be projected together through operating levers. Because of the comparatively shallow sashes each can be cleaned in safety from inside by hand.

Tilt and turn window

This window, designed as a bottom-hung opening-in window for normal operation with limited opening for ventilation, that can be converted to an inward-opening casement for cleaning, has been used in high buildings. The operation of the window is illustrated in Fig. 24. The frame and sash can be made in wood, metal or plastic with sections similar to an ordinary opening-in hinged window but fitted with a handle that locks the side hinge pins enabling the window to be opened for cleaning.

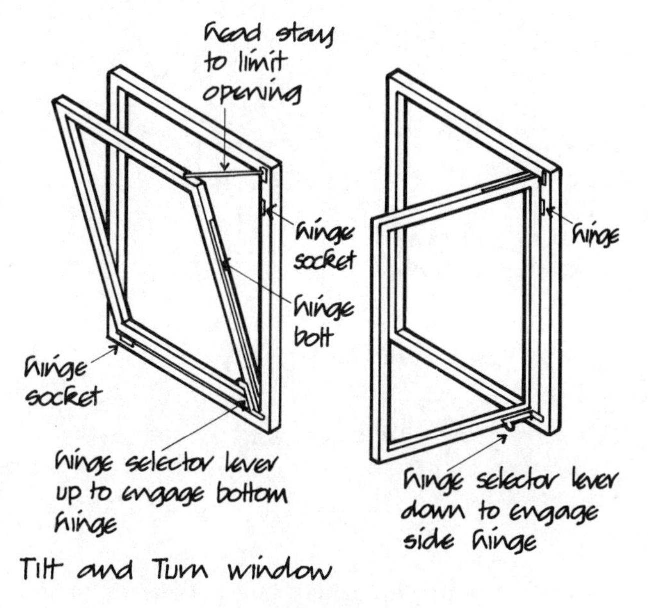

Fig. 24

Sliding folding windows

The sashes in this type of opening window are hinged to each other and fold horizontally in concertina fashion to one or both sides of the window to provide a clear unobstructed opening as illustrated in Fig. 25. This opening light system is used as either a horizontal window or fully glazed doors where indoor and outdoor areas can be combined in fine weather. Both steel and aluminium windows and doors are manufactured, the folding action being through a wheeled overhead track that carries the weight of the windows and a bottom wheeled track to guide the opening movement.

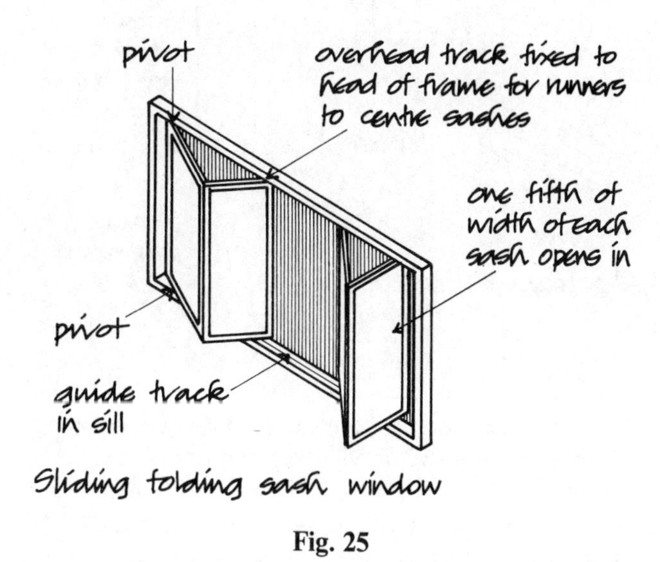

Fig. 25

WINDOW FRAMING

Wood casement windows

To provide adequate strength and stiffness in the frame, casements and ventlights of casement windows, and to accommodate rebates for casements and ventlights and for glazing, timber of adequate section has to be used and joined. The traditional joint used is the mortice and tenon joint in which a protruding tenon, cut on the end of one section, fits into a matching mortice on the other, the joint being made secure with glue and wedges as illustrated in Fig. 26. This traditional wood jointing technique which was, and still is to a considerable extent, formed with hand tools makes a strong joint adequate for frames and casements. More recently the casements and frames of mass-produced wood windows are joined with the combed joint illustrated in Fig. 27, which consists of interlocking tongues cut on the ends of members which are put together, glued and pinned. With the use of modern glue techniques this joint is as strong as a mortice and tenon joint. The combed joint is used in mass-produced windows as it can more rapidly be cut and assembled by woodworking and assembly machines than a mortice and tenon joint.

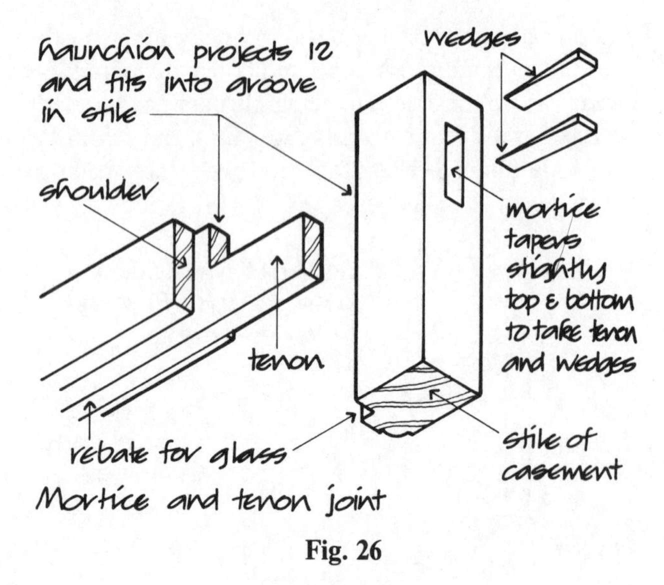

Fig. 26

Joinery The craft of accurately cutting and joining the timber members of windows and doors is termed joinery and those who practise it are called joiners. For centuries the joiner's craft was executed with hand tools used for preparing, cutting and assembling timbers. A mortice and tenon joint can readily be cut and made by skilled joiners and as it very rigidly joins timbers it was the joint always used in framing the members of windows and doors up to some seventy years ago.

During the present century woodworking machinery has been increasingly used to prepare, cut and assemble windows and doors so that today standard windows and doors are machine made.

The skilled joiner can quickly cut and assemble a mortice and tenon joint but the time taken by machinery to cut and assemble this joint is greater than that required to cut and assemble a combed or a dowelled joint. In consequence mortice and tenon joints are used less than they were.

It is usual to specify the sizes of timber for joinery for windows, doors and frames as being ex. 100 × 75, for example. The description 'ex.' denotes that the member is to be cut from a rough sawn timber size 100 × 75, which after being planed on all four faces would be about 95 × 70 finished size. This sytem of specifying the sawn sizes of members is used when joinery is to be prepared by hand operated tools or machines so that the member may be wrought or planed down to a good surface finish without limitation of a precise finished size, yet maintaining the specified size of window. Where joinery is wrought or planed by machine it is practice to specify the precise finished size of each member as this is the dimension the operator needs to

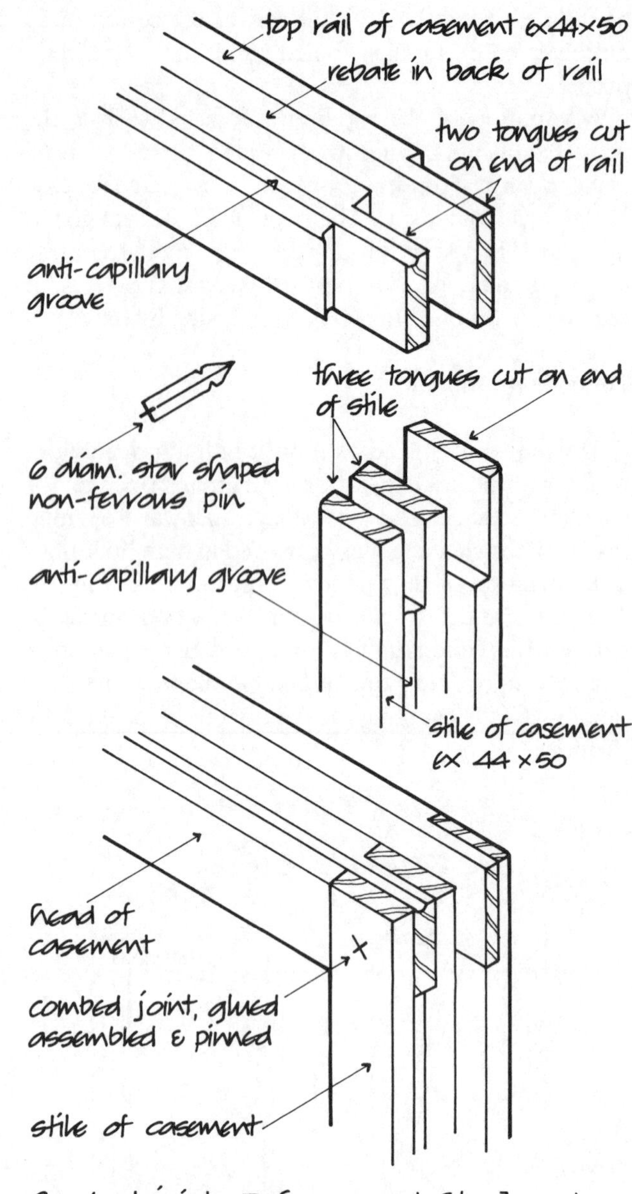

Fig. 27

know when setting up the machine and it is up to him to select the size of sawn timber to be used to produce the finished size.

Fig. 28 illustrates the arrangement of the parts of a wood casement window, the members of the frames, casements and ventlights being joined with mortice and tenon joints. It will be seen that casements and ventlights fit into rebates cut in the members of the frame. These rebates, which are usually 13 deep, serve as a check to wind and rain in normal positions of exposure.

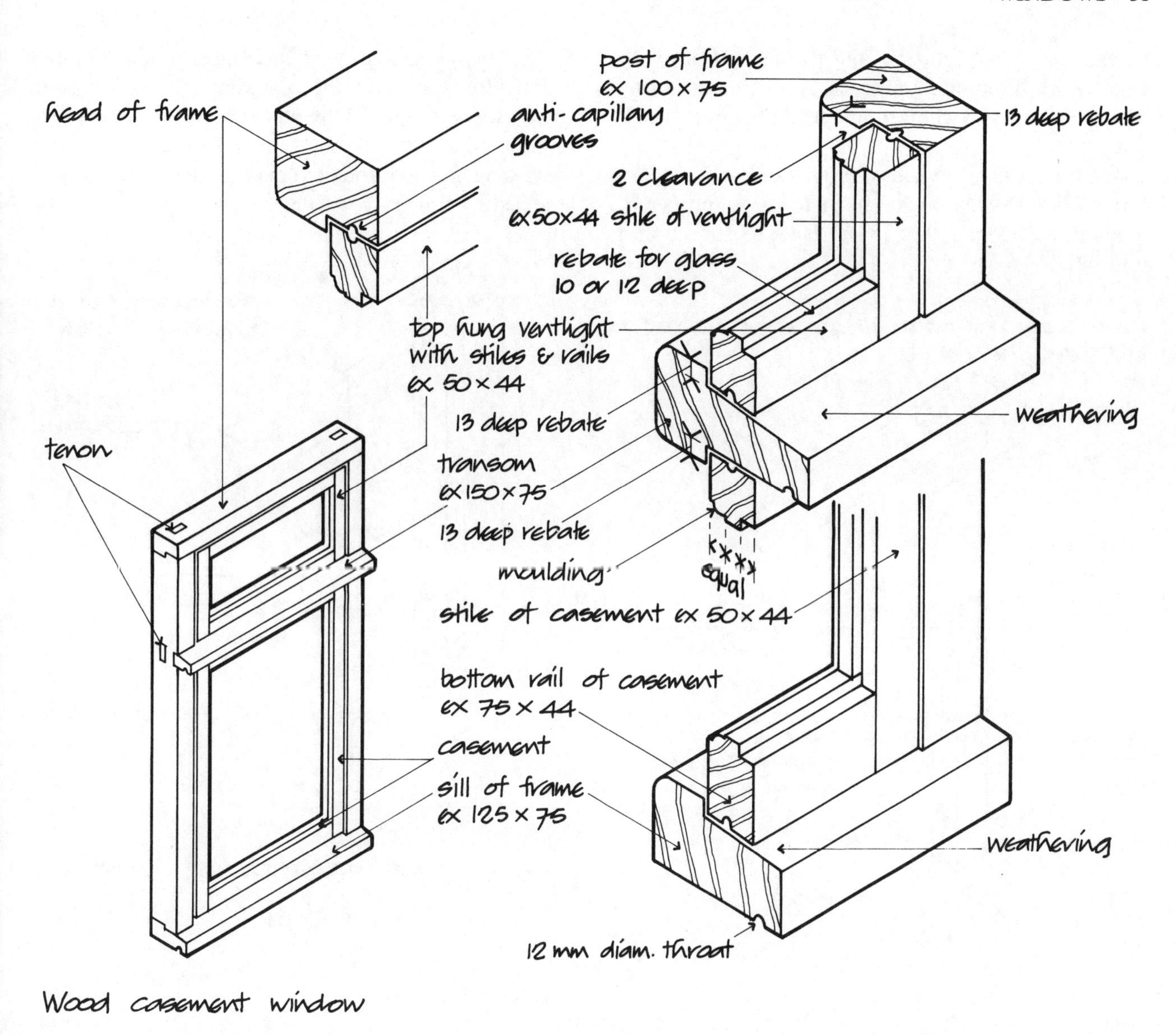

Fig. 28

Window frame

A casement window frame consists of a head, two posts (or jambs) and a sill joined with mortice and tenon joints, together with one or more mullions and a transom, depending on the number of casements and ventlights.

The members of the frame are joined with wedged mortice and tenon joints as illustrated in Fig. 29. The posts (jambs) of the frame are tenoned to the head and sill with the ends of the sill and head projecting some 40 or more each side of the frame as horns. These projecting horns can be built into the wall in the jambs of openings or they may be cut off on site if the frame is built in flush with the outside of the wall. The reason for using a haunched tenon joint between posts and head is so that when the horn is cut off there will still be a complete mortice and tenon left.

It will be seen from Fig. 29 that one face of the tenon is cut in line with the rebate for the casement. It is usual practice in joinery to cut one or both faces of tenons in line with rebates or mouldings to keep the number of faces cut across the grain to a minimum. The mortice and tenon joints are put together in glue, cramped up and wedged.

When there is a transom in the frame it is joined to the posts by means of tenons fitted and wedged to mortices. Mullions are joined to head and sill with tenons wedged to mortices and to the transom with stub tenons fitted into a mortice. A stub tenon is one which does not go right through the timber in to which it is fitted.

The members of a wood window frame are cut from 100 × 75 or 75 × 50 sawn timbers for head, posts and mullion and from 150 × 75 or 100 × 63 for sill and transom.

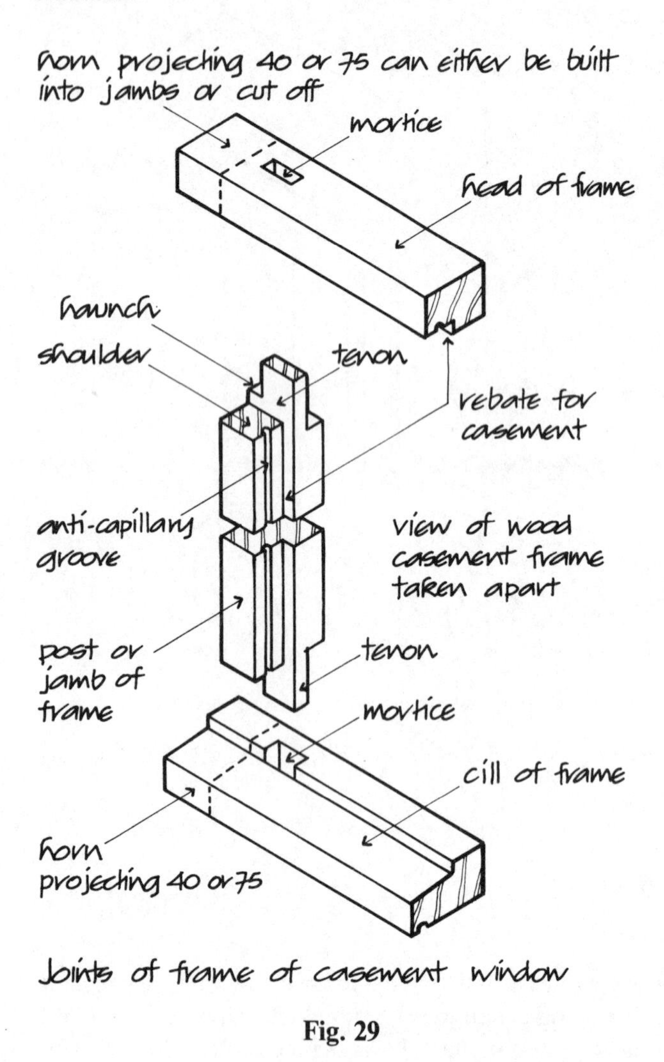

Fig. 29

Casement

The four members of the casement are two stiles, top rail and bottom rail. The stiles and top rail are cut from 50 × 44 timber and the bottom rail from 75 × 44 timber. The stiles and rails are rebated for glass and rounded or moulded on their inside edges for appearance sake. The rails are tenoned to mortices in the stiles and put together in glue, cramped up and wedged. Fig. 30 is an illustration of the joints taken apart.

The tenons are cut with their faces in line with the rebate for glass and the moulding, for the reasons previously explained. Obviously the tenons on the rails cannot be as deep as the rails if they are to fit into enclosing mortices and their depth is usually about half the depth of the rails on which they are cut.

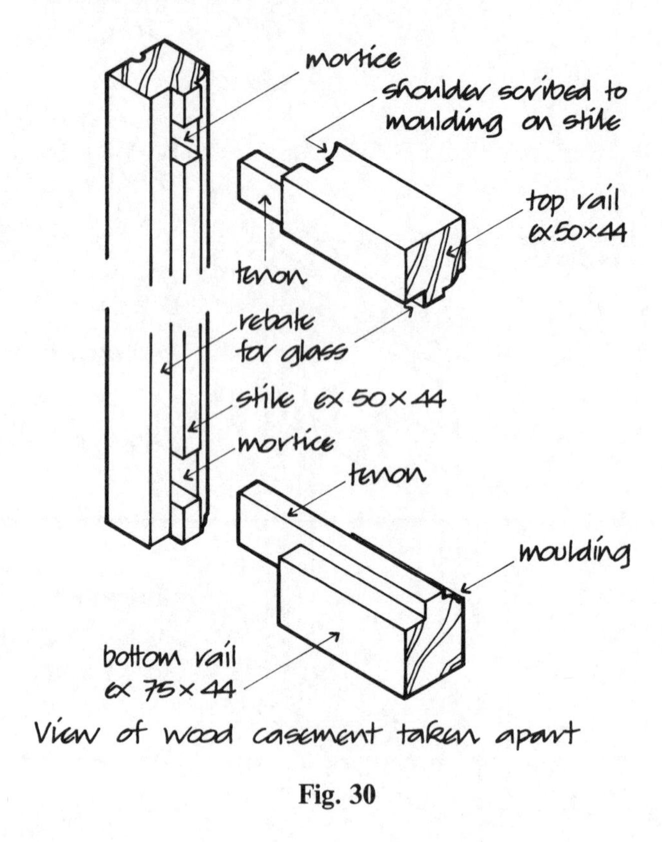

Fig. 30

Ventlight

The four members of the ventlight are cut from timbers the same size as the stiles of the casement and are rebated, moulded and joined in the same way as for the casement.

Standard wood casement

The manufacturers of wood windows produce a range of standard windows; standard sizes and designs are offered, the advantage being the economy of mass production. In line with the move to dimensionally co-ordinate building components and assemblies the standard range of windows is designed to fit basic spaces with such allowances for tolerances and joints as appropriate. The purpose of dimensional co-ordination is to rationalise the production of building components and assemblies through the standardisation of sizes within a framework of basic spaces into which the standard components and assemblies may fit. The difficulty has been to adapt this factory assembly technique to conditions on the average building site, without recourse to cutting components on site to fit and the use of gap-filling or gap-covering materials. This difficulty has yet to be overcome because of the deep-rooted tradition in building of roughly putting together and cutting and filling to make a fit.

The casements and ventlights are cut so that their edges lip over the outside faces of the frame by means of a rebate in their edges, as illustrated in Fig. 31. These lipped edges are in addition to the rebate in the frame so that there are two checks to the entry of wind and rain between opening lights and the frame. The members of the frame and of the opening lights may be joined with mortice and tenon or combed joints. A combed joint is commonly used to join the members of standard wood window opening lights as illustrated in Fig. 27. The joint is put together with the faces of the combs or tongues coated with glue and a metal or wood dowel

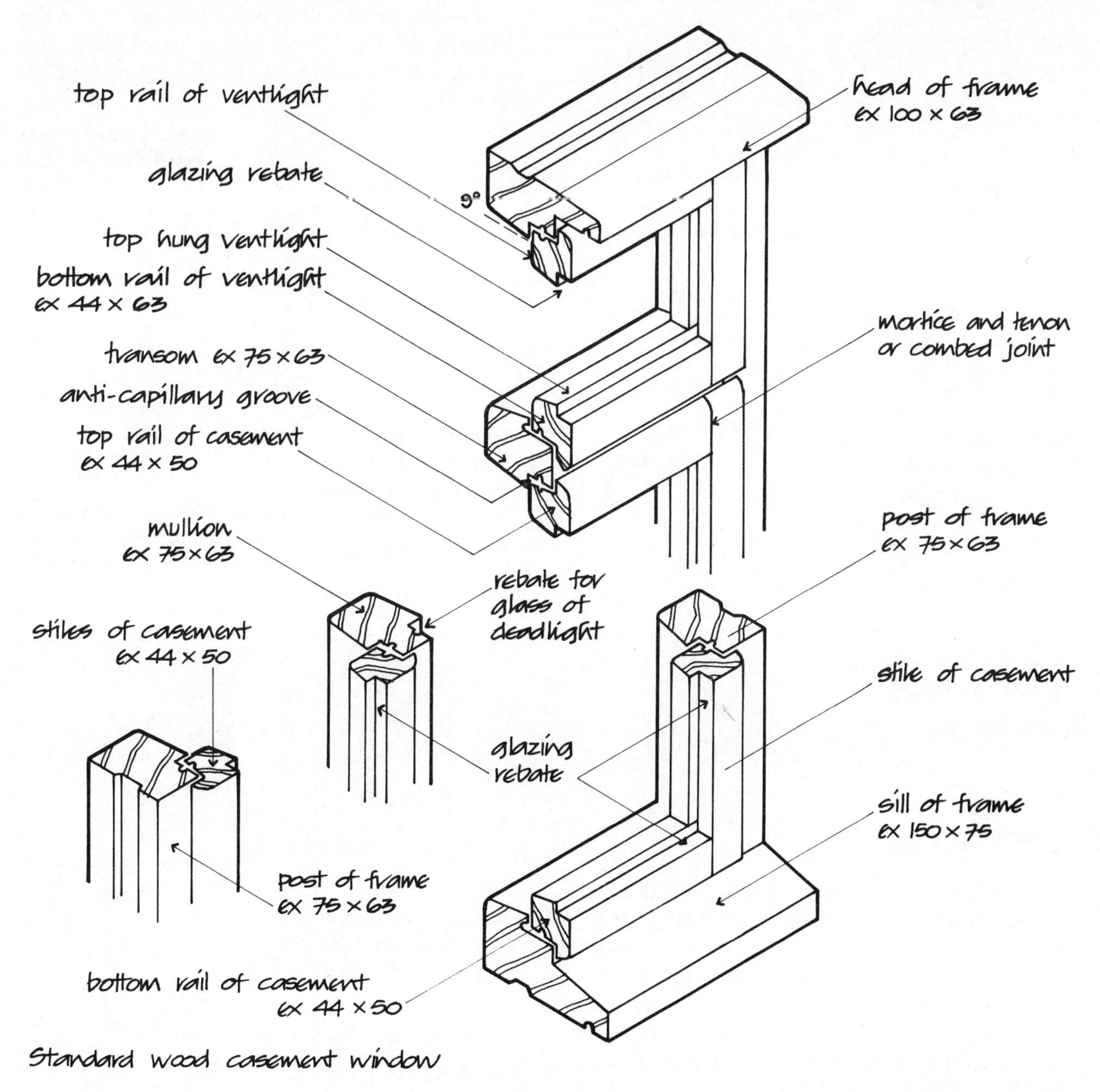

Fig. 31

pin is driven through the joint. If accurately cut and properly put together this joint is adequate for the opening lights and frame of casement windows.

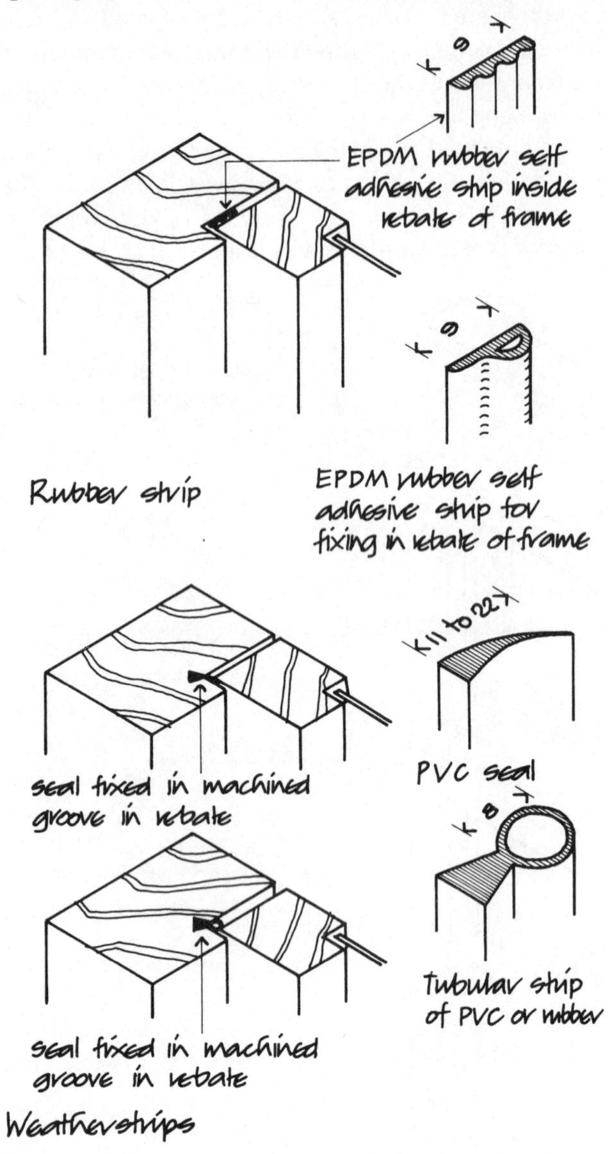

Fig. 32

Weatherstripping

In sheltered positions the outward facing rebate in the frame into which opening lights close will generally prevent rain penetration. The rebate will not, however, prevent draughts of cold air being blown through the clearance gaps around opening lights by wind pressure. To minimise cold draughts and to act as a seal against wind driven rain in all positions of exposure, it is practice today to fit weatherstripping to all opening lights of new and old windows. The two forms of weatherstripping that are commonly used are a flexible bulb or strip of rubber, synthetic rubber or plastic which is compressed between the frame and opening light, or a strip of nylon filament pile between the frame and opening light.

For maximum effect these seals should be fitted or fixed on the back face of the rebate or the inner face of the frame so that the rebate acts as a first defence against driven wind and rain.

The elastomeric seals illustrated in Fig. 32 are stuck or fitted to a groove ready-cut in the rebate of the frame of new windows, and the plastic strip illustrated in Fig. 32 is tacked to the frame of existing windows. The nylon filament illustrated in Fig. 33 is for use with old windows.

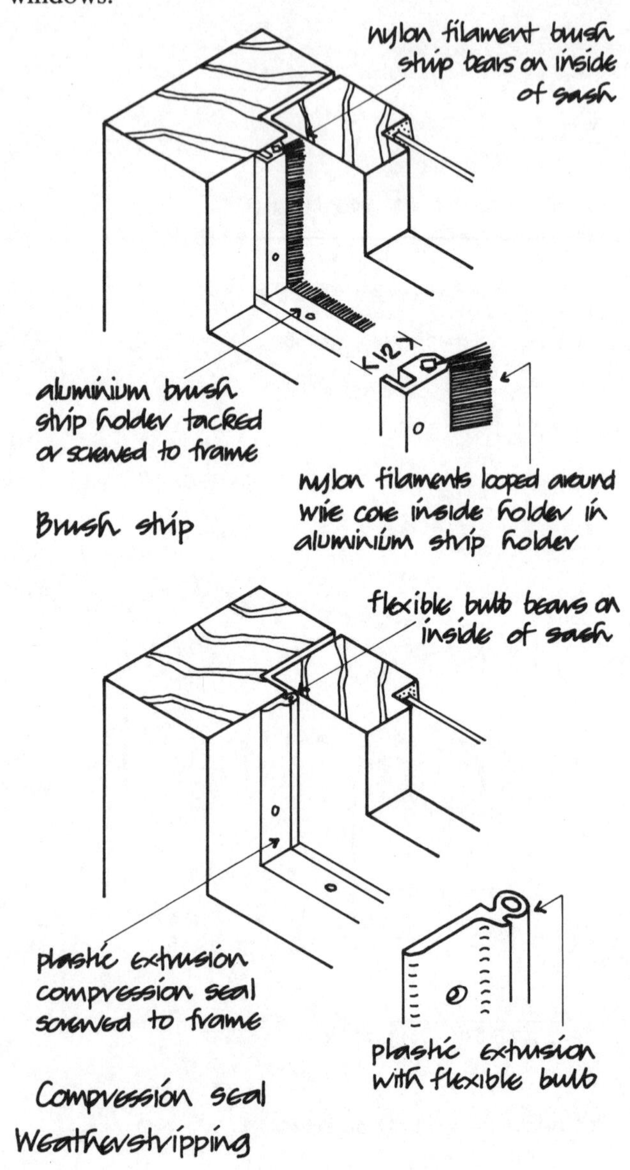

Fig. 33

Hardware

Hinges, fasteners and stays Wood casements, ventlights and sashes are hung on a pair of pressed steel butt hinges similar to those used for doors. To inhibit rusting the hinges are galvanised and finished with a lacquer coating. As an alternative, metal offset hinges may be used for casements, in which the pin is offset outside the casement so that when the casement is open there is a gap between the hinged edge of the casement and the frame sufficient to allow for cleaning the outside of glass from inside the building.

To close casements and sashes a casement or window fastener or latch is fitted halfway up the height of the opening light. The handle of these fasteners is raised to release the latch to open the window. As a security these fasteners are supplied with a loose key to operate a lock to lock the opening light shut. Fasteners are made of cast zinc, aluminium or steel, usually finished with a protective coating of anodising, powder or liquid organic coating. Fig. 34 shows a lockable casement fastener.

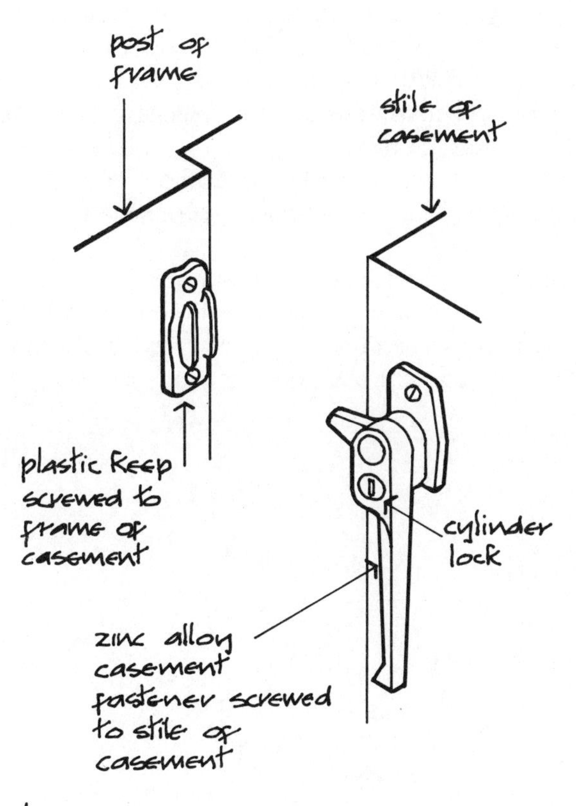

Fig. 34

To maintain casements and ventlights in an open position, casement stays are fitted to the bottom rail of casements and ventlights (Fig. 35). To open the light the stay is unclipped from the catch and engaged with a peg that maintains the light in a predetermined number of open positions. For security these stays are supplied with a loose key that can lock the stay shut in the closed position.

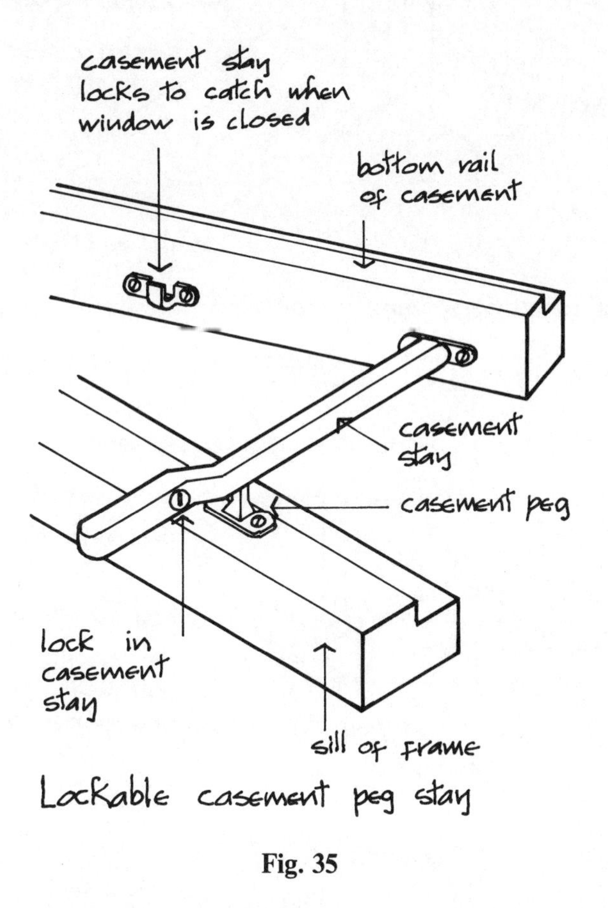

Fig. 35

Fixing windows

Wood window frames are usually 'built-in' to solid walls as the walls are raised. The term 'built-in' describes the operation of fixing the window in position and then building the wall around it. In this way there is a good fit of the wall to the window. The alternative method is to 'fix' the window in position after the wall is built which requires some care in building to ensure that there is sufficient clearance for the fitting of the window in position. Most softwood joinery windows are built-in as the window will be masked by subsequent painting. Hardwood windows are often fixed-in to avoid damage to the wood surface, which will not be covered by paint.

Wood window frames are secured in position in solid walls by means of galvanised steel cramps or lugs that are screwed to the back of the frame and built into horizontal brick or block courses as the wall is raised, as illustrated in Fig. 36. Where the back of the frame coincides with a cavity in a wall, a combined cavity closer and ties may be used, as illustrated in Figs. 36 and 37. One cramp, lug or tie is used for each 300 or 450 of height of window each side of the frame. Where wood frames are fixed-in, it is necessary to either leave pockets in the jambs of the wall into which the window lugs can be fitted and the walling then made up in the jambs of the opening, or to screw the frame to plugs in the jambs of the opening.

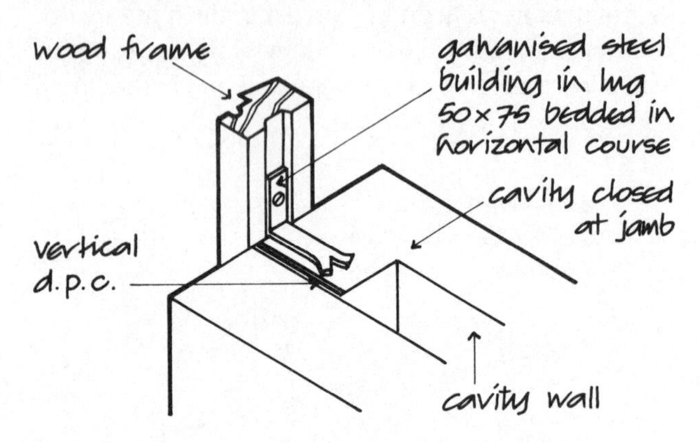

Fixing wood window or door frame

Fig. 36

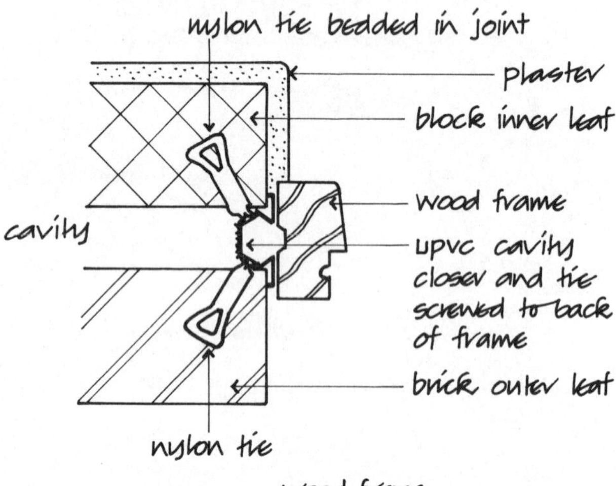

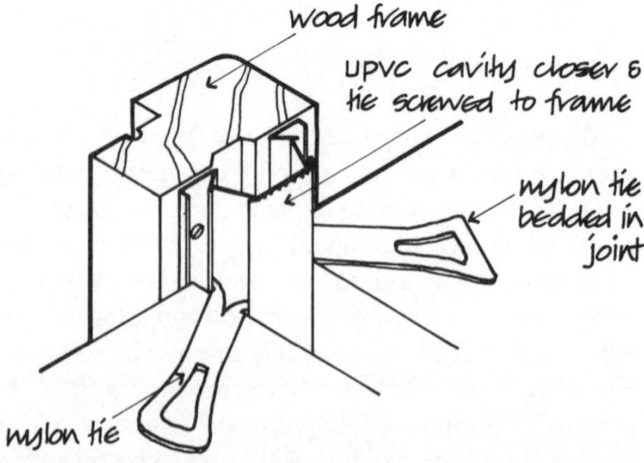

Upvc cavity closer and ties for wood window or door frames

Fig. 37

Perimeter sealing to wood windows Most wood window frames are bedded in mortar as the frame is built-in, with the mortar pointed as perimeter seal.

Steel casement windows

The steel section window has lost favour principally because of the ill repute it gained from rapid deterioration by rusting before the introduction of hot-dip galvanising in the 1940s. The strong slender sections of this type of window were at one time considered its most attractive feature. Changes in fashion mean that the steel window does not have the popularity it enjoyed between 1930 and 1950. The disadvantages of the steel window are that the small section will not comfortably accommodate the thickness of double glazing, and the material being a good conductor of heat acts as a cold bridge to transfer of heat and encourages condensation which in turn may encourage rusting.

Steel casement windows are made either of the standard Z section hot-rolled steel illustrated in Fig. 38, or the universal section illustrated in Fig. 39. The latter section is made with channels to take weatherstripping.

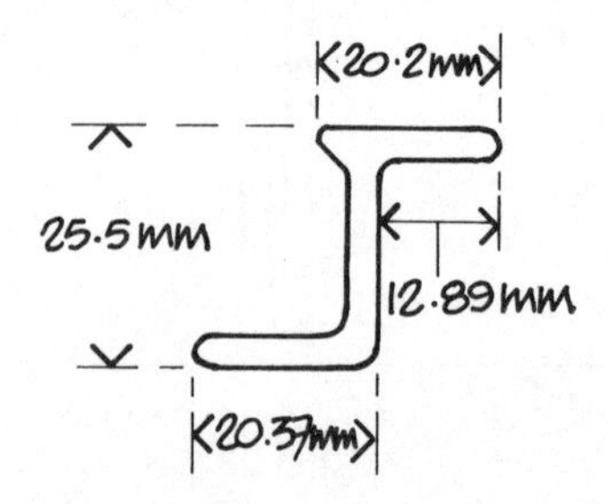

Standard section for frame, casements and ventlights of standard metal windows

Fig. 38

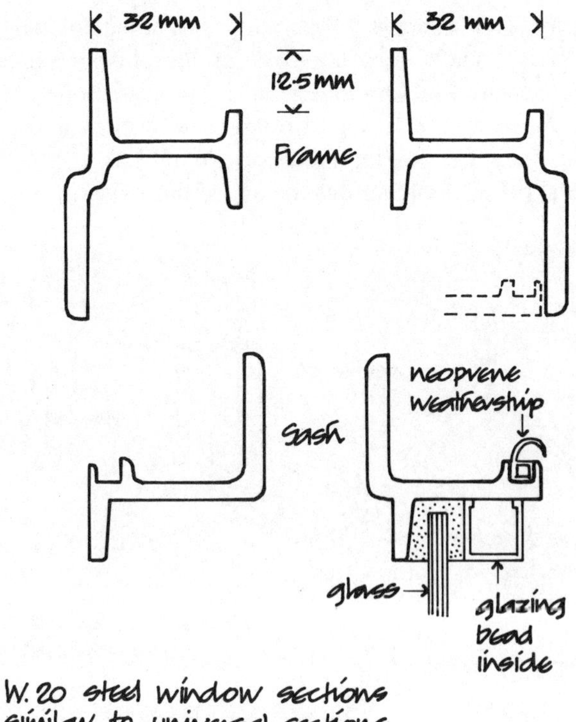

Fig. 39

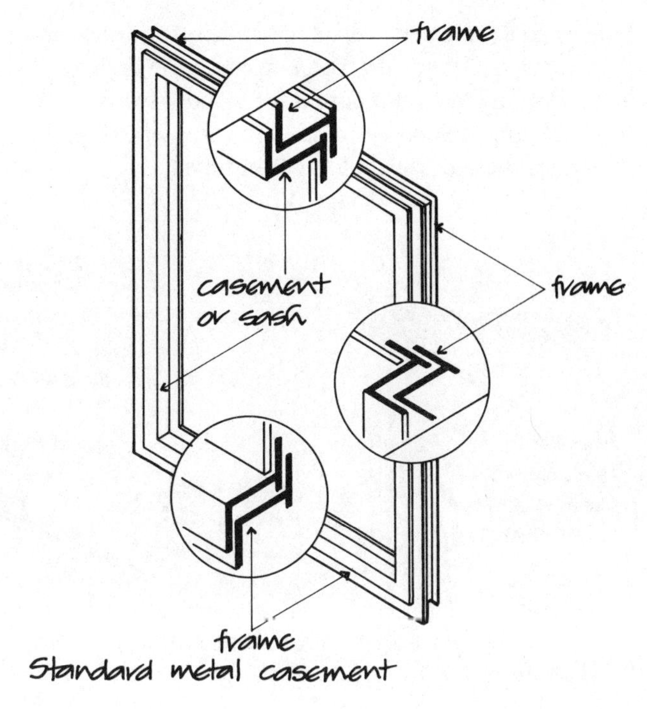

Fig. 40

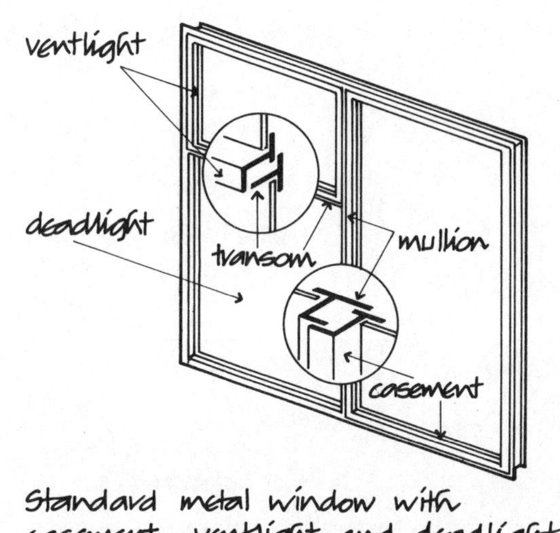

Fig. 41

Standard steel casements are made from the hot-rolled steel Z section which is used both for the frame, casements and ventlights. The section is cut to length and mitred and welded at the corners. The assembled and cleaned parts of the window are then 'rustproofed' by the hot-dip galvanising process in which the window parts are dipped in a bath of molten zinc. The zinc adheres strongly to the steel in the form of a thin coating which protects the steel from rusting. This protective coating will be effective for many years. The casement and frame sections fit together as illustrated in Fig. 40. Where there are two casements, or a casement with a dead light, a mullion section is welded into the frame, and with ventlights a transom section is welded into the frame as illustrated in Fig. 41.

As an alternative to the standard Z section the universal or W20 steel section is often used. These heavier sections, illustrated in Fig. 39, may be used for casements and are generally used for pivoted windows where the greater variety of section available is of advantage in making up the more complicated requirements of frame and sash sections, for the fixing of weatherstripping and to accommodate double glazing.

Steel section windows can be finished with a polyester powder coating to provide additional protection against corrosion and as a decorative finish. The galvanised windows are cleaned, chromated and electrostatically coated with polyester powder which is then stoved at a temperature of 200°C. The usual colour of this coating is white, but black, red and Van Dyke brown are also used. The powder coating should not require painting for many years.

Hinges and fasteners Steel casement windows are fitted with steel butt or offset hinges and lever catches and stays similar to those used for wood windows, the fittings being welded to frame and casement. Fig. 42 illustrates these hinges, catches and stays.

Fixing steel windows Standard steel casement windows are usually built in to openings in solid walls and secured with building-in lugs or ties that are bolted to the back of the fitting through a slot that allows adjustment for building into horizontal brick or block courses. Fig. 43 illustrates the use of these fittings.

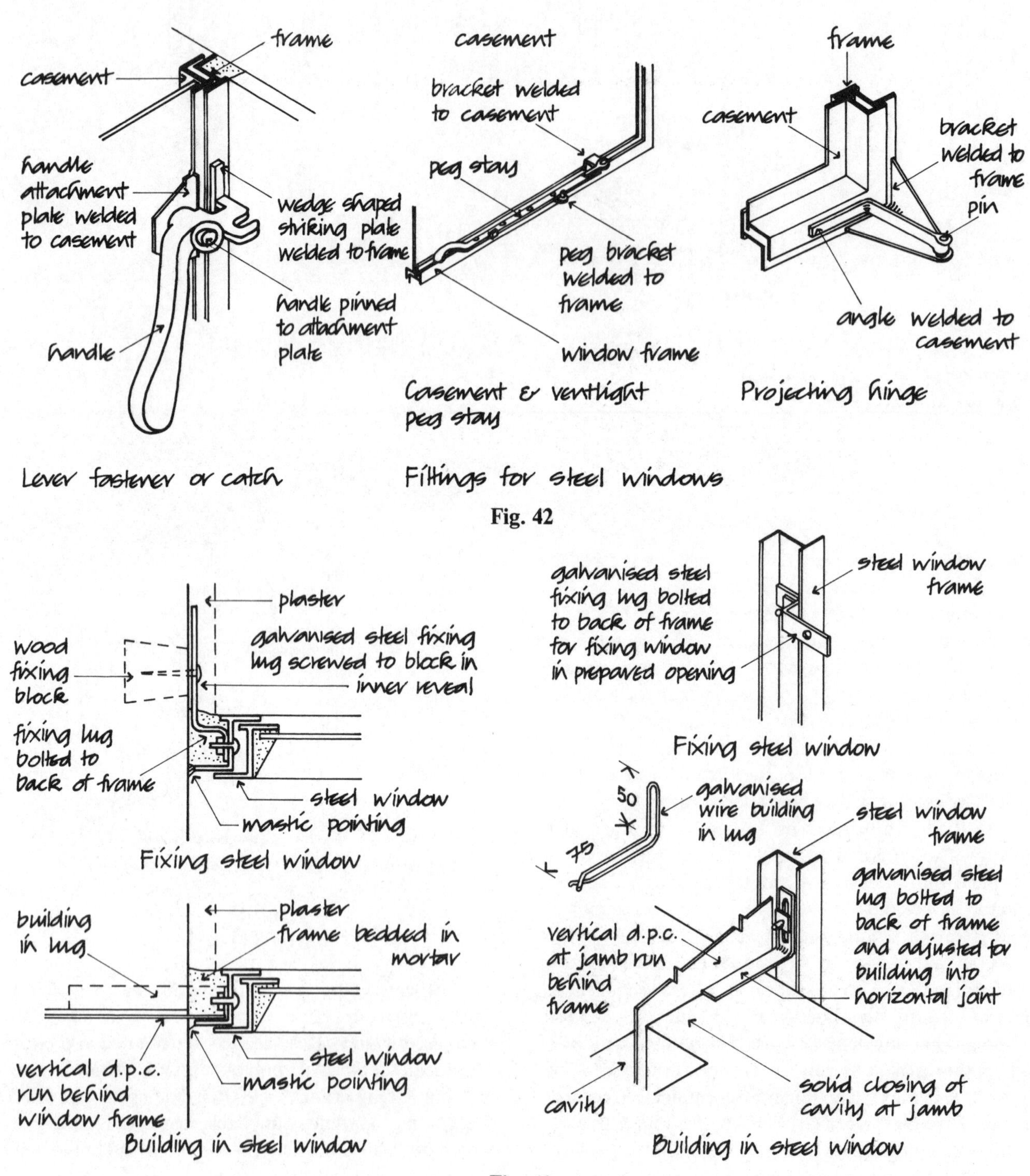

Fig. 42

Fig. 43

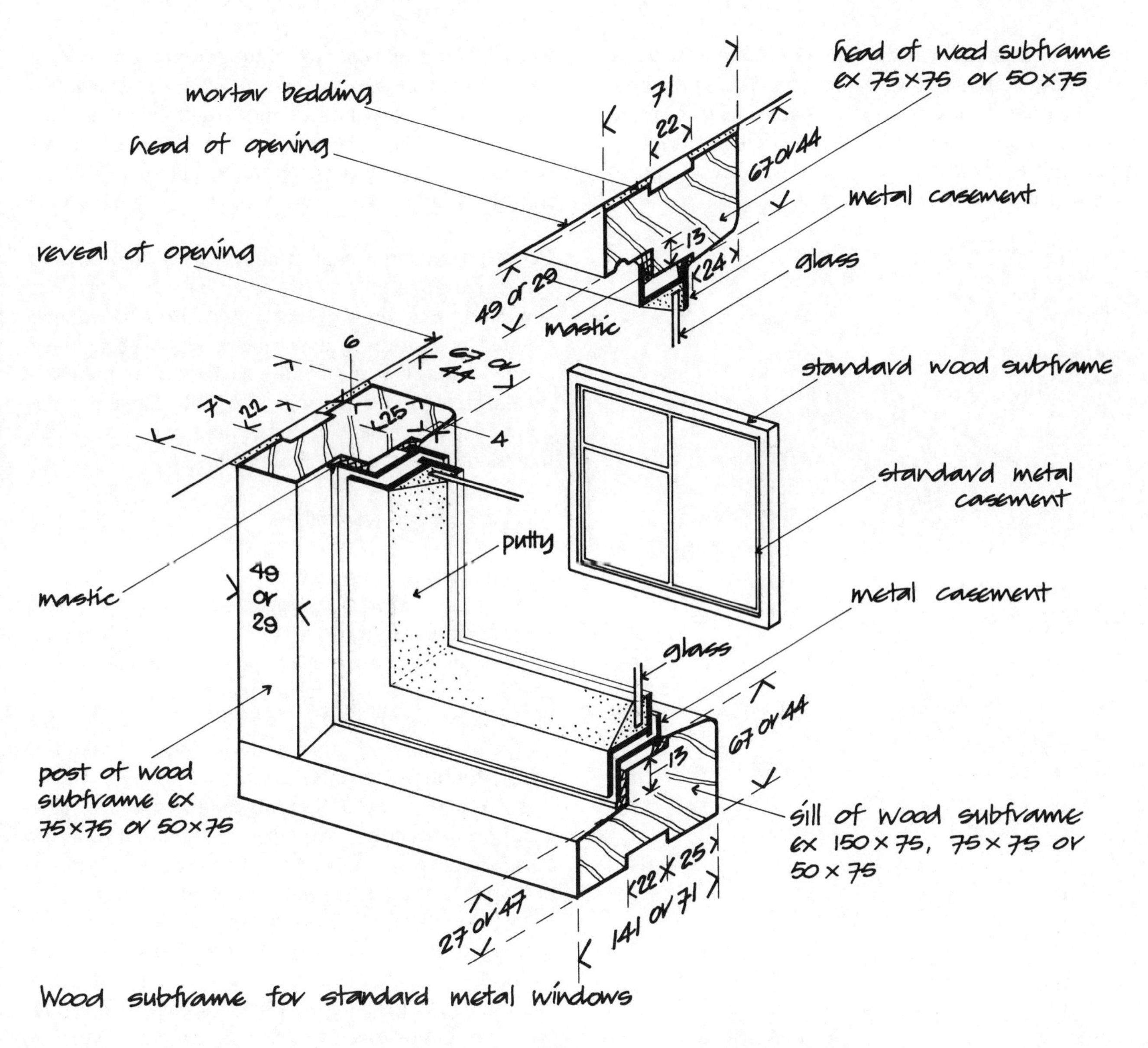

Fig. 44

Timber sub-frame for steel windows The sections used for steel windows are slender and may be damaged in transit or during handling on site. The manufacturers of these windows will supply wood surrounds for their steel windows to give added strength and rigidity or to provide a more substantial surround to the window for appearance sake. The wood surrounds are usually cut from 75 × 75 or 50 × 75 softwood timber which is wrought (planed smooth), rebated and joined with mortice and tenon or dowelled joints solidly glued. Fig. 44 is an illustration of the section of a wood surround with the steel window in position. The steel window is secured to the wood surround with countersunk headed wood screws driven through holes in the steel frame into the wood sub-frame. Mastic is packed between the steel frame and the timber sub-frame to exclude rain. The two rebates cut in the wood sub-frame are so spaced that they accommodate the flanges of the steel frame. The wood surround is secured with L-shaped lugs or ties that are screwed to the back of the frame and built into horizontal joints of brick or blockwork.

uPVC sub-frames for steel windows As an alternative to wood sub-frames, hollow section uPVC sub-frames may be used. The hollow sections are similar in size to standard wood sub-frames. The uPVC sub-frames are fixed with straps at head and sill and by direct fixing to jambs as illustrated in Fig. 45.

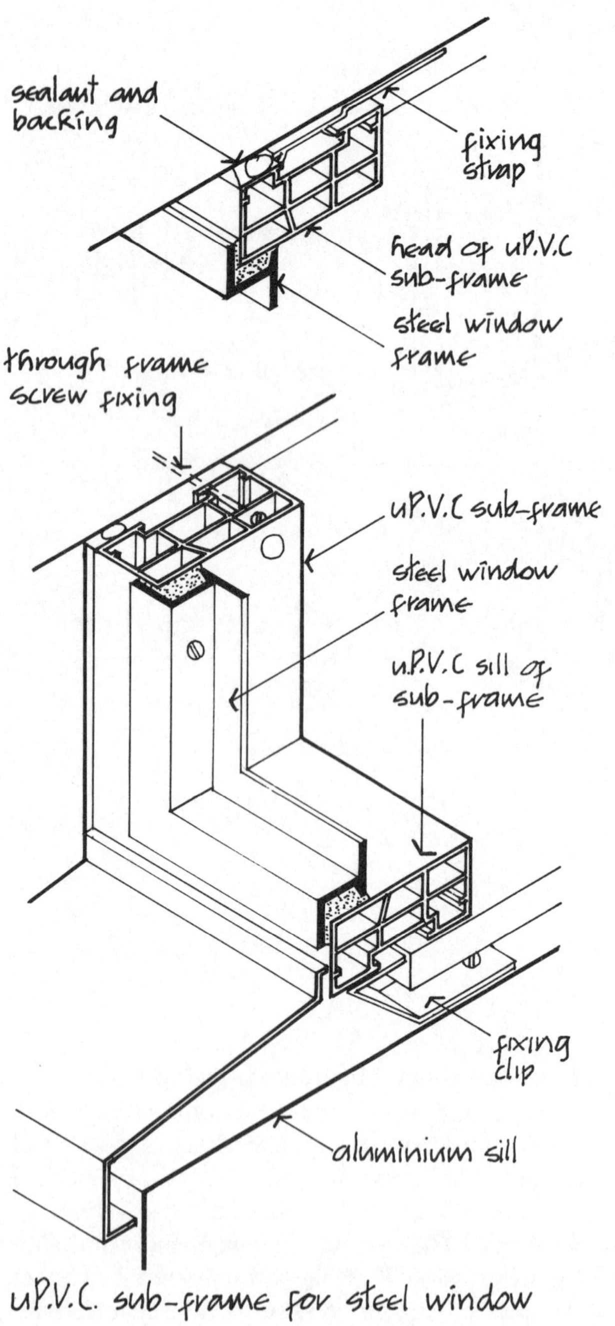

Fig. 45

Sub-sills for steel windows Steel, aluminium or PVC sub-sills may be used with steel windows. The galvanised, pressed steel and extruded aluminium sill sections, which may be organic powder coated, are designed to overlap the external wall face as a drip, as illustrated in Fig. 45.

Perimeter sealing around frames Steel windows and sub-frames for steel windows are usually built-in as the surrounding walls are raised and are bedded and pointed in mortar as a perimeter seal. Where these windows and their sub-frames are fixed-in to prepared openings, so that coatings are not damaged, the perimeter seal is made with a sealant in the same way that aluminium windows are finished.

Aluminium casement windows

Aluminium windows were originally made as a substitute for hot rolled steel section windows, in small sections similar to those of steel. Aluminium windows of small solid sections are less used now than they were, partly due to changing fashion and more particularly because the small section, which acts as a thermal bridge to encourage condensation, does not take the wider, doubled glazed, IG units in use today.

The majority of aluminium windows that are made today are of sections extruded from aluminium alloy in a wide range of channel and box sections with grooves and lips for weatherstripping and double glazing.

In the extrusion process through which these window sections are formed, molten metal is forced through a die as thin sections of material. The extrusion process allows for a wide range of sections which are more straightforward to vary than comparable uPVC sections. This is a particular advantage of aluminium as a window material where special sections are required.

The sections are mitre cut and mechanically cleated or screwed at joints which are sealed against entry of water as illustrated in Fig. 46. The fabricated frames and opening lights are then given a protective coating by anodising, polyester powder or liquid organic coating.

Anodised finish An aluminium oxide coating is formed by the sulphuric acid electrolytic method. Metal oxides are deposited on the surface of the aluminium as an alternating current is applied across the bath of sulphuric acid. A limited range of colours can be produced by various electrolytes. The usual range of colours is from silver grey through bronze to black. The

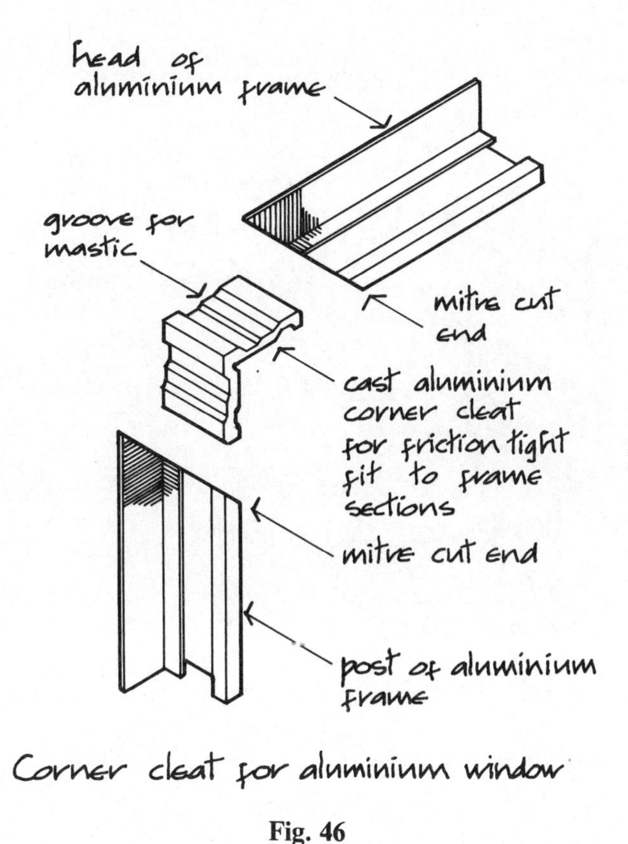

Fig. 46

anodised finish is covered with a clear, thin coating of lacquer to protect it against alkaline materials used in building operations.

In the past some anodised finishes have failed due to the finish being damaged by building operations or other causes, and unsightly corrosion of the aluminium below has occurred and spread. Because of this and the limited colour range, anodised finish is less popular.

Polyester powder organic coatings After the window frames and opening lights have been cleaned, etched and chromate conversion coat treated, the polyester powder is sprayed on to the windows. The coating is then stoved at 200°C when the powder flows and fuses to form a durable coating.

Liquid organic, acrylic coating After pre-treatment the same as that for the powder coating, the acrylic is applied in liquid form by electrophoretic dip for white or by electrostatic spray for colours. The liquid finish is then stoved at 200°C to form a hard, smooth durable coating.

A range of bright colours is practical with both the polyester powder and the liquid acrylic coatings, from white through blue, green, red and black. These organic finishes provide a decorative, protective, durable coating that requires only occasional cleaning with water to remove grime. In time the stronger colours may bleach due to the effect of ultra-violet light.

A disadvantage of aluminium as a window material is that it is a good conductor of heat and in consequence moisture vapour in warm air will condense to water on the cold inner surfaces of these window sections in periods of cold outside temperature. In most rooms this condensation on the comparatively slim sections will be merely a nuisance. In rooms such as kitchens, where air will be heavily saturated with moisture vapour, the condensation may spoil decorations and affect the seal to double glazing. Here it is advantageous to use the 'thermal break' window construction illustrated in Fig. 47. Separate aluminium window sections are mechanically linked to the main window sections through plastic thermal break sections. The IG, double glazed units are secured with aluminium beads and the window is weatherstripped with preformed synthetic rubber seals.

Hardware of hinges, lockable casement fasteners and stays is made of anodised finish, cast aluminium or die-cast zinc alloy, chromium plated.

Fixing The aluminium frame is secured to the surrounding wall by aluminium lugs that clip to the back of the frame at centres of up to 600 and also adjacent to hinges and fasteners, with the lugs screwed to plugs in the wall.

Perimeter sealing around frames Aluminium frames may be built-in as walling is raised and bedded and pointed in mortar, but usually these windows are fixed-in to prepared openings so that the window and its finish are not damaged while walls are being built. To make a weathertight joint around the perimeter of the window frame it is necessary to use a sealant.

A sealant is a material that is initially sufficiently liquid or plastic for application and which cures or changes to a material that will adhere to surrounding surfaces, retain its shape and accommodate some small movement without loss of seal against wind and rain.

Sealants used for sealing perimeter joints around window frames are classed as plastoelastic, elastoplastic or elastic. Plastoelastic sealants, which have some elastic property, remain predominantly plastic and can be moulded. Elastoplastic sealants, which develop predominantly elastic properties as they cure, will

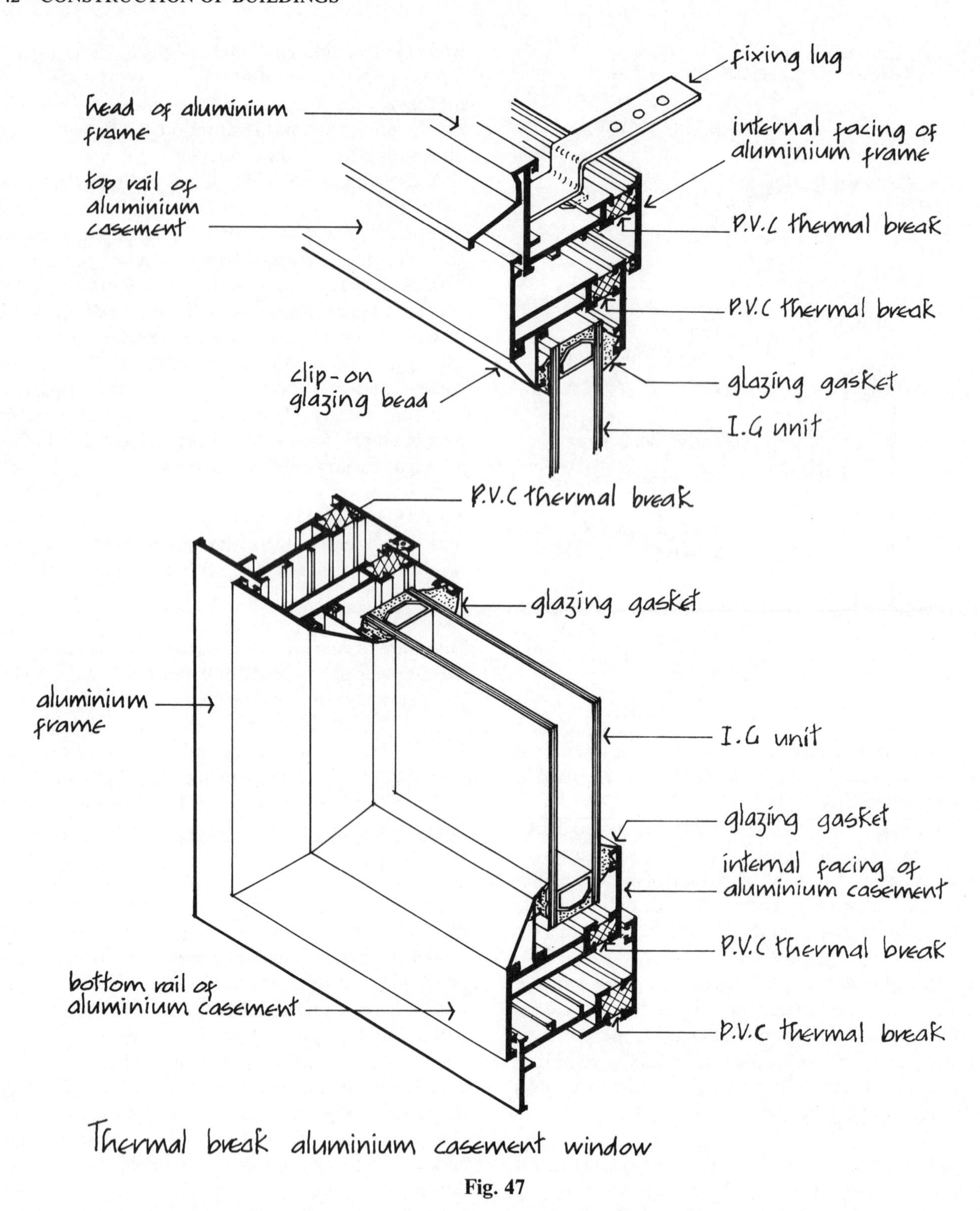

Fig. 47

return to their former shape when stress is removed and also retain some plastic property when stressed over long periods. Elastic sealants will, after curing, have predominantly elastic properties in that they will continue to resume their former shape once stress is removed, during the anticipated useful life of the material.

The materials that may be used for perimeter sealing around window and door frames are acrylic, polysulphide, polyurethane and silicone. Of these acrylic is classed as plastoelastic, polysulphide as elastoplastic and polyurethane and silicone as elastic. In general the plastoelastic material is easier to use because of its predominantly plastic nature, but it will not form so tough and elastic a surface as elastoplastic materials that have some plastic property. The elastic materials need some experience and skill in use for successful application.

Polysulphide and polyurethane sealants are produced as either one part sealants ready to use, or as two part sealants which have to be mixed before use. The one part sealants are more straightforward to use as there is no mixing and the material cures or loses plasticity fairly slowly, allowing adequate time for running into joints and compacting by tooling. The two part sealants require careful, thorough mixing and as they cure fairly rapidly require skill in application. The advantage of the two part sealants is that as they cure fairly rapidly they are less likely to slump and lose shape and adhesion than the more slow curing one-part sealants. Silicone sealants which cure fairly rapidly to form a tough, elastic material require rapid application and tooling for compaction.

As the prime function of a sealant to perimeter gaps around window frames in traditional walling is as a filler to exclude wind and rain, it should adhere strongly to enclosing surfaces, be resistant to the scouring action of weather and sufficiently elastic to accommodate small thermal movements for the anticipated life of the material. The expected useful life of sealants, after which they should be renewed, is up to 15 years for acrylic and up to 20 years for polysulphide, polyurethane and silicone. For appearance, the sealant should not be too obvious.

To ensure maximum adhesion, the surfaces on to which a sealant is run should be clean, dry and free from dust, dirt and grease. Rough surfaces such as open textured brick, textured rendering, scratched finishes and textured masonry paints are unsatisfactory as a base; the sealant will not readily adhere and would produce an unsightly appearance. Some window surfaces such as aluminium and plastic will require solvent cleaning to remove oil, grease and other coatings if the sealant is to make satisfactory adhesion. Sealants are usually run into joints from a gun operated by hand pressure or air pump.

The form of sealant joint used depends on the width of the perimeter gap between window frame and surrounding wall and whether the frame is set in a rebate. The types of joint used are butt joint, lap joint and fillet seal. The most commonly used is the butt joint (Fig. 48), formed between the back of the frame and the reveal of the opening. Foamed polyethylene is first run into the gap as a backing for the sealant. The sealant is then run into the joint and tooled with a spatula to compact the material and make good adhesion to the two surfaces. It is finished with a slight concave finish up to the edge of the window frame.

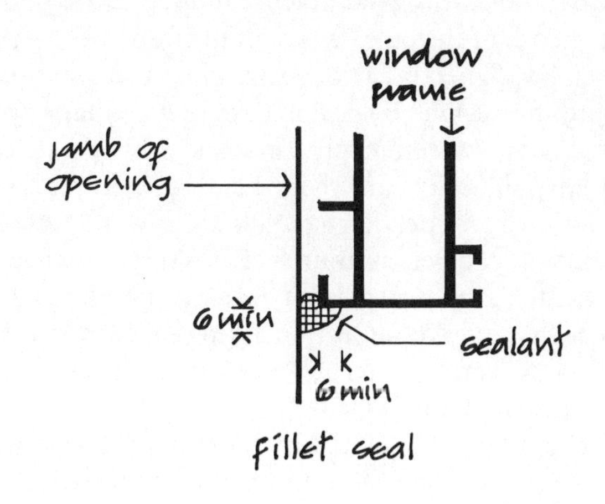

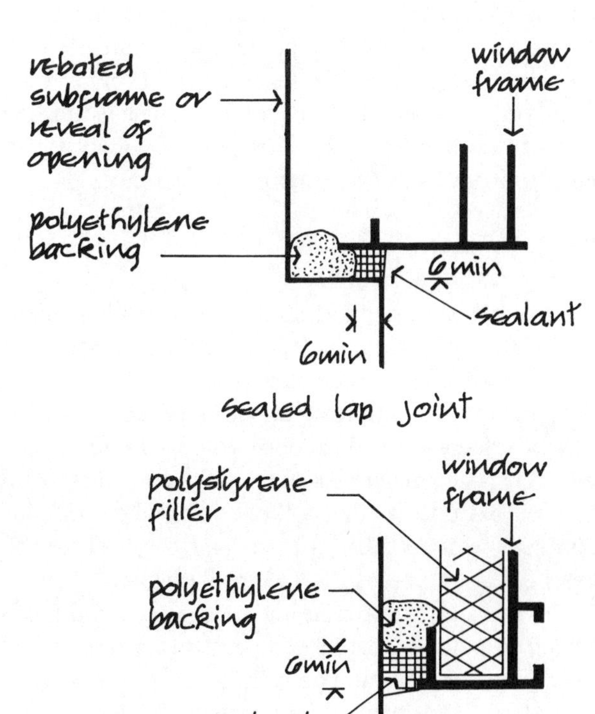

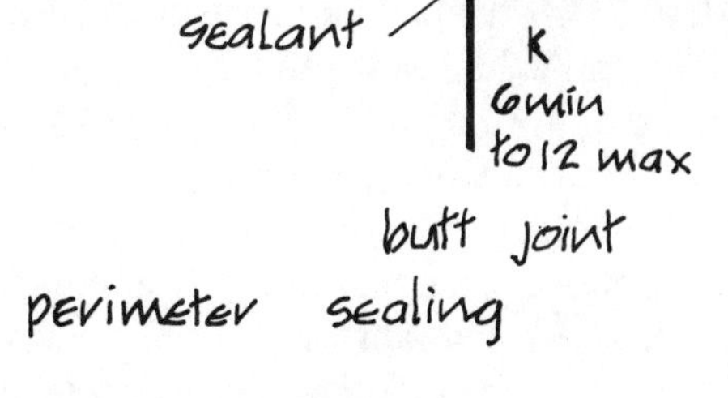

Fig. 48

The best gap width for this joint is from 6 to 12, which is wide enough for application of the sealant, small enough to contain the sealant and not too obvious. Butt joints up to 25 are practical with the depth of the sealant being half that of the gap. These wider joints tend to look somewhat unsightly. To prevent the sealant adhering to the outside face of the window frame it is good practice to use masking tape up the edge of the outside face of the frame. Once the sealant is sufficiently cured the masking tape is stripped towards the sealant. Polysulphide or polyurethane two-part sealants are commonly used by skilled operatives.

A lap point is formed where the window frame is set behind an accurately formed rebate in wood, metal, masonry or concrete surrounds (Fig. 48). The sealant is run into the gap over a polyethylene backing and tooled to a slight concave finish to masking tape. It is more difficult to form or renew this joint, which is less obvious than a butt joint.

A gap of less than 6 between the window frame and the opening in the wall is too narrow for gunned in sealant. Here a fillet seal is used, which is formed to adhere to the outside face of the frame and the wall opening (Fig. 48). The fillet seal is run as a convex fillet to provide sufficient depth of sealant, which is finished as it comes from the gun. This type of sealant finish tends to have a somewhat untidy appearance.

uPVC casement windows

These windows are fabricated from extruded, high-impact strength, white, uPVC (unplasticised polyvinyl chloride). Modifiers, such as acrylic, are added to the PVC material to improve impact strength. Pigment may be added to produce body coloured uPVC. The heated, plastic material is forced through dies from which it extrudes as thin-walled, hollow box sections complete with rebates, grooves and nibs for beads, weatherseals, glazing seals and for fixing hardware.

The wide range of sections are made in multi-cell form for rigidity, with one main central cell and two or more outer or surrounding cells. Main wall thickness varies from 3 for the bulkier German section, 2.8 mm for British and 2.2 mm for some French sections.

Tests for uniformity of section, colour and freedom from visible distortion of surface are carried out by reputable extruders. The dozen or so extruders in this country produce some hundreds of differing profiles. Some extruders also fabricate uPVC windows. The majority of the 6000–7000 fabricators take their sections from the handful of extruders.

The extruded sections are mitre cut to length, metal reinforcement is fitted and secured inside the main central cell, and the corner joints are welded together by an electrically heated plate that melts the end material, with the ends then brought together to fuse weld. The process of cutting and welding is fully automated, which makes it a comparatively simple operation to set the machine to make one-off sizes of windows for the replacement window market. This advantage is singular to uPVC windows.

At the mitred, welded corner joints a rough, curled edge of weld material protrudes from the face of the sections. This excess material is cut away for appearance, either flush with the external faces or more usually as a shallow groove, the material around the joint being routed out to a depth of about 0.25 mm. This latter finish tends to mask the slight difference in texture at the joint more than the flush finish.

Reinforcement is fixed inside the hollows of the uPVC cells to provide rigidity to the sections that might otherwise distort due to thermal movement, handling, fixing and in use as opening lights. Reinforcement should be fitted to all frames more than 1500 long and all opening lights more than 900 long. For fixing frames to surrounding walls and for secure fixing of hardware it is advantageous to use reinforcement to all window sections.

Reinforcement is either of galvanised, rolled steel or extruded aluminium sections, aluminium having the advantage that it does not destructively corrode and expand where water may find its way into the hollow sections.

In use coloured uPVC material, particularly dark colours, may bleach in an irregular, unsightly manner after some years due to ultra-violet light, so white or off-white uPVC is recommended.

In fire uPVC, which does not readily ignite, will only burn when the source of heat is close to the material and will not appreciably contribute to the spread of flame. The rate of generation of smoke and fumes produced when uPVC is subject to fire is no greater than that of other combustible materials used in building. Recent reports from Germany of the dangers of toxic fumes from these windows in fires were given publicity to the extent that some German authorities did for a time ban their use. In the event it has been established that PVC gives no more toxic fumes than other materials that may ignite in fires and is thus but a part of the hazard of fire.

Other than occasional washing with water to remove dirt these windows require no maintenance and this is

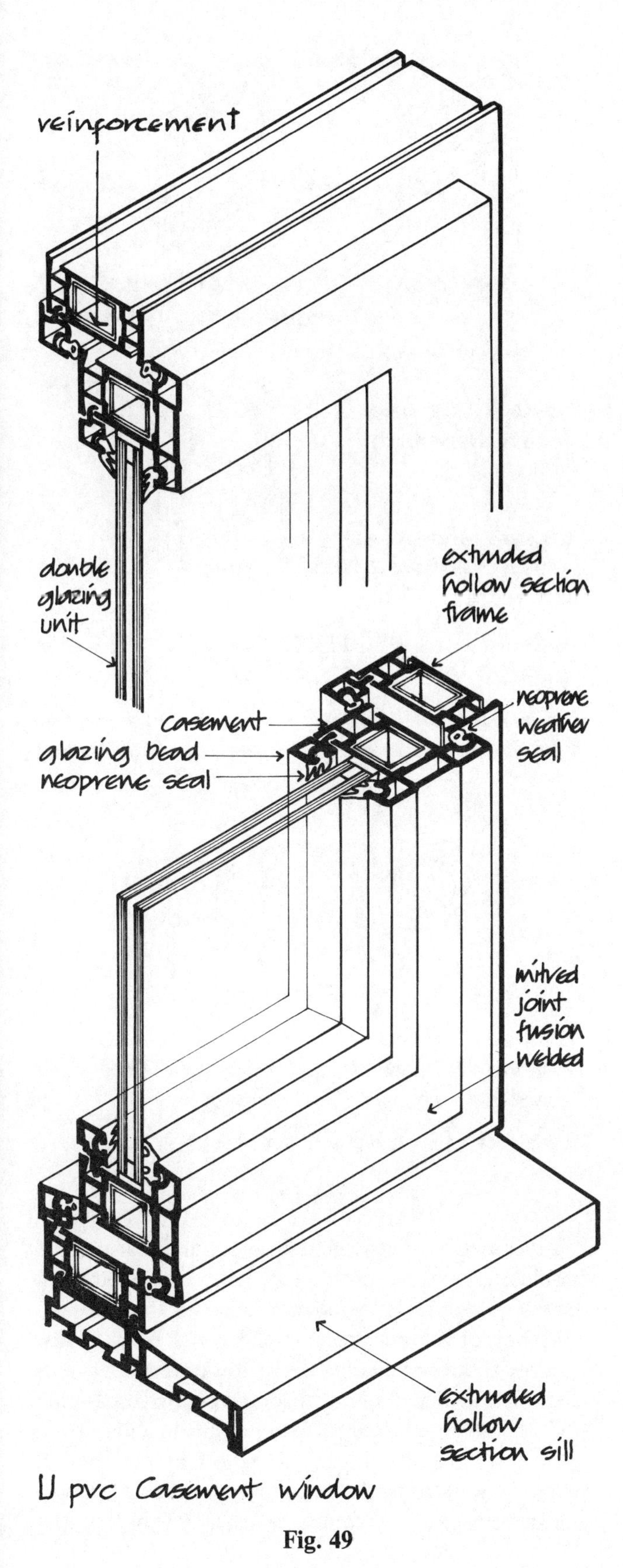

Fig. 49

their principal advantage.

The uPVC casement illustrated in Fig. 49 is glazed with an IG double glazed unit set in synthetic rubber seals and fitted with weatherstripping and reinforcement of galvanised steel or aluminium sections.

Because most uPVC sections are bulky they are not suited for use in the comparatively small casement window. The extruders of uPVC sections make a less bulky section specifically for use in casement windows.

Hardware is made from cast aluminium alloy and diecast zinc alloy with anodised, powder or organic liquid coatings for lockable fasteners and stays that are screwed through the outer wall of the uPVC sections into the reinforcement.

Fixing uPVC frames To avoid damage to the frames during building operations these windows are usually fixed in position after the surrounding walls have been built. Fixing is usually by driving strong screws through holes in the frame and reinforcement into surrounding walls or by means of lugs bolted to the back of frames which are screwed to plugs in walls. Fixings are at 250 to 600 centres and from 150 to 250 from corners.

Perimeter seals to uPVC windows The gap between the window and the surrounding wall is sealed with silicone or polyurethane sealant with backing of foamed, compressible, pre-formed strips or gunned in expanded, adhesive foam for joints more than 6 wide.

Encapsulated unbonded timber core windows

Preservative treated timber or laminated frame, casement and ventlight sections are sheathed in a rigid uPVC extrusion, minimum 0.85 mm thick, that covers the surface of the timber and is mitred and sealed at angles to corner spigots by heat or solvent welding. The sections include rebates, a drainage channel and weather seal between opening lights and frame. The uPVC encapsulation or sheath has a white finish that requires no maintenance other than occasional washing. The sections are made to take either single or double glazing.

These windows, which combine the solidity, strength and stiffness of wood with the maintenance free surface of uPVC, are used as both replacement windows and in new work. Fig. 50 illustrates a typical casement window.

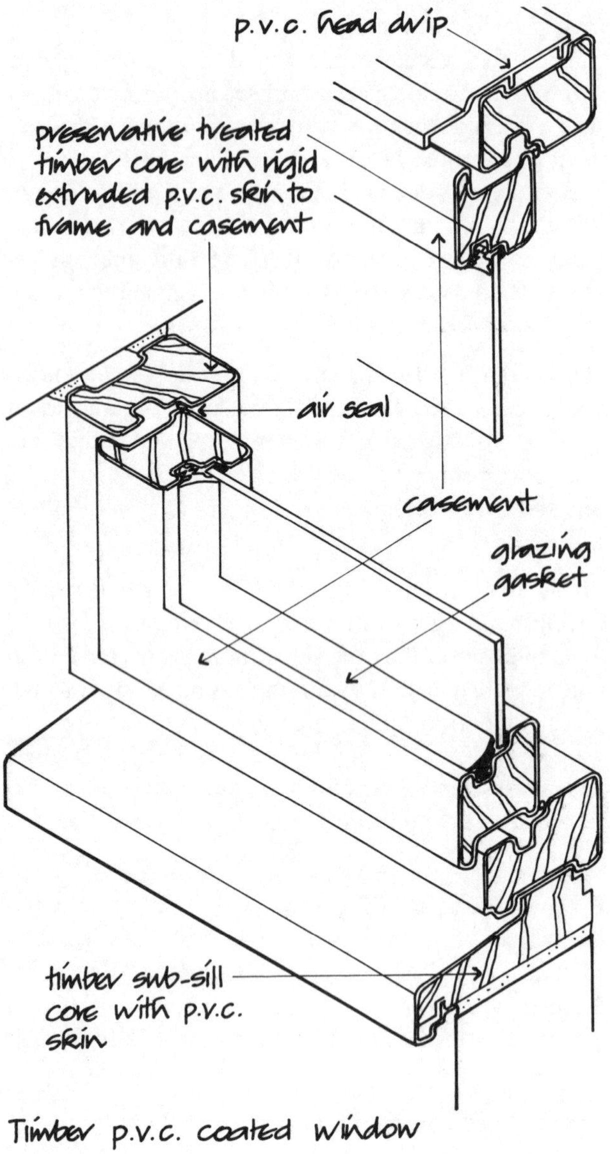

Timber p.v.c. coated window

Fig. 50

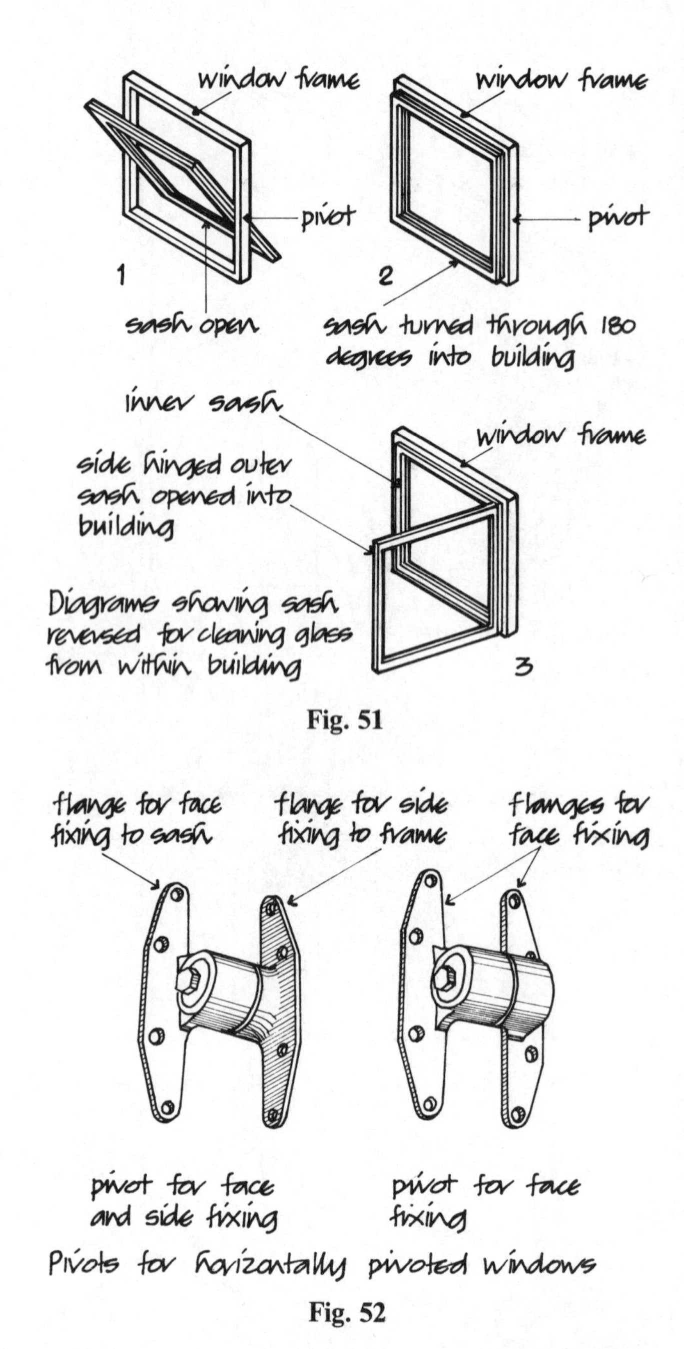

Diagrams showing sash reversed for cleaning glass from within building

Fig. 51

Pivots for horizontally pivoted windows

Fig. 52

Pivoted windows

Horizontally-pivoted wood window Fig. 53 is an illustration of a double-glazed horizontally-pivoted wood window with a reversible sash for cleaning the glass both sides from inside. The frame is solid and rebated for a stop bead that is fixed to the sash above and the frame below the pivots. The sash is made with separate inner and outer sashes, each of which is glazed, and the sashes are normally locked together. The inner sash is pivoted to the frame and the outer sash is hinged to the inner sash so that when the sash is fully reversed through 180° the inner sash, now facing into the building, may be unlocked and hinged to open in for cleaning the glass (Fig. 51).

The air space between the sashes and the glass is ventilated from the outside through the gaps between the sashes, to prevent condensation, which reduces the efficiency of the air space as a thermal barrier. The window is opened against the action of friction pivots (Fig. 52) and locked shut with lever-operated espanolite bolts that secure the sash at four points top and bottom against weatherstripping. The action of the friction pivots is not strong enough to resist gusts of wind in exposed positions, so some form of stay is necessary.

Vertically-pivoted steel windows Fig. 54 is an illustration of a steel section window fabricated from

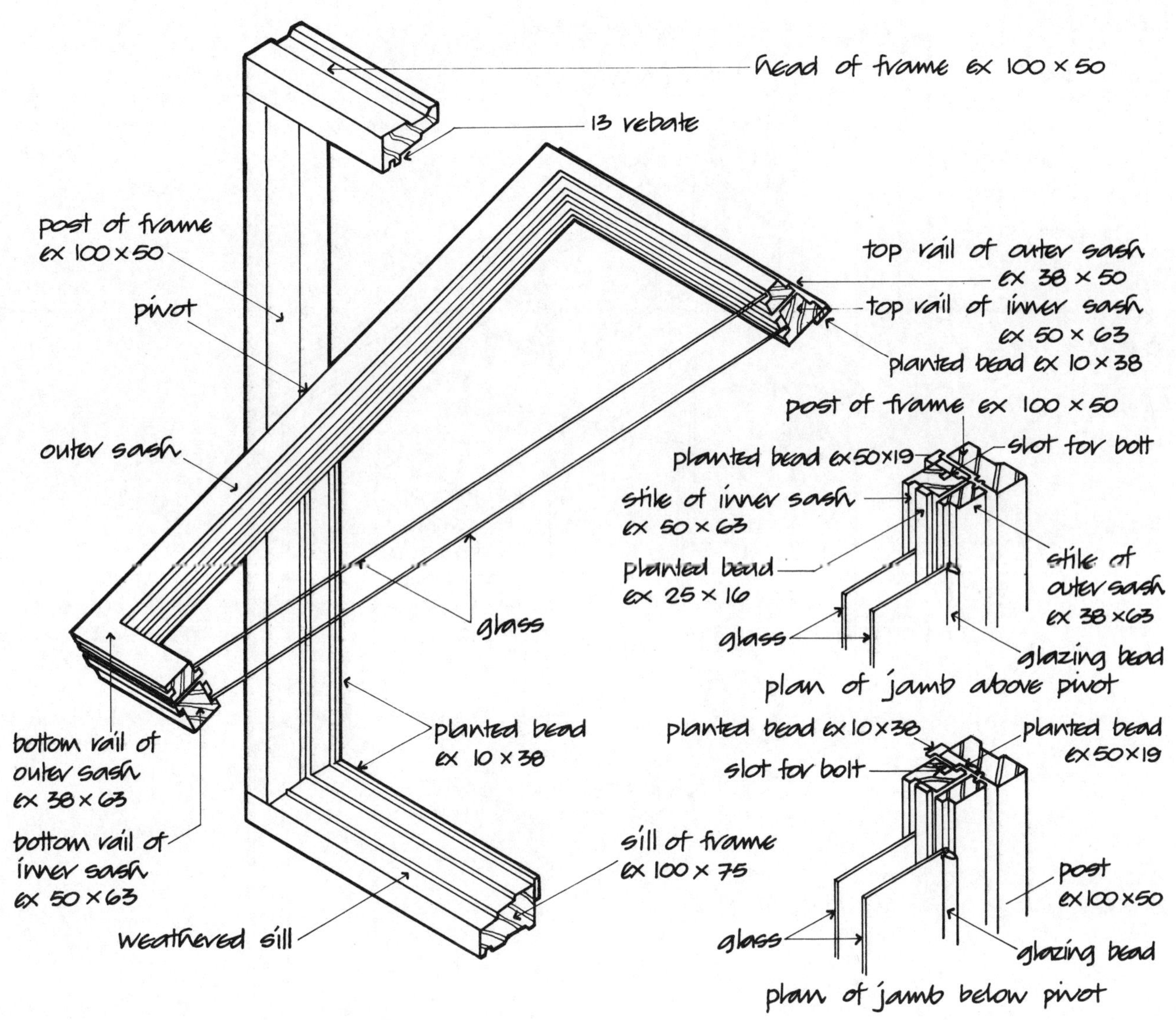

Fig. 53

universal sections, hot-dip galvanised, fitted with weatherstripping and single glazed. The sash is pivoted so that two thirds of the width opens out and one third in, to minimise internal obstruction. This type of window is used for warehouses and other unheated or partially heated buildings where the disadvantage of the cold bridge to the transfer of heat of the frame is of little account and the low initial cost is attractive.

Sliding windows

Vertically sliding wood sash window (double-hung sash) This traditional window is framed from thin section timbers to form a box or cased frame inside which the counterbalance weights are suspended to support the sliding sashes which are suspended on cords that run over pulleys fixed to the frame.

The jambs of the frame are cased from three thin members tongued, grooved and glued together as illustrated in Fig. 55. The pulley stile, in which the pulley supporting the sash cords is fixed, is usually 7 thicker than the outer and inner linings because it carries the weight of the sashes and the weights. A thin strip of wood, the parting slip, is suspended inside the cased jambs to separate the weights of the sashes. A strip of plywood or hardboard is nailed across the back

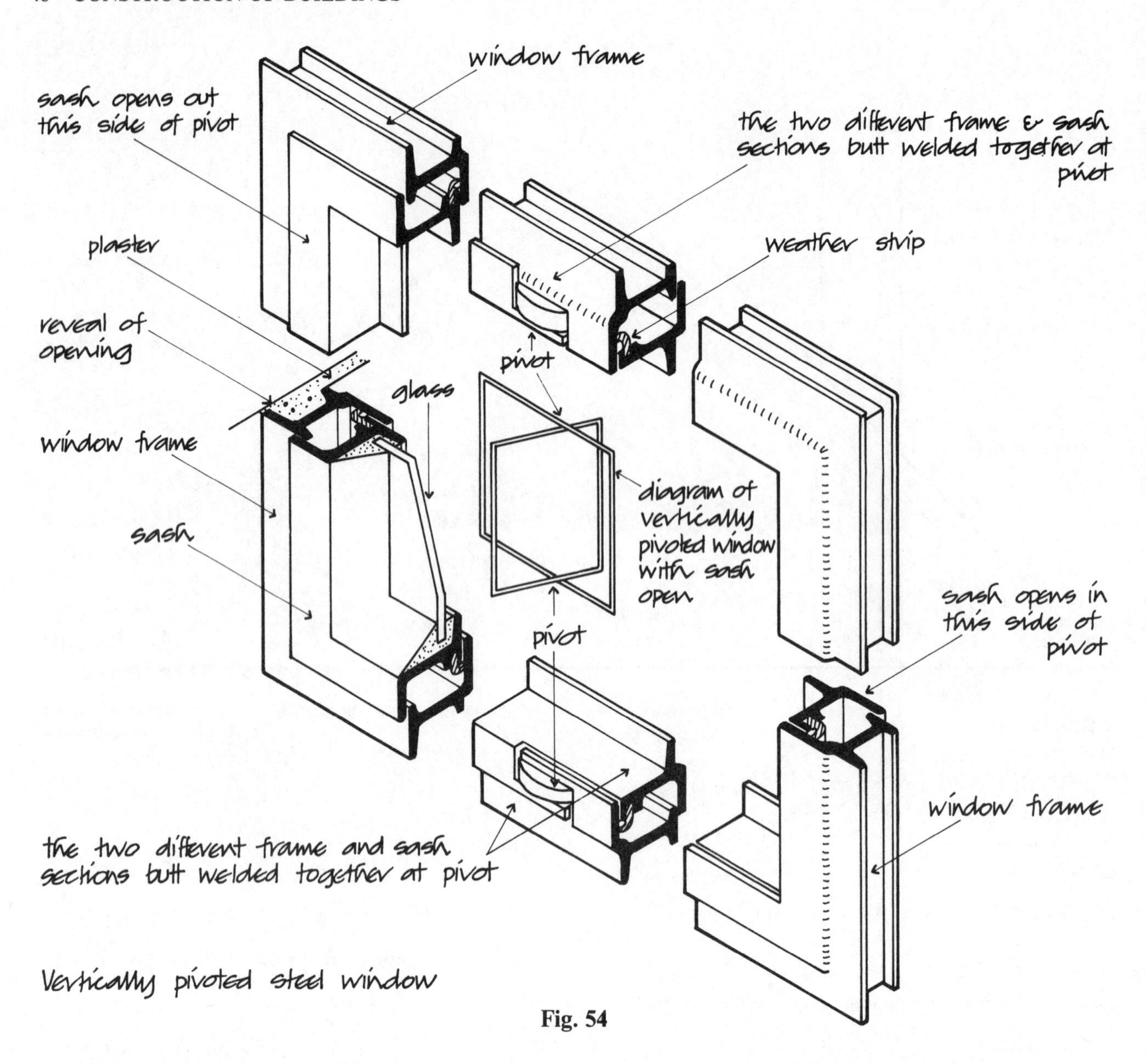

Fig. 54

of the linings as a back lining, to prevent the weights catching the reveals of the opening.

The size of the cased jambs depends on the thickness of the sashes. If the sashes are cut from 38 thick timbers the inside of the cased jamb is usually 85 wide and 50 deep, and with sashes cut from 44 thick timbers, 105 wide and 50 deep. The head of the cased frame is usually constructed from three thin members put together with glued tongued-and-grooved joints, or a solid rectangular section of timber 38 deep may be used.

The sill of the frame is cut from a solid section of hardwood, such as oak, 75 deep and as wide as the cased frame overall. The sill is weathered and sunk on its top surface. The word weathered denotes that the top surface of the sill is cut to slope outwards to throw rainwater off. The sinking, which is a shallow rebate some 6 deep in line with the face of the lower sash, serves to prevent rainwater being blown between the sill and the sash. A 12 wide semi-circular groove on the underside of the sill forms a throat-and-drip edge and the rectangular groove takes a 25 × 3 galvanised steel or wrought iron water bar which is bedded in mastic in the oak sill and in cement in the stone sub-sill. The head of the frame, which is cut to fit between the linings of the jambs and the pulley stile, is tongued to a groove in the soffit lining. Similarly the sill is cut to fit between the jamb linings, and the pulley stile is wedged into a groove in the sill. The outer linings of the jambs and head project 13 beyond the face of the pulley stiles to act

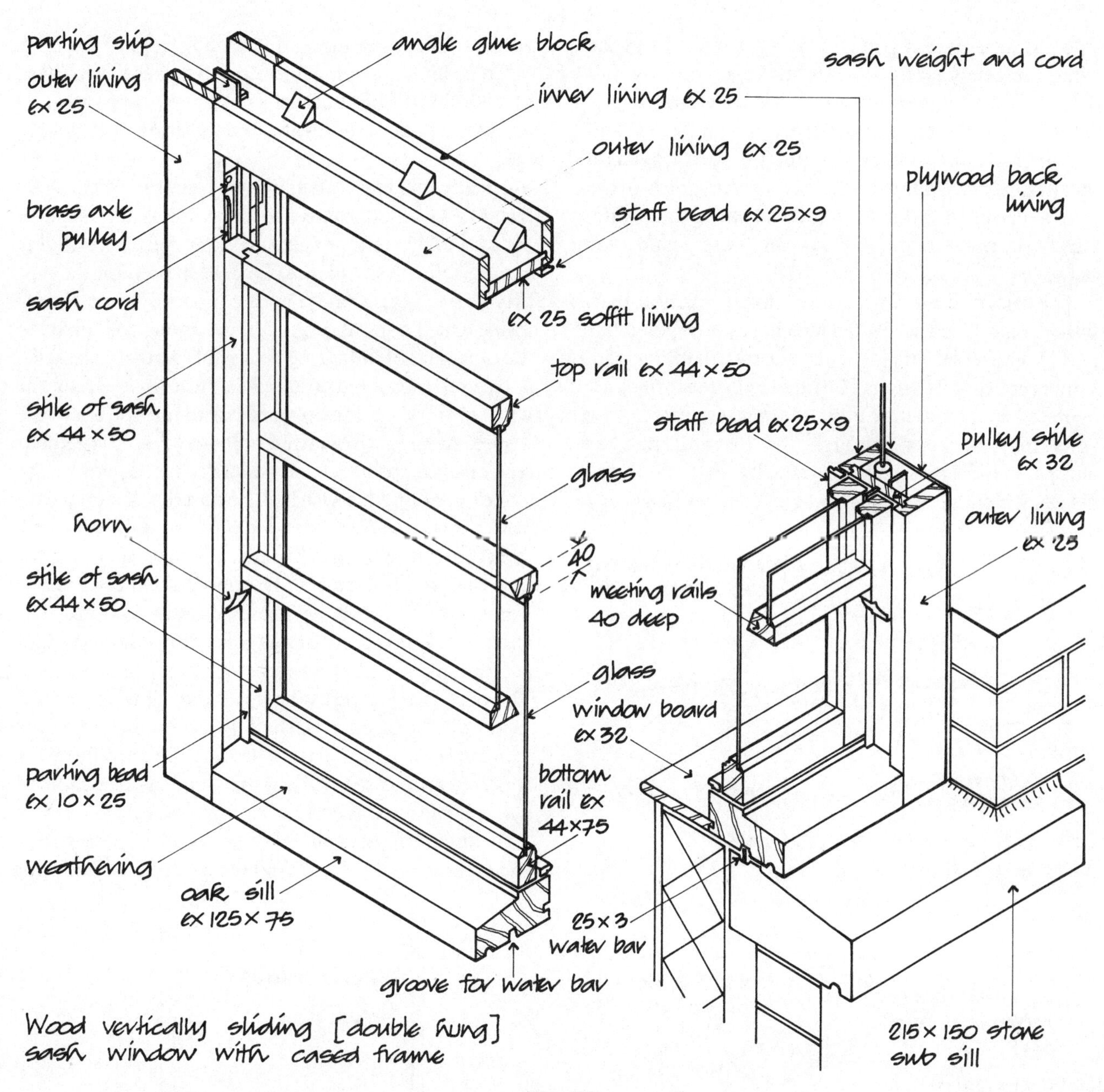

Fig. 55

as guides for the top sash. Parting beads 10 × 32 are set into grooves in the pulley stiles to separate and act as guides to the sashes, and a staff bead is screwed to the edge of the inner linings to act as a removable guide for the lower sash. Brass axle pulleys, two to each sash, are fixed in the pulley stiles as illustrated in Fig. 55.

To renew sash cords, the staff beads and parting beads are removed and the sashes are lifted out into the room. Small traps cut in the bottom of the pulley stiles, called pockets, are taken out so that the new sash cords can be attached to the weights inside the cased frames.

Sashes are usually of the same depth so that the meeting rails come together in the middle of the height of the window.

The stiles and top rail of sashes are cut from 38 or 44 thick × 50 deep timbers, rebated for glass and moulded inside. Meeting rails are cut from 63 × 38 timbers rebated and splayed to meet and rebated for glass. The bottom rail of bottom sashes is cut from timber 38 or 44 thick × 63 or 75 deep, rebated for glass and moulded. The top rail of top sashes and bottom rail of bottom sashes are tenoned, wedged and glued to mortices in the stiles, and the meeting rails, which are tenoned to the stiles, are extended as horns to make a tenon the full depth of the thin meeting rails as illustrated in Fig. 55.

Sash cords are made from twisted or braided flax and cotton cord some 6 thick. In time the cord frays and breaks and needs fairly frequent renewal. Cords made from a mixture of nylon fibre and flax have a longer life. Heavy sashes are often hung on brass chains for durability.

The sashes are secured in the closed position by a sash fastener fixed across the meeting rails. Pivoted bar or fitch fasteners are used as illustrated in Fig. 56. Either type of fastener may be fitted with a security lock operated by a loose key. For additional security, screw bolts may be fixed through the meeting rails as illustrated in Fig. 57 or a lock attachment fixed to the sashes, which allows them to be opened sufficient for ventilation and no more, operated by a loose key.

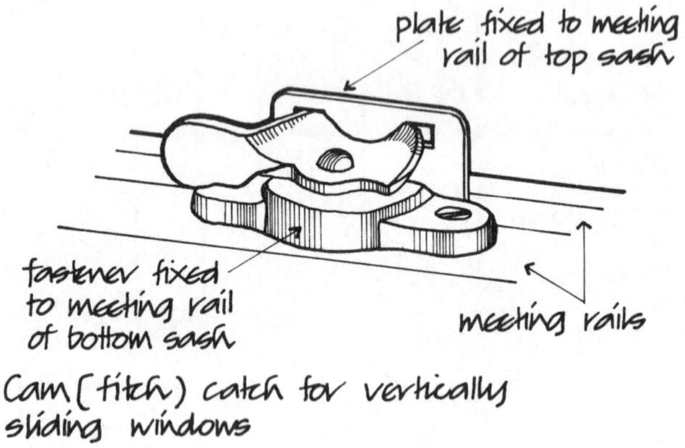

To facilitate opening and closing sashes it is usual to fit sash lifts to the bottom rail of the lower sash and cords and pulleys for raising and lowering the top sash.

plate fixed to meeting rail of top sash

fastener fixed to meeting rail of bottom sash

meeting rails

Cam (fitch) catch for vertically sliding windows

Fig. 56

Vertically sliding wood sash window with solid frame As an alternative to the traditional system of cords, pulleys and weights to hang vertically sliding sashes, spiral sash balances have been used for the past fifty years. The spiral balance consists of a metal tube inside which a spiral spring is fixed at one end. Fixed to the other end of the spring is a metal cap through which a twisted metal bar runs. The tube is fixed to the window frame and the twisted bar to the bottom of the sashes. As the sash is raised or lowered the twisted bar tensions the spring which supports the weight of the sashes, enabling the sashes to be raised or lowered with little effort. Fig. 59 shows one of these sash balances.

Because of the sash balance there is no need for hollow cased frames to take counterbalances and the frame members can be made of solid sections. The window frame is constructed from four solid rectangular sections of timber, two posts (jambs), head and sill. The posts are joined to the head and sill with combed joints glued and pinned, similar to those described for standard casements. The sashes are similar to those for windows with cased frames, the members being joined with mortice and tenon or combed joints.

A range of vertically sliding wood windows with solid frames cut from standard sections and put

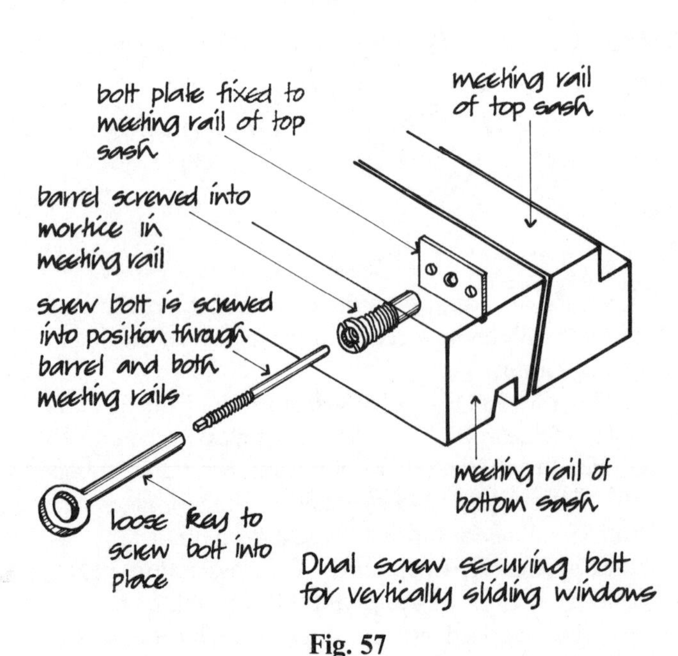

Dual screw securing bolt for vertically sliding windows

Fig. 57

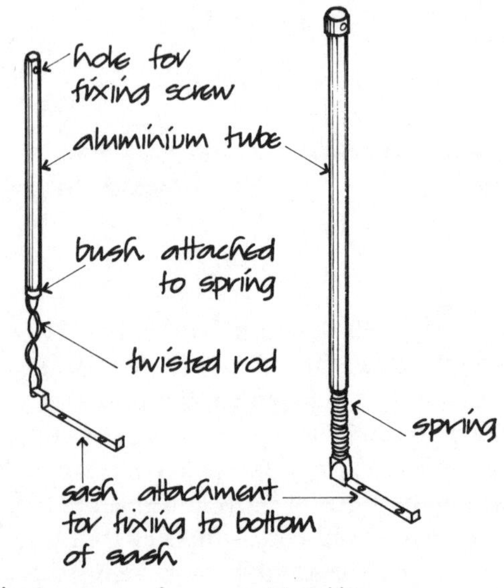

Sash balances for vertically sliding windows

Fig. 58

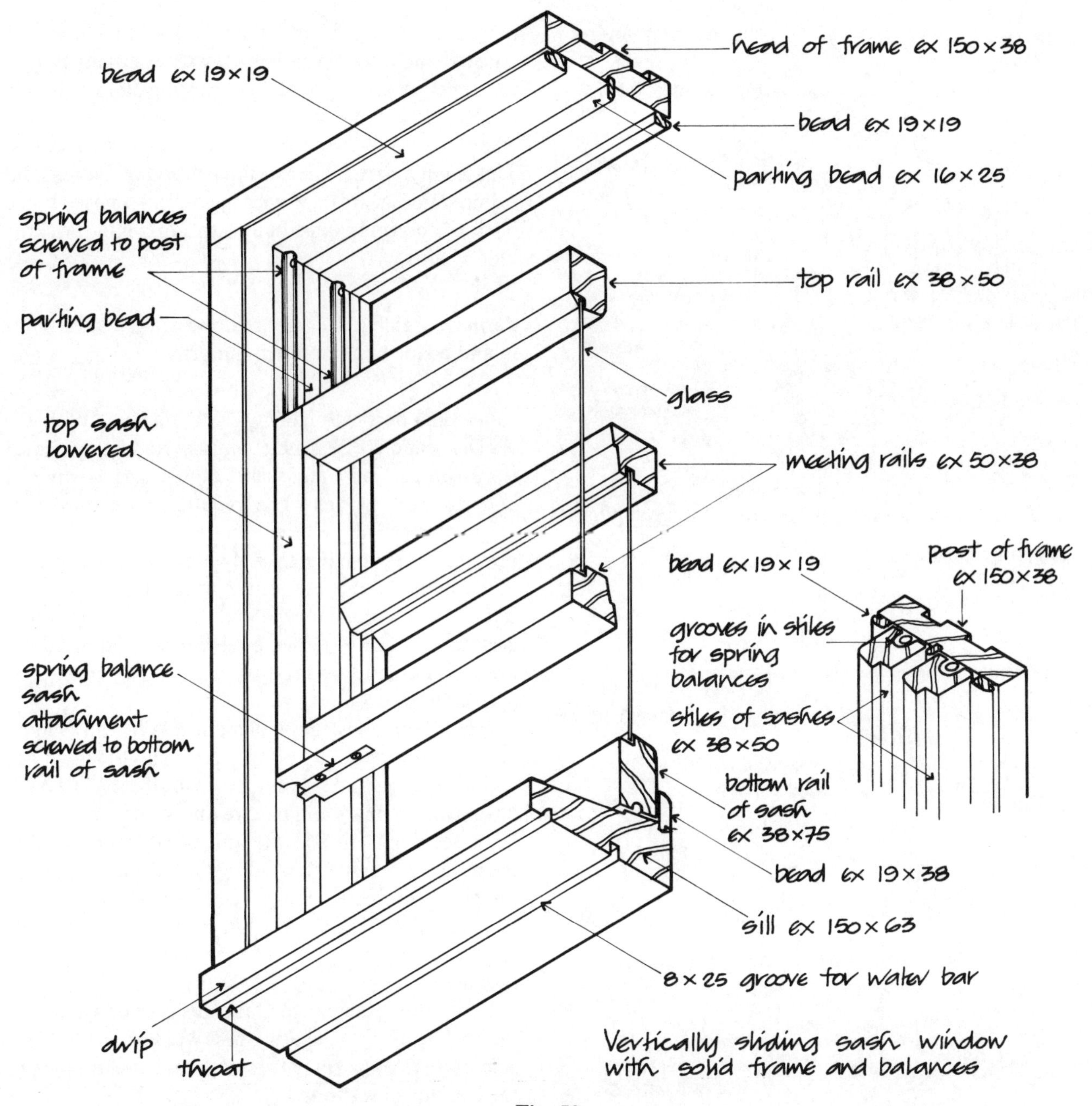

Fig. 59

together in a range of standard sizes is manufactured with the sashes being grooved for and hung on spring balances (Fig. 58).

Weatherstripping To allow free movement of the sashes there cannot be a close fit of the sashes to the frame. To exclude wind it is necessary to fit some form of weatherstripping. The neoprene seals illustrated in Fig. 60 are held in aluminium sections that are fixed to the frame and sashes.

An alternative method of weatherstripping consists of uPVC parting beads that are screwed to the frame instead of wood beads. Nylon filament and plastic blade weatherstrips are fixed in the plastic parting beads to wipe and seal to both inner and outer sashes. uPVC strips fitted to a rebate in the staff beads hold similar nylon filament seals that wipe and seal on the inside faces of sashes and a similar seal is fitted to a groove in one of the meeting rails so that it seals between the meeting rails when the window is closed.

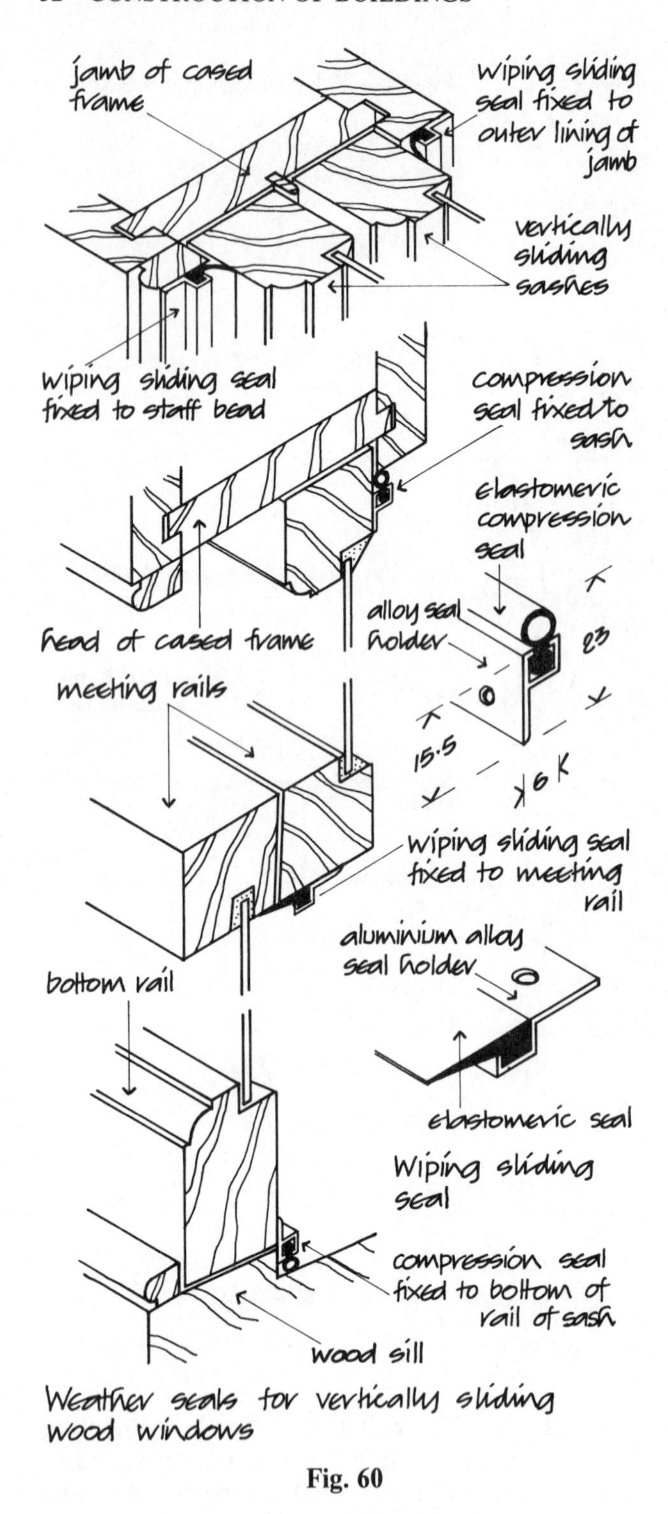

Fig. 60

Fixing vertically sliding sash windows

It used to be common to fix vertically sliding cased frame windows in rebated jambs of solid masonry walls behind a ½*B* deep rebate so that most of the frame was covered and only the sashes showed externally to give the appearance of a window consisting almost entirely of glass. The frame was wedged in position with wood wedges driven between the frame and the brick jambs. This is not a particularly secure method of fixing. Solid frame windows are fixed with lugs or ties screwed to the back of the frame and built into horizontal brick or block joints.

Perimeter sealing These windows are generally built-in and bedded and pointed in mortar.

Aluminium vertically sliding sash window With the slender sections of extruded aluminium alloy practical for use in the frame and sashes of this type of window, and because the material requires no painting at frequent intervals, aluminium vertically sliding windows have become increasingly popular. The extruded aluminium sections are joined with screw nailed butt joints, mechanical mortice and tenons or mitred and cleated joints with stainless steel screws. The windows are mill finished, anodised or have a stoved powder or acrylic finish.

Fig. 61 is an illustration of a typical vertically sliding aluminium window. The jamb section houses the spiral spring balances and acts as a guide for the sashes which have nylon runners and pile weatherstrips. The hollow head section is fitted with pile weatherstrip and a neoprene seal and the sill with weatherstrip. The slender sash sections take either single or double glazing, have integral sash lifts and the meeting rails close to a neoprene seal and are secured with a fitch fastener.

As it is not practical to make this type of window in thermal break construction, these windows are not best suited where warm, moist air may condense to water on the cold metal sections.

uPVC vertically sliding sash window The uPVC vertically sliding sash window illustrated in Fig. 62 is double glazed, suspended in spiral balances and weather-stripped. These windows can be designed so that both sashes can be tilted into the building to facilitate cleaning glass both sides from inside.

The comparatively bulky frame and slim sash sections are specifically designed to match the sections of wood windows so that these uPVC windows can be used as replacement windows without obvious change of appearance.

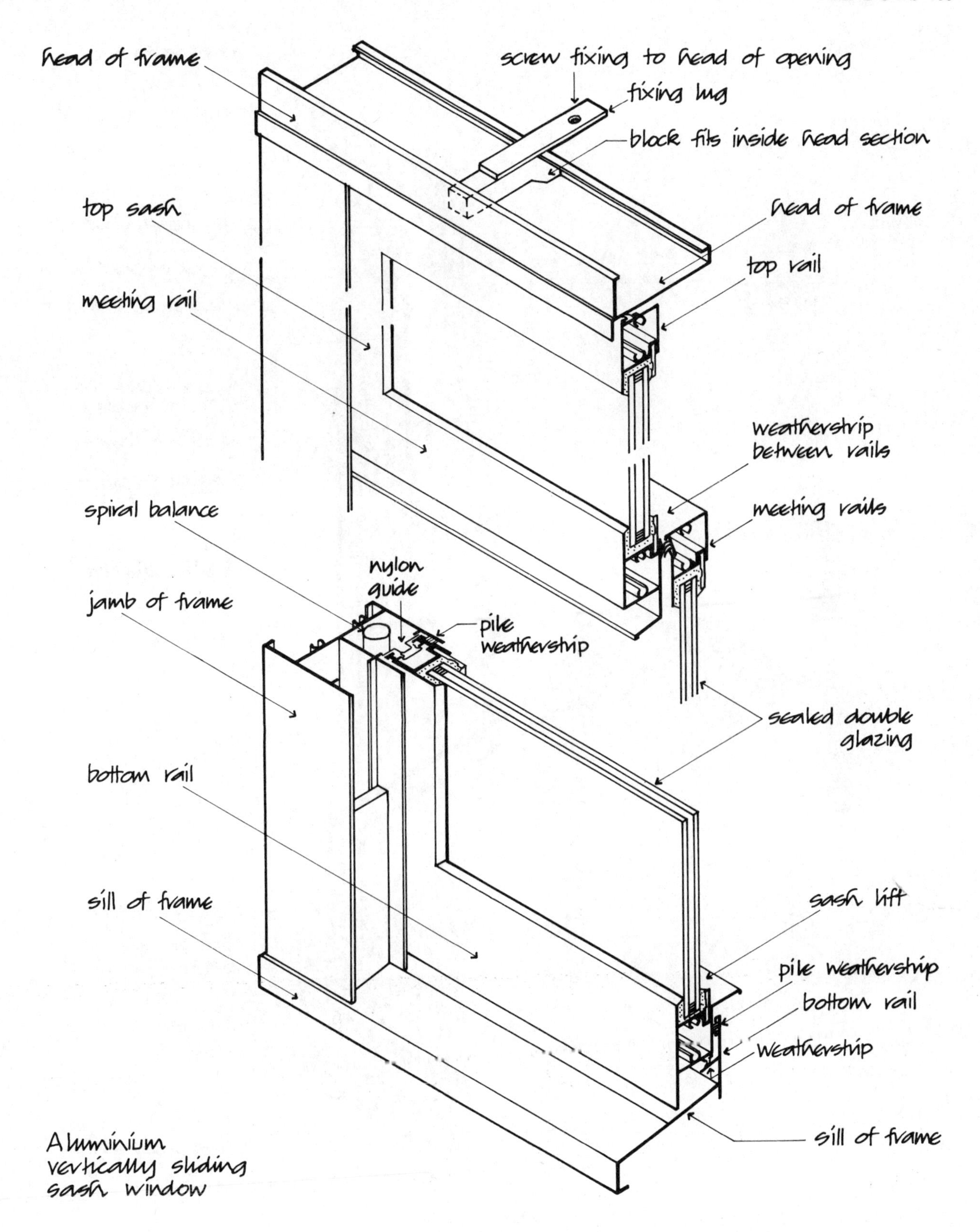
head of frame
screw fixing to head of opening
fixing lug
block fits inside head section
top sash
head of frame
top rail
meeting rail
weatherstrip
between rails
spiral balance
meeting rails
nylon
guide
pile
weatherstrip
jamb of frame
sealed double
glazing
bottom rail
sill of frame
sash lift
pile weatherstrip
bottom rail
weatherstrip
sill of frame
Aluminium
vertically sliding
sash window

Fig. 61

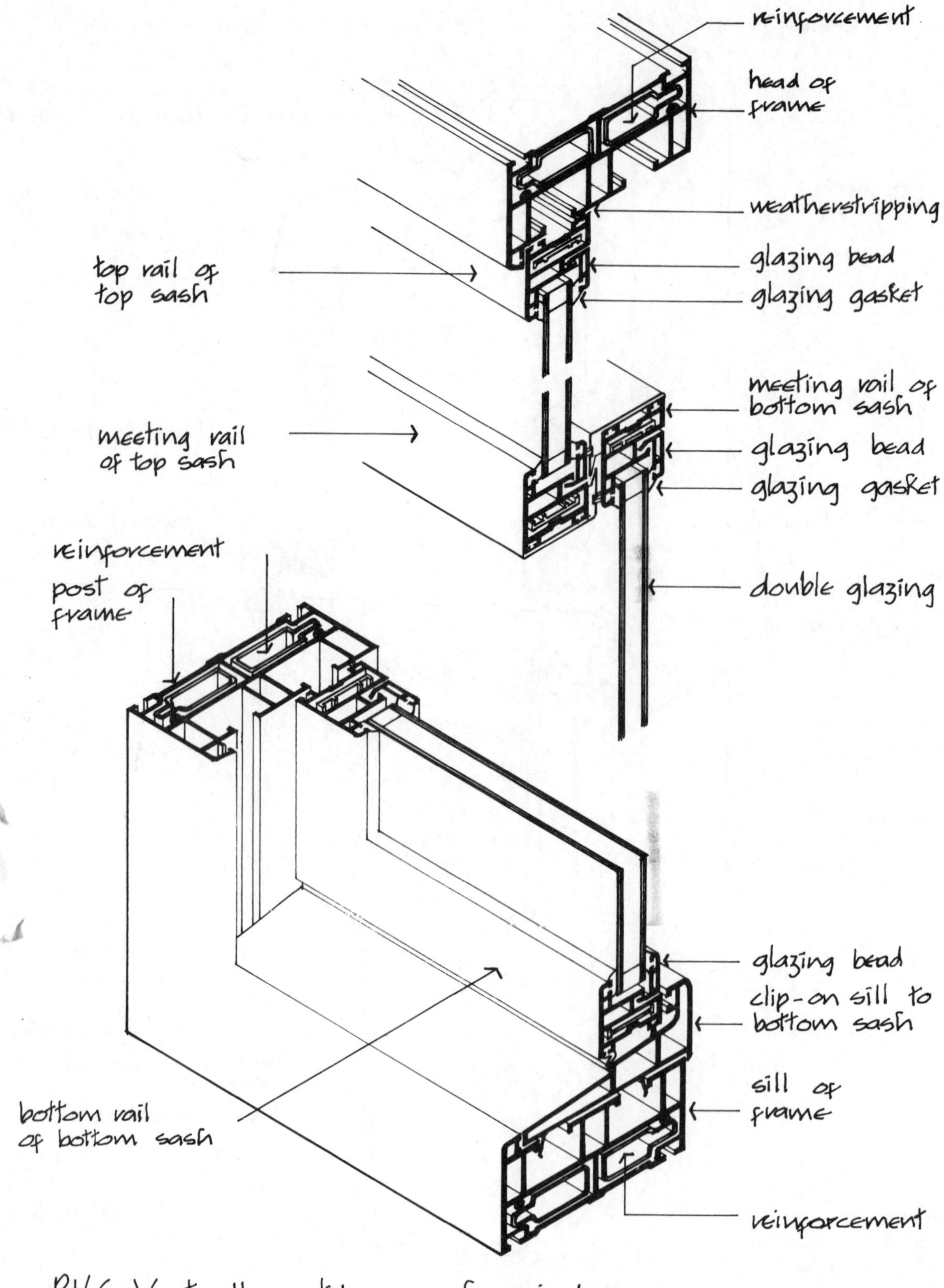
reinforcement
head of frame
weatherstripping
glazing bead
glazing gasket
top rail of top sash
meeting rail of bottom sash
glazing bead
glazing gasket
meeting rail of top sash
reinforcement
post of frame
double glazing
glazing bead
clip-on sill to bottom sash
sill of frame
bottom rail of bottom sash
reinforcement
uP.V.C Vertically sliding sash window

Fig. 62

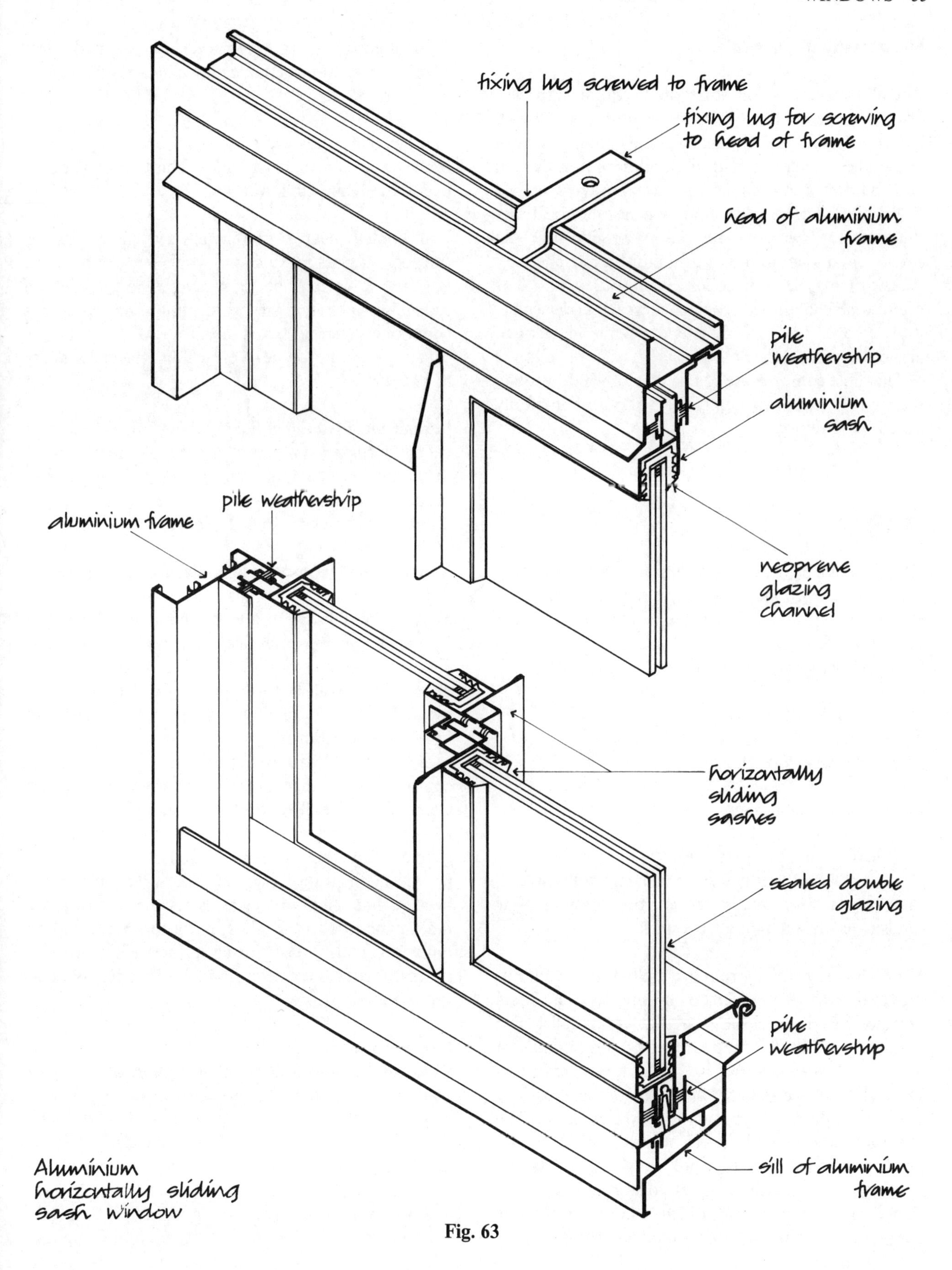
fixing lug screwed to frame
fixing lug for screwing to head of frame
head of aluminium frame
pile weatherstrip
aluminium sash
pile weatherstrip
aluminium frame
neoprene glazing channel
horizontally sliding sashes
sealed double glazing
pile weatherstrip
sill of aluminium frame
Aluminium horizontally sliding sash window

Fig. 63

Horizontally sliding windows

The aluminium section horizontally sliding sash window illustrated in Fig. 63 slides on a bottom track with nylon filament, pile weatherstripping acting as weatherseal and as guide to both top and bottom rails, and with pile weatherstripping to stiles. To avoid having over bulky sections, this window is not designed as a thermal break window and will in consequence be a source of condensation in warm moist atmospheres.

As grit may in time collect around the track, it is often somewhat difficult to open these windows which, if forced open or closed, may tend to jamb on the track and so be more difficult to open.

The most common use of this type of window today is as fully glazed horizontally sliding doors, commonly called patio doors.

GLASS

Glass is made by heating soda, lime and silica (sand) to a temperature at which they melt and fuse. Molten glass is either drawn, cast, rolled or run on to a bed of molten tin to form flat glass.

Glass may be classified into three groups:

(a) Annealed flat glasses.
(b) Processed flat glasses.
(c) Miscellaneous glasses.

(a) Annealed flat glasses

(1) Float or polished plate glass

Float or polished plate glass is transparent with surfaces that are flat and parallel so that they provide clear undistorted vision and reflection.

Float glass is made by running molten glass continuously on to a bed of molten tin on which the glass floats and flows until the surfaces are flat and parallel. The continuous ribbon of molten glass is then run into an annealing lehr or chamber in which the temperature is gradually reduced to avoid distortion of the glass. The glass gradually solidifies and the solid glass is cut. The natural thickness of the sheet of glass is 6.5 mm. To produce thinner glass the molten ribbon of glass is cooled and stretched between rollers and to make thicker glass the spread of the molten ribbon is restricted to produce the required thickness.

Float glass has largely replaced sheet and plate glass in this country, and is made in thicknesses of 3, 4, 5, 6, 10, 12, 15, 19, and 25.

Polished plate glass is made by grinding and polishing both surfaces of rough cast glass. This type of glass has been superseded by float glass.

Body tinted float glass or polished plate glass is transparent glass in which the whole body of the glass is tinted. This type of glass reduces solar radiation transmission by increased absorption. This material is commonly termed 'solar control glass'. Tints are usually green, grey, blue or bronze and thicknesses 4, 5, 6, 10 and 12.

Surface modified tinted float glass is transparent glass which during manufacture has a coloured layer of metal ions injected on to the glass. Solar control properties are provided by an increase in reflection and absorption. Thicknesses 6, 10 and 12.

Surface coated float glass (reflective float glass) is transparent glass which has a reflective surface layer applied either during or after manufacture. The reflective layer may be on a clear or a body tinted glass. Transmission of solar radiation is reduced by increase in reflexion and absorption and the glass has a coloured metallic appearance. Colours are silver, blue and bronze by reflection.

Surface modified and surface coated glasses are solar control glasses that are also referred to as low-emissivity glasses. The effect of the surface coating is to reflect back into the building the long wave energy generated by heating, lighting and occupants, while permitting the transmission of short wave solar energy from outside. This type of glass is designed for use in the inner pane or sheet of glass in sealed double glazing units where the greater inside surface temperature of the glass reduces condensation and the effect of 'cold spot' discomfort.

(2) Sheet glass

Clear sheet (drawn sheet) glass is transparent glass manufactured by the flat drawn process in which a continuous sheet is drawn from a bath of molten glass. The continuous sheet is gradually cooled to minimise distortion and then cut into sheets as it solidifies. The drawn sheet is not exactly flat or uniform in thickness and will cause some distortion of vision.

Sheet glass is no longer manufactured in England.

Body tinted sheet glass is transparent glass in which the whole body of the glass is tinted to give solar control properties. Tints are usually green, grey or bronze and thicknesses 3, 4, 5 and 6.

(3) Cast glass (also known as patterned glass)
Clear cast glass is translucent glass made by the rolling process. The deeper the pattern the greater the obscuration and diffusion.

Body tinted cast glass is similar, with the whole of the glass tinted for solar control and decorative purposes.

Wired glass is cast or rolled with wire completely embedded in it. One type of wire is available: Georgian 13 square mesh. Cast wired glass is translucent with a cast or patterned surface. Polished wired glass is transparent, through grinding and polishing.

(b) Processed flat glasses

(1) Toughened (tempered) glass is made by heating annealed glass and then rapidly cooling it to cause high compression in the surfaces and compensating tension in the centre of the thickness of the glass. This is a safety glazing material that is less liable to break on impact, and when broken it fragments into comparatively harmless small pieces.

Clearfloat, sheet, polished plate and solar control glass may be toughened.

(2) Laminated glass is made of two or more sheets (panes) of glass with an interlayer of reinforcing material between the sheets. The interlayers are permanently bonded to the enclosing sheets of glass. This glass is resistant to impact shock and when broken the reinforcing layer prevents extensive spalling of fragments. The reinforcing interlayer is usually in sheet form and made of polyvinyl butyral. This type of glass is specified as three ply, that is two sheets of glass and one of reinforcement and similarly five ply with three sheets of glass and two of reinforcement.

This type of glass is often described as safety or security glass.

Safety glass Toughened glass and laminated glass are described as flat safety glass which, on breaking, result in a small clear opening by disintegration into small detached fragments that are neither sharp nor pointed and are unlikely to cause cutting or piercing injuries.

The practical guidance in Approved Document N to the Building Regulations 1991 defines critical locations where safety glass should be used. These critical locations are in glazed panels in internal walls and partitions between floor and 800 above that level and in glazed doors and door side panels, between floor and 1500 above that level.

Polycarbonate sheet Flat plastic sheets made of polycarbonate are manufactured as transparent, translucent and colour tinted sheets for use as safety glazing. The sheets are 2, 3, 4, 5, 6, 8, 9.5 and 12 mm thick.

The principal characteristic of this material is its high impact resistance to breakage. These sheets do not have the lustrous, fire glazed finish of glass nor are they as resistant to abrasion scratching and defacing. A special abrasion resistant grade is produced.

Polycarbonate sheet, which is about half the weight of a comparable glass sheet, has a high coefficient of thermal expansion. To allow for this, deeper rebates and greater edge clearance are recommended than for glass. To allow for the flexibility of the material and thermal expansion, one of the silicone compounds is recommended for use with solid bedding.

(c) Miscellaneous glasses

In this group of glasses is included roof and pavement lens lights, copper lights, leaded lights and hollow glass blocks.

Insulating glass (IG) units, double glazing

Because of increased demand for thermal comfort in buildings, together with the requirements of the Building Regulations 1991 for energy conservation, and the continually increasing cost of fuel, it has become common for some years to fit double glazing to the majority of new and replacement windows. The term double glazing describes the use of two sheets or panes of glass in a window or door. The type of double glazing most commonly used today is the insulating glass unit (IG unit) which comprises two sheets or panes of glass spaced some 10, 12, 16 or 20 apart with a perimeter seal so that the air or gas trapped between the glass serves as thermal insulation to reduce transfer of heat through windows.

The U value (thermal transmittance) of a single sheet of 6 thick glass is 5.4 W/m^{2}K and that of an IG unit with two sheets of 6 thick glass spaced 10 apart is 3.1 W/m^{2}K and spaced 20 apart is 2.8 W/m^{2}K.

The advantages of insulating glass (IG) units are that

there is some reduction of heat loss or gain as compared to single glazing and that because of the lower *U* value, that is better insulation against transfer of heat, a larger area of glass in windows may be used in complying with the requirements of the Building Regulations for conservation of energy. There is in addition some small reduction in airborne sound transmission.

Because of the greater thermal insulation of double glazing as compared to single glazing there may well be a noticeable reduction of 'cold spots' near large areas of glass. 'Cold spots' is the term used to describe the sensation of cold, experienced close to large areas of cold surface in a room, caused by the automatic response of the body in radiating heat towards the cold surface in an attempt to maintain normal skin temperature. Similarly there is reduced condensation of moisture vapour from air because of the higher temperature on the inside of double glazing as compared to single glazing, particularly in bathrooms and kitchens.

The principal causes of heat loss from rooms are by transfer of heat through glass in windows and doors and by draughts of cold air being forced by pressure through gaps around opening parts of windows and doors, into rooms. Double glazing which will appreciably reduce transfer of heat through glazing will, by itself, effect no reduction in draughts of cold air into rooms. Draught stripping or sealing around opening parts of windows and doors will substantially reduce draughts of cold air irrespective of whether the glazing is single or double.

The considerable initial cost of double glazing and the subsequent cost of replacing double glazed units will at best be covered by savings in fuel costs over very many years, 25 or more. The comparatively small cost of weatherstripping around opening parts of windows and doors provides a much better return on capital than the installation of double glazing. The majority of double glazed windows include effective weatherstripping or seals to all opening windows and doors.

Insulating glass (IG) units, sealed double glazing units

The terms double glazing, sealed double glazing and insulating glass are generally interchangeable. The term double glazing embraces all systems of double glazing whether the glazing is unsealed as in double windows or sealed as in insulating glass units. The term sealed double glazing more precisely describes the system of glazing than the term favoured by the trade, insulating glass units.

Insulating glass units are made up from two sheets (panes or squares) of glass that are hermetically sealed to a continuous spacer in the perimeter of the unit. This spacer maintains the space between the two sheets of glass and supports the sealant. The usual space between the two sheets of glass is 10, 12, 16 or 20.

Spacer tube or bar The spacer is either a hollow aluminium section or a butyl-based bar with an integral aluminium strip, as illustrated in Fig. 64.

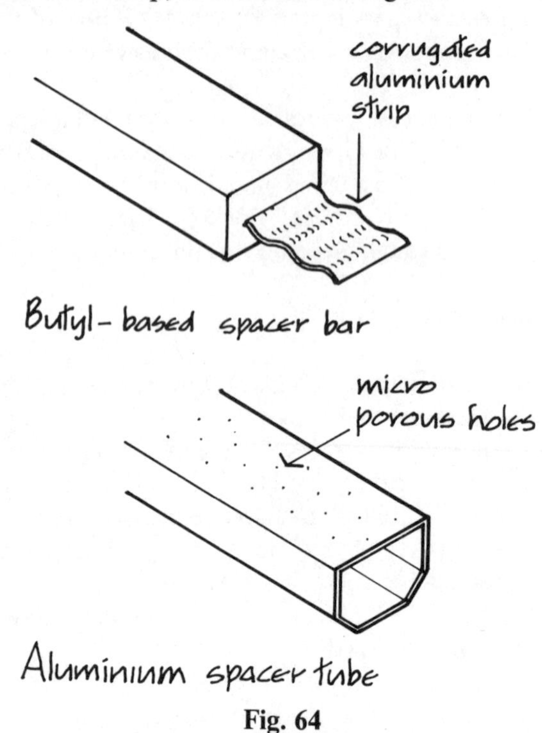

Fig. 64

Aluminium spacer tubes are either cut to length and connected with metal or nylon corner keys, or the spacer tube is bent to the required shape and the ends of the tube welded together to form a continuous spacer. The usual width of spacer tubes is from 5.5 mm to 19.5 mm to provide a space between sheets of glass from 6 to 20. The narrow widths are made for use in narrow sections of timber and steel windows and the wider for the wider section of uPVC and aluminium windows.

The butyl-based bar with integral aluminium strip consists of a preformed section with a corrugated, integral aluminium strip. The aluminium strip serves as reinforcement and in part as a barrier to moisture. A dessicant is embedded in the surface of the bar. The advantages of the bar, commonly known as 'Swiggle strip' are that it is continuous and being self adhesive makes fitting more rapid than using a spacer tube and sealant.

The useful life of a sealed double glazing unit depends on the dehydrated air or gas sealed between the sheets of glass remaining dry. The spacer tube, which serves as a significant barrier to moisture vapour, also serves to support and control the depth of the sealant used.

Dessicant To ensure that the air or gas in the sealed cavity between the sheets of glass remains dry, it is necessary to fill the hollow of the spacer tubes with a dessicant which will absorb moisture vapour which might otherwise condense to water on the inside faces of the glass of the unit. The visible face of the spacer tubes is perforated with micro-porous holes to facilitate absorption of moisture vapour by the dessicant. The butyl-based spacer bar has dessicant in the face of the bar.

The sealed space between the two sheets of glass is usually filled with dehydrated air. To provide somewhat better thermal insulation the space may be filled with Argon gas and for better sound insulation it can be filled with SF gas.

Edge seals To hermetically seal the two sheets of glass to the spacer tube it is necessary to run some material around the outside of the spacer tube that will adhere strongly to both the spacer tube and the sheets of glass. To hold the spacer tube in place between the two sheets of glass as an aid to assembly, it is common to use double-sided tape. The sealant in liquid form is then gunned into the space between the two sheets of glass and the spacer tube. The sealants commonly used are polysulphide, polyeurethane, silicone, epoxy-polysulphide or hot melt butyl. As the double-sided tape does not act as a permanent sealant, this type of seal is described as a single seal, Fig. 65. As an alternative to double-sided tape a hot-melt bead of butyl or polyisobutylene is applied to the sides of the spacer on to which the sheets of glass are pressed to keep them in place and then the seal is run in the space between the sheets of glass and the spacer tube. Properly applied the hot-melt butyl will act as a primary seal and the edge sealant as a secondary seal. This dual seal method may provide a longer useful life than a single seal. Fig. 65 is an illustration of edge seals to IG units.

As an alternative to hollow spacer tubes an extruded butyl-based bar with integral aluminium strip may be used, as illustrated in Fig. 65. The bar is self adhesive and at once serves to keep the two sheets of glass in place and acts as edge seal by virtue of the aluminium strip and the butyl bar. This much facilitates the fabrication of IG units and is extensively used.

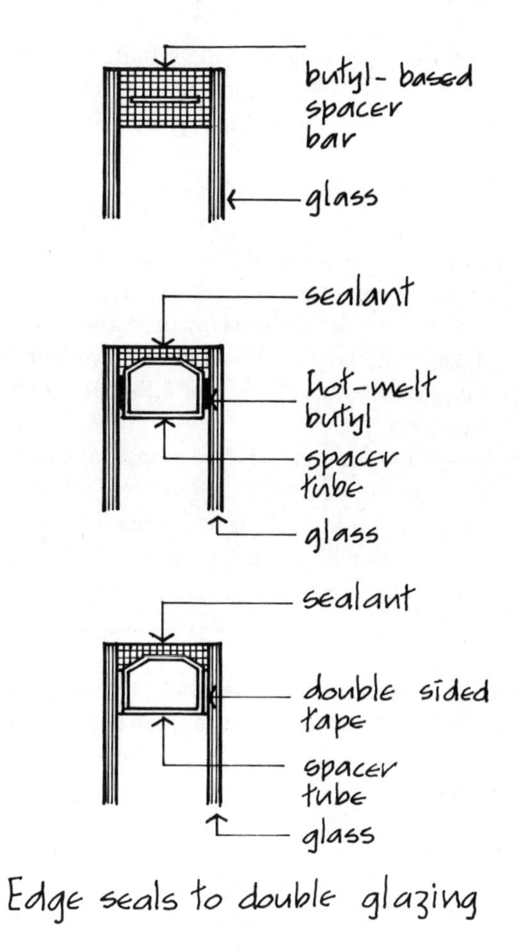

Fig. 65

For appearance, the overall depth of spacer tube and edge sealant and depth of spacer bar should be such that the edges of the face of the spacer that are visible through the glass – the sight faces – should not be visible from outside the glazed unit when it is fixed in position in the window. Likewise for appearance, the sight faces of spacer tubes can be finished, colour coated or anodised.

The useful life of a double glazed unit depends mainly on the integrity of the seal or seals as a barrier to moisture vapour penetrating the space between the sheets of glass, to the extent that the dessicant is unable to absorb most of the moisture vapour that will in consequence condense to water on the cold inside face of the glass. Double seal units generally have a somewhat longer life than single seal units.

It is generally accepted that the useful life of sealed double glazed units is at best up to 20 years. Some manufacturers give conditional guarantees of up to 15 years, conditional on workmanship in handling and fixing the units and on the glazing materials used.

Setting blocks, location blocks

To accommodate movement between glass and window frames, casements or sashes, due to different thermal and mechanical movements, a minimum clearance must be allowed all round IG units of from 3 to 5 depending on the size of the unit and 6 at sill level for drained glazing.

Setting blocks are used between the bottom edge of IG units and the window to support and centralise the unit in place and prevent compression of mastic sealants. Location blocks are used between the edges of IG units, other than the bottom edge, to prevent movement of the unit within the window, as illustrated in Fig. 66.

Setting blocks and location blocks should be of some resilient, non-absorbent material such as sealed teak or mahogany, hammered lead, extruded uPVC, plasticised PVC or neoprene. The width of setting blocks should be equal to the thickness of the IG unit plus the backface clearance and at least 25 long. Location blocks should be 3 wider than the IG unit and at least 25 long.

Fig. 66

Distance pieces are used to prevent displacement of glazing compounds or sealants by wind pressure on the glass, by retaining the IG unit firmly in the window. Distance pieces should be used except where load bearing tapes or putty are used for stepped units. Distance pieces should be the same thickness as face clearance and made of a resilient, non-absorbent material similar to that for setting blocks.

Sight size, sight line Sight size is the actual size of the opening that admits daylight and the sight line is the perimeter of the sight size. The sight line is determined by the edges of the frame, casement or cash and the exposed edge of the glazing material. Practice is to fix or apply glazing material so that the sight line on the inside of glass coincides with that on the outside and just above the exposed face of the spacer tube or bar in the IG units, for appearance sake.

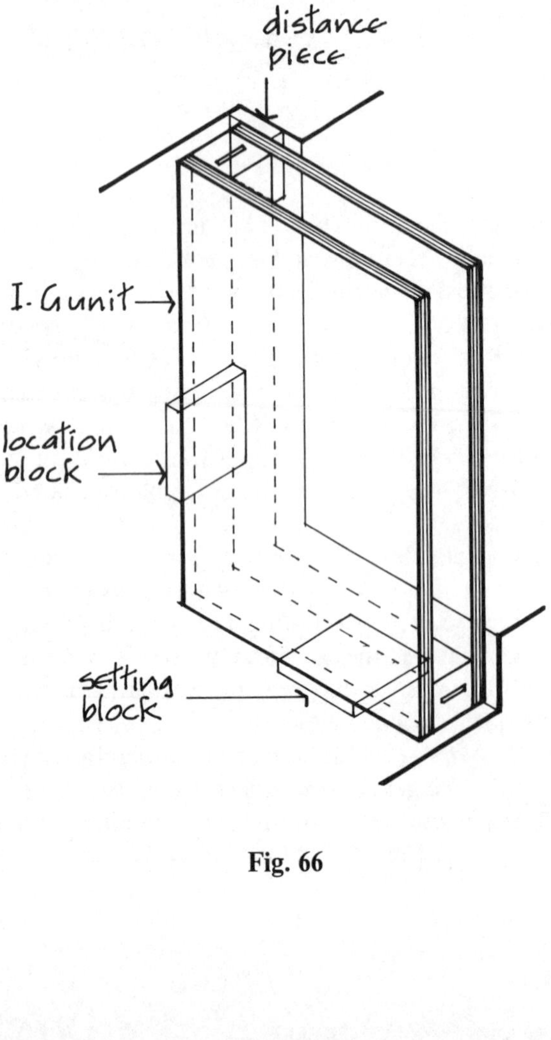

GLAZING

Single glazing

The operation of fixing glass in windows, doors and openings is termed glazing.

Glass must be accurately cut to size to provide an edge clearance between the edges of the glass and the bed of the rebate to allow for variations in the sash or frame and of the glass and to facilitate setting the glass in position. An edge clearance of 2 for putty glazing and 3 for other methods of glazing for glass up to 6 thick and up to 5 for thicker glass must be made. To secure glass in the glazing rebates with the requisite edge clearance all round, setting blocks are placed below the glass as illustrated in Fig. 67, and also at the sides and top as necessary. These setting blocks are of PVC, hammered lead, hard nylon or hardwood from 25 to 150 long and of the same thickness as the edge clearance.

The two common methods of glazing are putty and bead glazing.

Glazing with putties For glazing to absorbent wood sashes, putty made from linseed oil and fillers is used.

Putty is spread in the primed glazing rebate and setting blocks are pressed into the putty; the glass is then placed in the rebate and pressed on to the putty until there is about 1.5 mm of back putty between the glass and the back of the rebate. The glass is then secured in position with metal sprigs that are tapped into the rebate. Sprigs are small, cut, headless nails. Putty is then spread in the rebate around the glass and finished off at an angle as illustrated in Fig. 67. The sloping outside putty which is weathered should be painted as soon as the putty is firm enough, to protect the putty from drying and cracking.

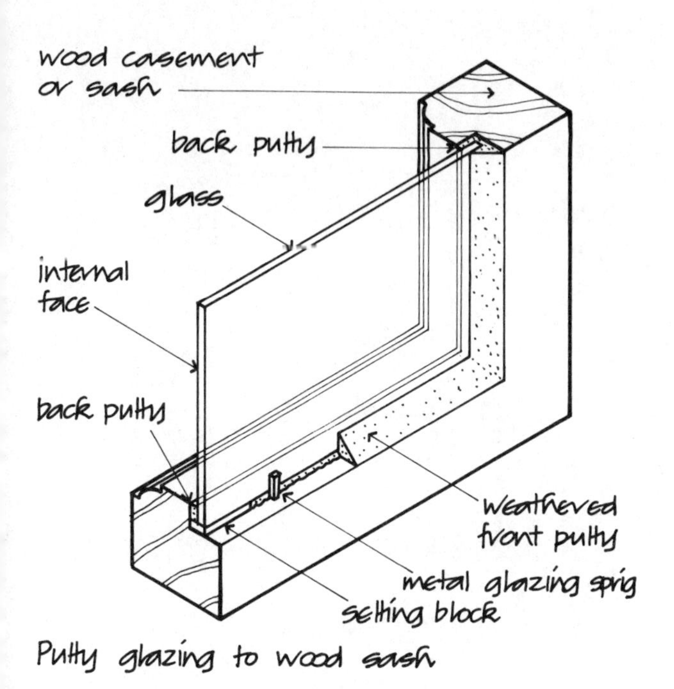

Fig. 67

For glazing to non-absorbent galvanised steel sections, putty made from blended vegetable oils and fillers which adheres to non-absorbent surfaces and sets hard, is used. The glazing procedure is the same as that for glazing to wood except that metal clips, set in holes in the metal frame are used to secure the glass in position as illustrated in Fig. 68.

Glazing with beads With this system of glazing the glass is bedded in non-setting glazing compound or a strip sealant and secured in position with wood or metal beads that are screwed or clipped into position all round the glass. The beads may be fixed either from inside or outside the glass. Inside bead glazing is commonly used for softwood beads to protect them

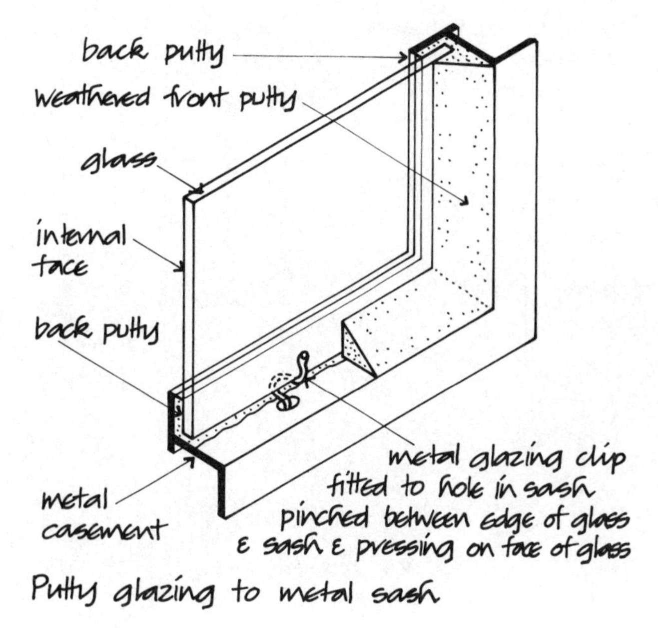

Fig. 68

from weather decay and with all types of bead to protect against the security risk of the beads being unscrewed or unclipped from the outside.

External bead glazing and internal bead glazing for large squares of glass are usually carried out with strip sealant for both wood and metal frames. The strip sealant is made in the form of a tape of butyl polymers in various widths and thicknesses.

A strip of load-bearing mastic tape is run all round the rebate upstand, setting blocks are put in position and the glass is pressed into position. A second strip of mastic tape is run round the edges of the glass and the beads are then pressed into the tape and secured in position. A sealant capping is applied by gun to the outside edges of the glass, tooled into place and finished with a smooth chamfer to shed water, as illustrated in Fig. 69.

Double glazing

Premature failure of the edge seal to double glazed units will occur if water is in contact with the seal for an extended period. The two methods of glazing that are used – vented or vented and drained method and the solid bedding method – are designed to prevent access of water to the seal either by ensuring that any water that penetrates to the seal is soon removed by ventilation or drainage or by solid bedding.

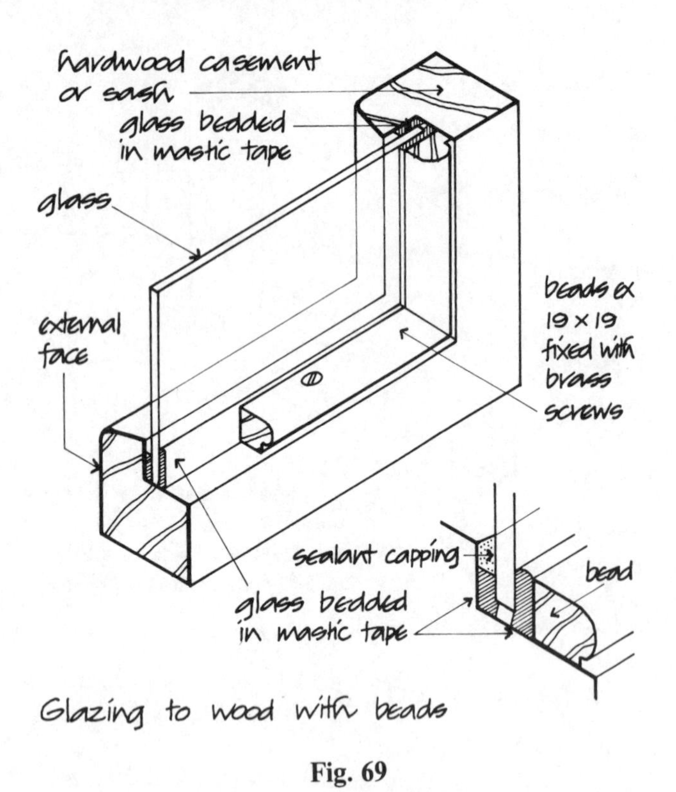

Fig. 69

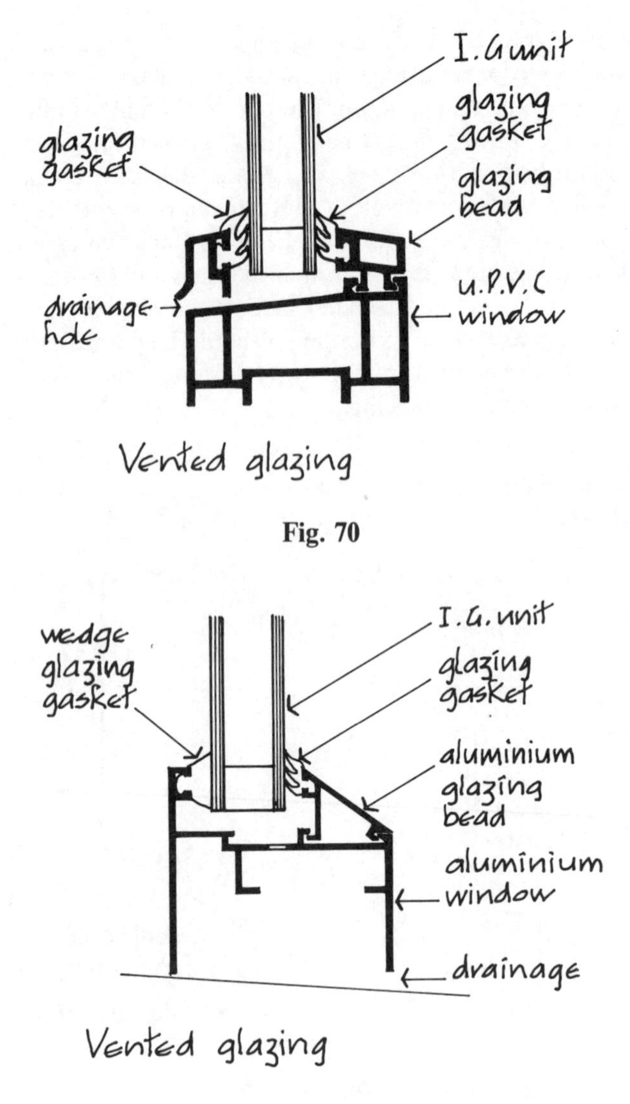

Fig. 70

Fig. 71

Vented and drained method The vented method of glazing is designed to rapidly remove any water that has penetrated to the edge seal of units, by drainage or ventilation or both. Drainage is achieved by sloping surfaces that conduct water to drainage slots. Drainage alone may not remove all water and the drainage slots are designed to provide ventilation for drying out. Slots 20 × 3 are generally adequate for both drainage and ventilation. To prevent wind driven rain being blown into the slots a hood is formed over each slot.

The Glass and Glazing Federation include vented, drained and vented and drained methods of glazing as being preferred to other methods of glazing for IG units because of the provisions for ventilation and drainage to remove any water that may have penetrated to the edge seal.

Insert gasket section glazing for aluminium and plastics frames with grooves and nibs for gaskets, PVC or synthetic rubber sections, profiled to fit into grooves, are cut overlength for tight fit and fitted into grooves as shown in Fig. 70.

As an alternative a wedge shaped gasket of PVC or synthetic rubber may be fitted internally as illustrated in Fig. 71. The advantage of the wedge shaped gasket is that it tends to wedge the IG unit between itself and the outer gasket.

Cellular adhesive strips or mastic tape glazing Profiled closed cell synthetic or closed cell PVC foam sections with self adhesive backing or pre-shimmed mastic tape are applied to the back rebate. The IG unit is set in position and profiled closed cell synthetic or closed cell PVC foam with self adhesive backing or pre-shimmed mastic tape is applied to the beads and the beads fixed in position up to the IG unit.

Load bearing tapes or synthetic rubber sections and sealant capping glazing The tapes or sections are applied to the rebate, the IG unit is fitted in position and the tapes or sections are fitted to the glazing beads that are fixed in position against the IG unit. The one or two part sealant capping is then applied both sides and finished with a smooth chamfer.

Solid bedding glazing

Solid bedding glazing depends on enclosing the double glazed unit in a sealant that is impermeable to liquid water and has a low moisture vapour transmission rate. The sealant used should be compatible with the material of the window, should be continuous all round the unit both inside and outside and should adhere to or make close contact with glass and window. The solid bedding sealant serves as a barrier to rain externally, condensation internally and as a barrier to water from washing, especially when a squeegee is used.

Silicone sealants are impermeable to liquid water but permeable to moisture vapour. Used as a sealant externally they provide an effective seal against water. Any water that might find its way past the seal, due to defects in the continuity of the seal, can to an extent evaporate through the seal as moisture vapour to outside air. A moisture vapour permeable (mvp) seal therefore acts as a ventilated seal depending on the area of the exposed surface of the seal and the exposure of the window. With mvp seals externally a seal with low moisture transmission rate, such as butyl mastic strip, is used internally against entry of moisture from inside.

For solid bedding sealants to be effective as a barrier to water they must be applied to clean, dry surfaces, applied with care and must be continuous, particularly at vulnerable corners. The advantage of factory glazing is that ease of access to frames, control of quality of workmanship and a clean, dry atmosphere are more likely than with on site glazing.

A preferred method of solid bedding, which is suitable for most window and door frames in low, medium or high exposure rating areas, utilises a moisture vapour permeable sealant, such as one component silicone, externally and a low permeability sealant, such as load bearing tape, polysulphide sealant or non-setting compound, internally, with an internal capping where severe condensation is likely. This method of solid bedding is illustrated in Fig. 72.

Other acceptable methods of solid bedding, recommended by the Glass and Glazing Federation, employ curing sealants or mastic tapes and sealant capping. One or two part curing sealant compounds such as silicone, polysulphide, polyurethane, polybutadiene or epoxy sulphide are run by gun application around the IG unit.

In the other method a non-setting sealant is run as a perimeter sealant, then load bearing tape or closed cell self-adhesive tape is applied to glass edges and the bedding is completed with a capping sealant of silicone, polysulphide, polyurethane or acrylic sealant, as illustrated in Fig. 72.

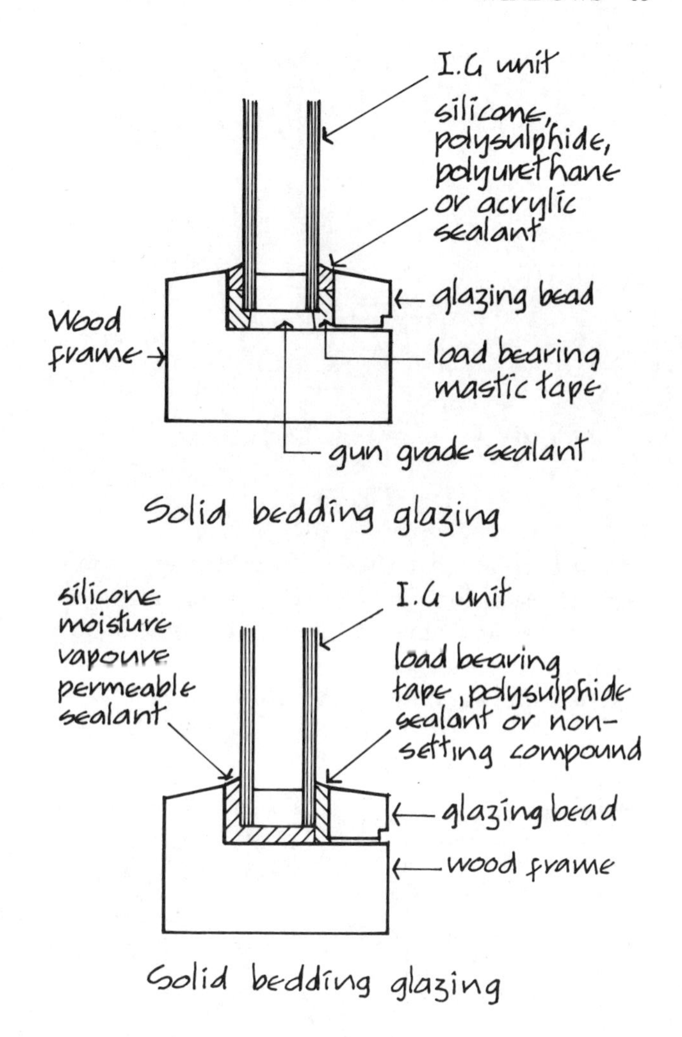

Fig. 72

Where there is no external access for glazing, a system of one part curing sealant as perimeter filling and heel and toe beads with back and front bedding of pre-shimmed tapes or self adhesive cellular sections, may be used.

For stepped insulating glazing units where exposure rating is low, solid bedding systems of linseed oil putty for softwood and metal casement putty for softwood pressure treated, hardwood and steel frames, are used, with the putty painted (Fig. 73).

Double glazing for sound insulation

The comparatively small space between the sheets of glass in IG units and contact between the sheets

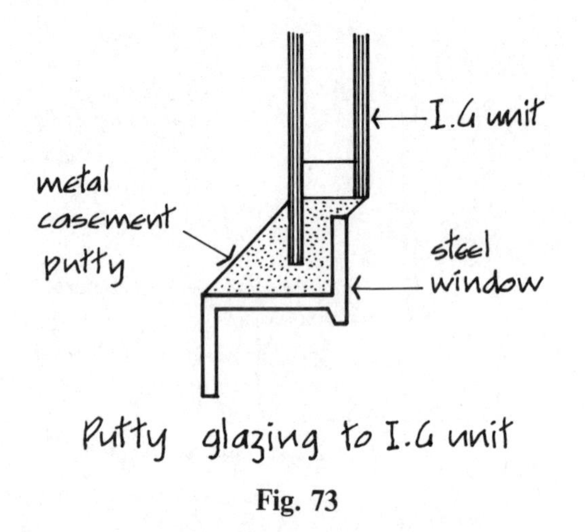

Fig. 73

through the spacer bar or tube affords little reduction in sound transmission. To increase sound insulation the sheets of glass in double glazing should be spaced some distance apart and not make contact through some rigid material, as illustrated in Fig. 74.

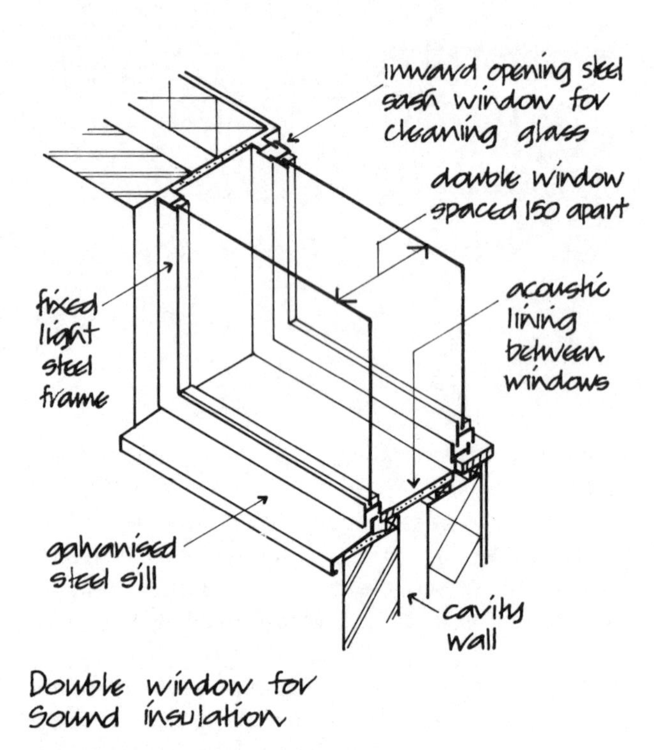

Fig. 74

WINDOW SILLS

It is good practice to set the outside face of windows back from the outside face of the wall in which they are set so that the reveals of the opening give some protection against driving rain. Wind-driven rain which will run down the impermeable surface of the window glass to the bottom of the window should be run out from the window by some form of sill. The function of an external sill is to conduct the water that runs down from windows away from the window, and to cover the wall below the window and exclude rain from the wall. The material from which the sill is made should be sufficiently impermeable and durable to perform its function during the life of the building.

External window sills are formed either as an integral part of the window frame, as an attachment to the underside of the window, or as a sub-sill, which is in effect a part of the wall designed to serve as a sill. Most materials used for external sills are pre-formed so that the dimensions of the sill determine the position of the window in relation to the face of the wall. As a component part of a wall external sills should serve to exclude wind and rain and provide adequate thermal insulation to the extent that the sill does not act as a cold bridge to heat transfer.

The internal sill of a window serves the purpose of a finish to cover the wall below the window inside the building, and as a stop for wall plaster. The material used for internal sills should be easy to clean and materials commonly used are painted softwood, plastic, metal, and clay tiles.

External sub-sills

These sills are constructed as a capping to a solid wall below windows, as a weathering to run water away from the window, to protect the wall below and as a finish between the wall and the window. The materials used are natural stone, cast stone, concrete, tile and brick. Natural sedimentary and igneous stone sills are little used due to the scarcity and cost of the material. Slate sills, which are readily available but comparatively expensive, are used to some extent.

Slate sills are cut from some fine-grained natural slate and finished in a range of standard and purpose-made sections to suit either timber or metal windows. The slate is finished with a sawn, sanded or fine-rubbed finish. The natural colour of the slate varies from light grey and green to blue and black. This dense, durable material is impermeable to water, requires no maintenance, is a poor thermal insulator and being brittle may crack due to movement of the building fabric. These sills should be cut and finished in one length to avoid the difficulty of making a weathertight joint in the material, a limitation which restricts their use to comparatively narrow windows.

The two standard sections illustrated in Fig. 75 are for use with standard steel windows which are screwed to the fillet fixed to the sill. Sills for use with wood windows are finished with a groove for a water bar. These sills are bedded in mortar on the wall below the window with the drip edge projecting some 38 beyond the face of the wall.

The sill shown in Fig 75 is cut in one length to fit between the jambs of the window opening, with the vertical joint between the end of the sill and the jamb filled with mortar and pointed with non-setting mastic compound. This is a satisfactory joint for normal exposure to driving rain. In positions of severe exposure it is wise to incorporate a lead d.p.c. tray under the sill, cut and shaped to extend some $\frac{1}{2}B$ each side of the opening, to exclude water.

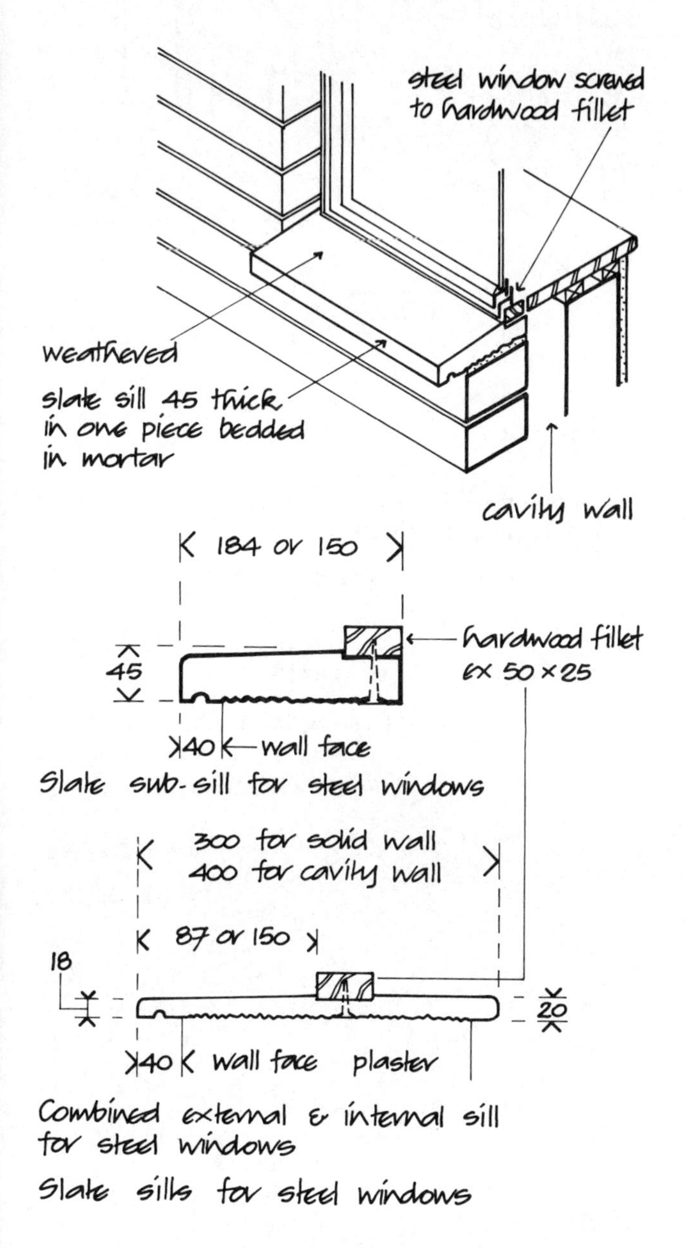

Fig. 75

These sills may be cut and finished with stooled ends for building into the jambs of openings. This costly end detail does not provide a satisfactory finish unless the sill is a full brick thickness, otherwise untidy cutting of brickwork is necessary. Where the ends of these comparatively thin sills are built in, differential settlement between the more heavy loaded walls at the jambs and the less heavily loaded walling under the window, may cause the sill to crack.

The combined external and internal sill shown in Fig. 75 acts as a cold bridge across the thickness of the wall and will encourage condensation on the top of the internal sill. Because it is comparatively thin the sill is liable to crack due to handling and slight movements in the wall and is used for only narrow window openings.

Precast concrete and cast stone sills Cast stone is made from a mix of cement and crushed natural stone as a finish to concrete to give the appearance of natural stone. The core of the stone is usually of natural aggregate concrete with an external facing of cast stone cast integrally with the concrete core. The cast stone finish, which may resemble the colour and texture of a natural stone initially, does not weather like the natural material and is liable to irregular and unsightly stains due to weathering.

Precast concrete sills are made from natural aggregate concrete, cast and compacted by vibration in moulds with either a natural concrete finish or an integrally cast finish of fine aggregate and cement with or without some colouring additive in the mix, or by the use of selected coloured aggregates. There is little to distinguish precast concrete from cast stone other than the finish of the latter that attempts to resemble a natural stone. These sills are cast in a range of standard sections or to the specifier's detail, with either square ends or stooled ends for building into jambs. The top of the sill is finished with a groove for a water bar for wood frames and a nib for steel windows.

The sill shown in Fig. 76 has square ends to fit inside the jambs of the opening with a d.p.c. tray under and behind the sill extending $\frac{1}{2}B$ each side of the sill to exclude rain. This under sill d.p.c. is particularly important in exposed positions and where the sill is in two or more lengths. The drip edge projects some 38

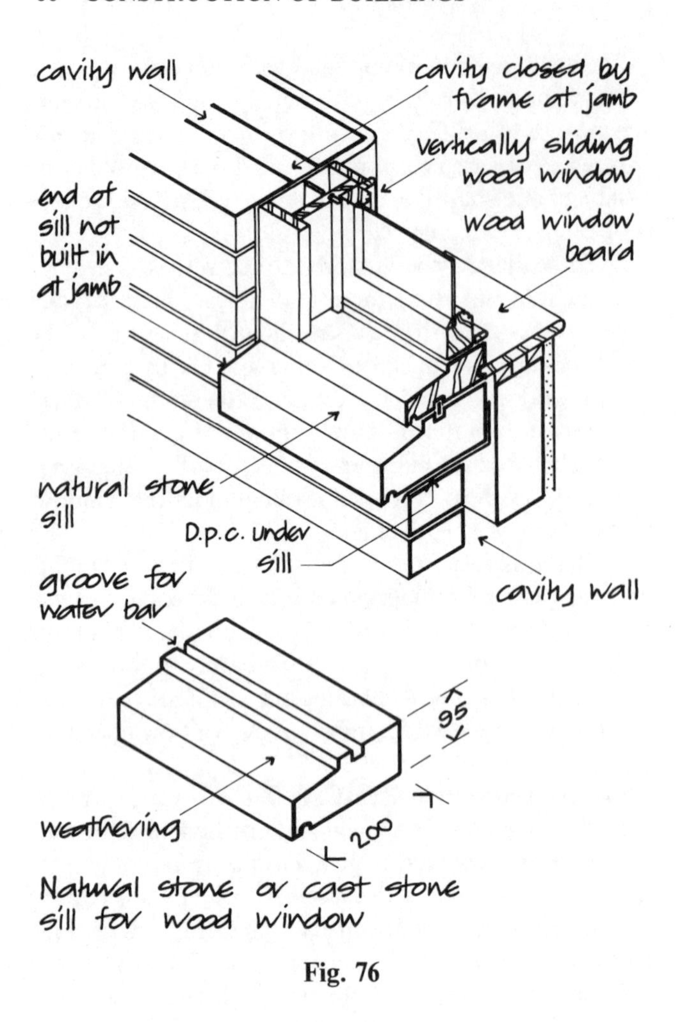

Fig. 76

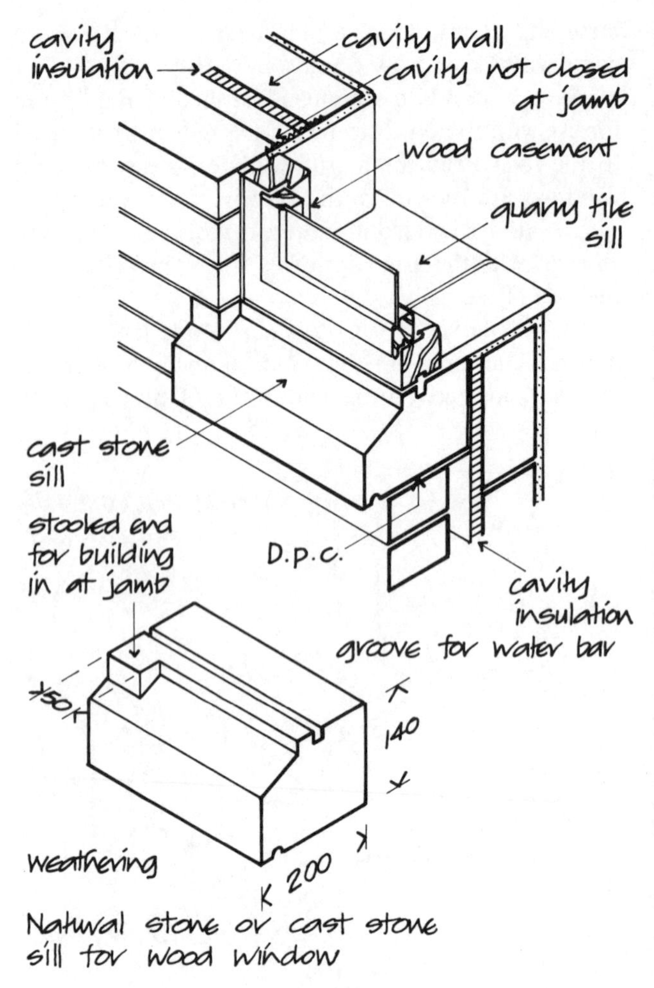

Fig. 77

from the face of the wall. The sill shown in Fig. 77 is of a depth to match the brick courses so that the stooled ends can be built in at the jambs without untidy cutting of bricks. In positions of moderate exposure, with sills in one length, it would be reasonable to omit the under sill d.p.c.

Natural stone sills are cut and used in the same way as precast concrete sills.

D.p.c. below sills Sub-sills that are constructed of small units such as brick and tile, sills that are jointed such as concrete and stone and sills that butt to the jambs of the window opening may not be entirely effective in excluding rain that may penetrate joints in the sill and, at the end of the sill, to the wall below. Particularly in positions of severe exposure to driving rain it is a wise precaution to build in some form of d.p.c. below sills. The material used for the d.p.c. is usually one of the non-ferrous sheet metals, lead being favoured for its ease of working. Fig. 78 shows a lead

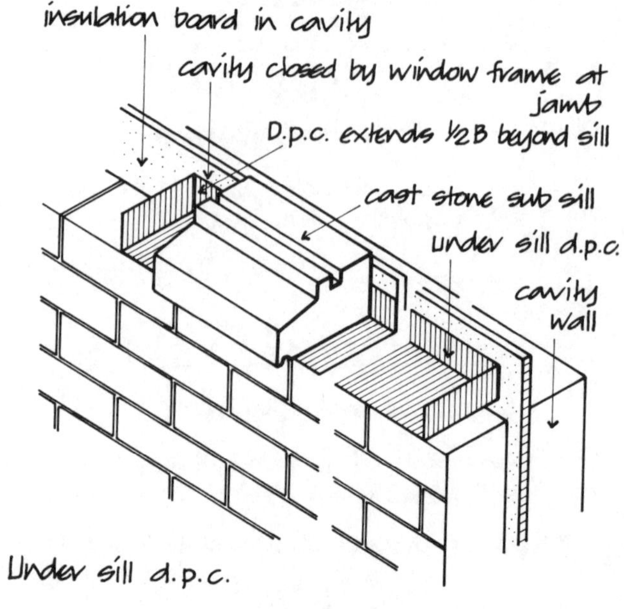

Fig. 78

d.p.c. that is cut and dressed to fit behind and below a cast stone sill. The d.p.c. is cut to extend ½*B* each side of the sill as a seal against the possibility of rain penetrating the butt end joints of the sill to the jambs of the opening. Similarly sheet lead or other suitable material should be bedded under sub-sills of other materials to exclude rain wherever there is a likelihood of rain penetrating the sill or joints between sill and jambs.

Tile sub-sills are formed by bedding clay or concrete roofing tiles in two courses breaking joint in cement mortar as illustrated in Fig. 79. The tiles should be laid with a slope or weathering not less than the minimum slope recommended for their use as roofing tiles. The tiles may be laid with their long axis parallel to the wall face to avoid cutting tiles or at right angles to the wall face and cut to suit the position of the window in the wall thickness. The tiles are laid to project some 38 from the wall face to act as a drip edge. The tiles may be finished square within the reveals of the opening or notched around the angle of the jambs. To protect the wall below the sill a d.p.c. should be built in and it should extend ½*B* each side of the opening in positions likely to suffer severe exposure. This is a comparatively cheap form of sub-sill.

Brick sub-sills A sub-sill of either standard or special bricks is formed by bedding standard bricks on edge or special bricks weathered outwards. Standard bricks, other than engineering bricks, laid on edge as a sill may in time disintegrate due to saturation by rain running down the window and the action of frost. In all but situations of moderate exposure it is wise to use engineering bricks or special bricks of engineering quality. Standard bricks may be laid on edge either flush with the wall or weathered to slope out and projecting to provide a drip edge. The special plinth bricks illustrated in Fig. 80 are bedded below the window as a sill and finished flush with the outside face of the wall. An under sill d.p.c. should be built in below all brick sills.

Metal sills Most metal window manufacturers provide standard section metal sills for fixing to the frame of their windows to give cover and protection to the wall below the window. The projection of the sill beyond the face of the wall is determined by the 25 width of the welded-on stop ends, which in turn determines the position of the window in the thickness of the wall. The joint between the ends of the sill and the

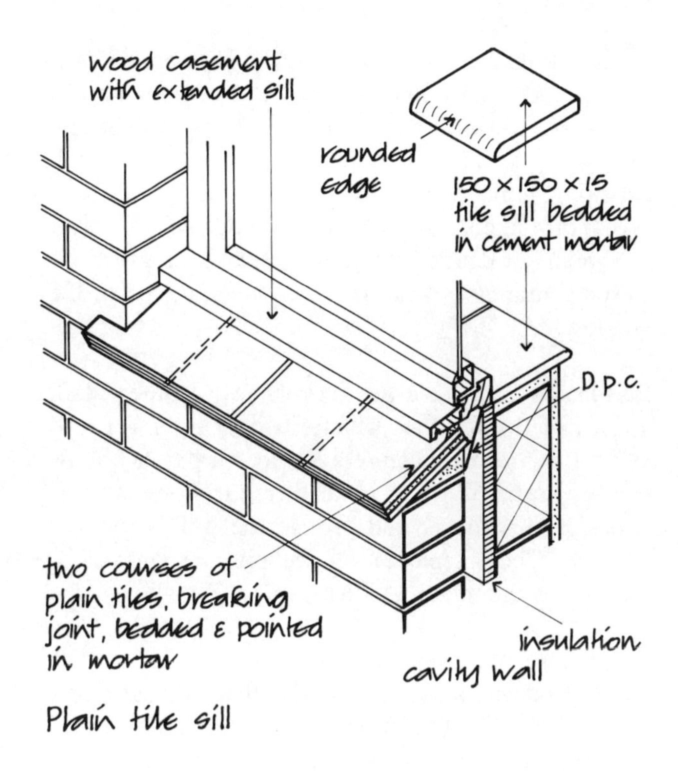

Fig. 79

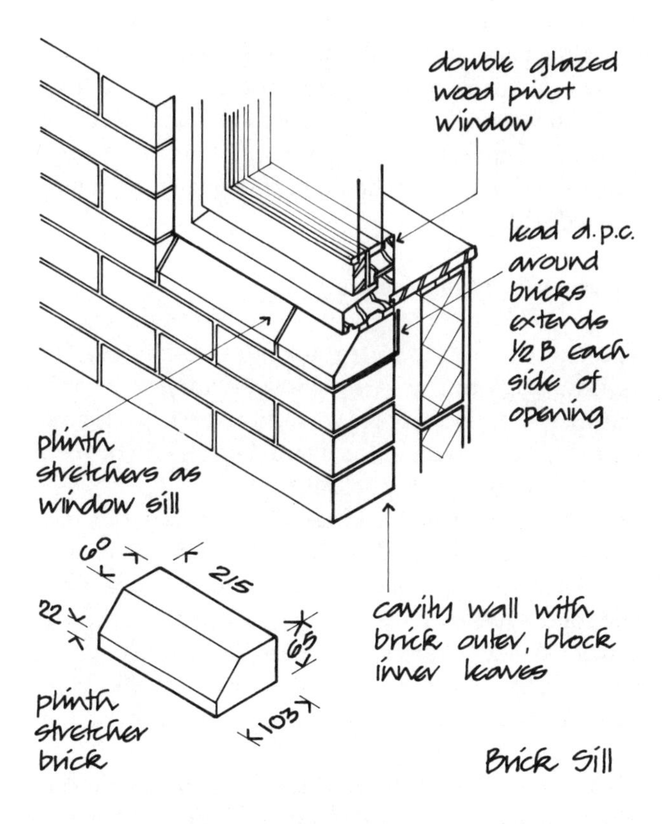

Fig. 80

jambs should be pointed with mastic. The steel sill itself will exclude rainwater but the end joints may be vulnerable to water particularly in positions of severe exposure. A d.p.c., in the course below the sill, extending either side of the opening, might be a wise precaution in conditions of severe exposure.

Similarly, extruded aluminium section sills are made to suit aluminium windows as an integral part of the window.

Plastic sills Most plastic windows have an integral sill as part of the window, which fits over some form of sub-sill. Some manufacturers provide a separate hollow section plastic sill which is weathered to slope out and is designed to cover and protect the wall below the window. These separate sill sections are clipped or screwed to the frame, as illustrated in Fig. 49.

Wood sills Most standard section wood windows can be supplied with a wood sill section that is tongued to a groove in the sill of the frame so that it projects beyond the window either to cover and protect the wall below or to overlap a sub-sill as illustrated in Fig. 79. The sill is designed to project some 25 to 38 from the face of the wall as a drip edge. Because softwood sills must be protected with a sound paint film which must be regularly maintained, otherwise rain penetration will cause rapid deterioration of the wood, it is sensible to use hardwood sills in positions of moderate and severe exposure. The sill fits between the jambs of the opening, and the end butt joints should be bedded in mortar and pointed with mastic.

Internal sills, windowboards

The internal sill surface to a window is covered for appearance with a painted softwood window board, clay tiles, slate, metal or plastic sections designed for the purpose. The surface of the internal sill should be such that it can easily be kept clean.

A common form of internal sill is a softwood board termed a window board cut from 19 or 25 boards and wrought smooth on one face and square or rounded on one edge. The board may be tongued to fit to a groove in wood window frames. The board is nailed to plugs or bearers nailed to the wall so that it projects some 25 or more from the finished face of plaster, as illustrated in Fig. 76.

Clay or concrete tiles may be used as an internal sill. The tiles are bedded in mortar on the wall and pointed in cement, as illustrated in Fig. 81. Rounded edge tiles

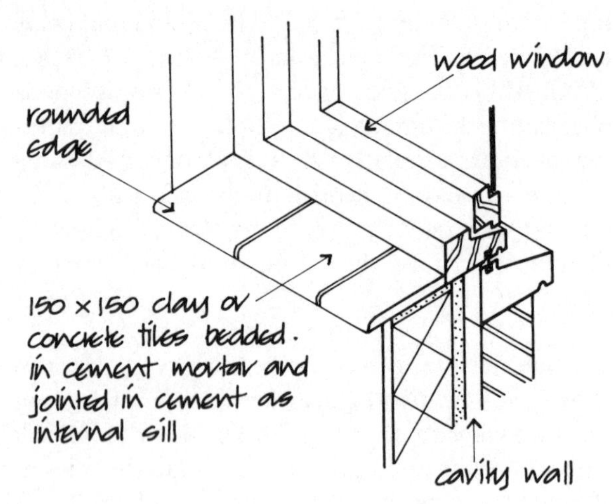

Fig. 81

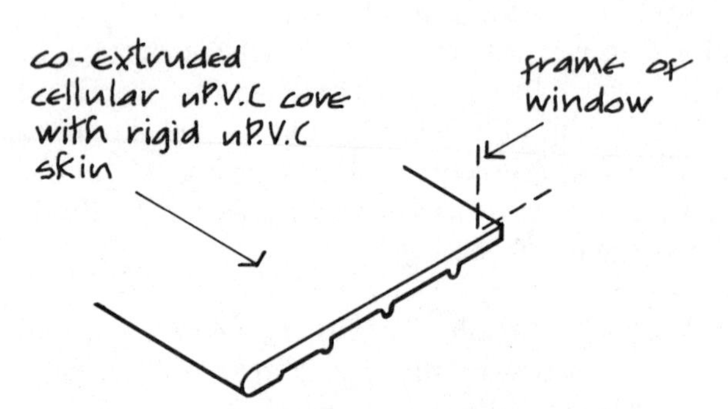

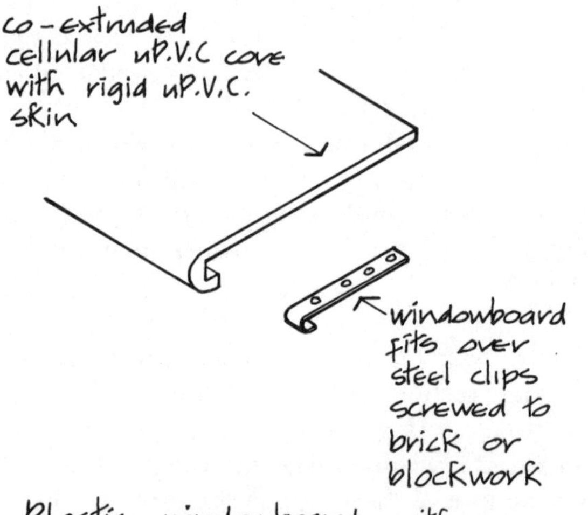

Fig. 82

are used and laid to project beyond the plaster face.

Various sections of plastic windowboard are made for use with uPVC and other windows and as replacement windowboards. The thin sections are of co-extruded uPVC with a closed cell, cellular core and an integral, impact modified uPVC skin. The thicker sections are of chipboard to which a uPVC finish is applied on exposed faces.

The advantage of these windowboards is that they do not require painting and are easily cleaned. The disadvantage is that they are fairly readily defaced by sharp objects and the damage cannot be made good.

Plastic windowboards are cut to length and width with woodworking tools and fixed with concealed fixing clips, mortar bedding, or silicone sealant adhesive or are nailed or screwed to prepared timber grounds with the nail or screw heads covered with plastic caps (Fig. 82).

CHAPTER TWO

DOORS

Since the first early settlements, wood has been the traditional material for doors. The ready availability of the natural material which can be cut, shaped and joined with simple hand tools, made it the obvious material for making doors. In the construction of doors, from the crude cottage door of boards nailed to battens to the larger, sophisticated, framed panelled and moulded doors, there is a wide variety of types of wood to choose from to suit utility and appearance. The grain and colour of the chosen wood may be used as a decorative feature by the application of oil, wax, polish or varnish.

Wood doors are still today commonly chosen for both internal and external use. For a time, early in the twentieth century, steel – framed as fully glazed doors – was in vogue to match the then fashionable steel window. Due to the progressive, destructive corrosion these suffered and due to change of fashion, the steel door soon lost favour. Aluminium, framed mainly as fully glazed doors, was and is today used as a substitute for steel. These metal doors suffer the disadvantage of being ready conductors of heat and are subject, therefore, to condensation.

During the last few years plastic doors have been made and used, principally as a substitute for wood doors to avoid the expense of the comparatively frequent painting that wood needs for maintenance. Fibre glass reinforced plastic and uPVC doors have been made to resemble the appearance of panelled wood doors. Doors made as a shell of fibre glass reinforced plastic with an insulating core have not for long stood up to the wear common to a door due to the variability of the moulding techniques and the brittle nature of the material. Plastic doors made as a frame of corner-welded, hollow extrusions of uPVC with moulded panels of fibre glass or acrylic, pressed to resemble the appearance of a framed, panelled wood door, do not look much like a wood door, offer poor security and do not for long stand up to normal use.

Wood remains the material best suited for making doors.

A door is a solid barrier to a doorway or opening, that can be opened for access and closed to deny access for privacy and security, and serves as a thermal, acoustic and fire barrier and as a weather barrier as part of an external wall.

A doorway is an opening in a wall or partition for access and a door frame or lining is the timber, metal or plastic frame or lining fixed in the doorway or opening to which the door closes on hinges, pivots or runners.

FUNCTIONAL REQUIREMENTS

The primary function of a door is:

Means of access

and the secondary function is:

Privacy.

The functional requirements of a door as a component part of a wall or partition are:

Structure – strength and stability
Resistance to weather
Durability and freedom from maintenance
Fire safety
Resistance to the passage of heat
Resistance to the passage of sound
Security.

For access alone, a doorway or opening in a wall or partition will suffice. A door is used as a hinged barrier which can be opened for access and closed for privacy, as a barrier to transfer of heat and sound and as a barrier to the spread of fire.

Before central or space heating of whole buildings was as common as it is today and when open fires and stoves were the common form of heating, doors served the very useful purpose of containing heat in separate rooms. Today this is no longer a need where whole buildings have space heating, and with changed patterns of living and working there has been a move to dispense with doors in both residential and office buildings. The so called, 'open plan' living and working arrangements are much used, with common living areas and office working spaces, and doors confined to use in toilets, bathrooms and private study or work places in houses and offices.

Means of access

A door opening should be sufficiently wide and high for reasonably comfortable access of people. An accepted standard width and height of 762 and 1981 (the metric equivalent of the former imperial sizes of 2′6″ × 6′6″) has been established for single leaf doors. A narrower width may be adequate for a single person yet not comfortable for a person carrying things. The standard height makes allowance for all but the few exceptionally tall people. For large spaces or rooms a greater width is often adopted for appearance. The standard width and height has been chosen as convenient for the majority of people. Double leaf doors are commonly used for access to grand, large spaces or rooms for appearance, and for convenience in busy corridors.

The side on which the door is hung by hinges or pivots, its hand or handing and whether it opens into or out of a space or room, are a matter of convenience in use. By convention doors usually open into the room of which they are part of the enclosure. When single rooms were separately heated it was common to hang doors so that they opened with the leaf of the door moving towards the centre of the room, to avoid too vigorous an inrush of colder outside air. This inconvenient arrangement is reversed with space heated buildings. There have been systems of describing the hand of doors by reference to opening in or out and as either left hand, right hand or clockwise, anti-clockwise. These are of very little use because of the difficulty of clearly defining what is outside and what inside.

Privacy

Doors should serve to maintain privacy inside rooms to the same extent as the enclosing walls or partitions. For visual privacy, doors should be as obscure as the walls or partitions. For acoustic privacy doors should offer the same reduction in sound as the surrounding walls or partitions and be close fitting to the door frame or lining and be fitted with flexible air seals all round. These seals should fit sufficiently to serve as an airborne sound barrier but not so tightly as to make opening and closing the door difficult.

Strength and stability

Whether it be side hinged, top and bottom pivoted, or on tracks to slide and fold, a door must have adequate strength to support its own weight and suffer knocks and minor abuses in service, as well as adequate shape stability for ease of opening and accuracy of closing to the frame or lining. Both strength and shape stability depend on the materials from which a door is made and the manner in which the materials are framed as a door.

British Standards Institution Draft for Development 171:1987 *Guide to specifying performance requirements for hinged or pivoted doors*, gives guidance on performance criteria and was published as a prelude to the publication of a Standard.

The tests suggested are applied in general to complete door assemblies of door leaf, frame or lining and hardware because the performance of a door is affected by the component parts of door assemblies. Provision is also made for performance requirements for door leaves in isolation, to assist those manufacturers – the majority – who make doors alone rather than door sets or assemblies.

To allow for the wide range of use of doors, categories of duty related to use are suggested, Table 10 from Light Duty (LD), through Medium Duty (MD) and Heavy Duty (HD) to Severe Duty (SD).

The functional requirements of doors are specified, relating to both the component parts and the whole of door sets or assemblies, as:

Strength
Operation, the ability to be operated
Stability or freedom from excessive distortion in use
Fire resistance
Sound insulation.

For external doors the additional requirements are:

Weather resistance
Thermal insulation
Security.

Strength

A door assembly should be strong enough to sustain the conditions of use set out in Table 10. The suggested tests are for resistance to damage by slamming shut or open, heavy body impact, hard body impact, torsion due to the leaf being stuck in the frame, resistance to jarring vibrations and misuse of doorhandles.

Operation

A door should be easy to open, close, fasten or unfasten and should stay closed when shut. Tests for the forces required to operate door assemblies by 95% of females in the specified age groups are defined.

Dimensional stability

A door should not bow, twist or deform in normal use to the extent that its appearance is unacceptable or it is difficult to open or close.

The dimensional stability of wood, metal and plastic doors is affected by temperature and humidity differences. Wood doors are affected mainly by temperature and humidity, metal doors by temperature, and plastic by thermal and hygrothermal movements.

Bow in doors is caused by differences in temperature and humidity on opposite faces, which may cause the door to bow with a curvature that is mainly in the height of the door and should not exceed 10.

Twist is caused, particularly in panelled wood doors, where movement of the spiral grain, due to changes of moisture content, causes one free corner to move away from the frame. This should not exceed 10.

Table 10. Categories of duty for doors and doorsets

Category	Description	Examples
LIGHT DUTY (LD)	Low frequency of use by those with a high incentive to exercise care, e.g. by private house owners – small chance of accident occurring or of misuse.	Internal doors in dwellings. External doors in dwellings providing secondary access to private areas.
MEDIUM DUTY (MD)	Medium frequency of use primarily by those with some incentive to exercise care – some chance of accident occurring or of misuse.	External doors of dwellings providing primary access. Office to designate public areas but not used by public or by people carrying or propelling bulky objects.
HEAVY DUTY (HD)	High frequency of use by public and other with little incentive to exercise care – high chance of accident occurring and of misuse.	Doors of shops, schools, hospitals and other buildings, which provide access to designated public areas and which are used by public and others frequently carrying or propelling bulky objects.
SEVERE DUTY (SD)	Subject to frequent violent usage.	Doors of stockrooms, etc. commonly opened by driving trolleys against them! Doors in educational establishments subject to frequent impact by people.

Taken from DD 171: Performance of doors

Durability and freedom from maintenance

In the introduction to DD 171 the authors state that because aspects of performance are related to appearance, which is a subjective criterion, and the aspect of durability includes a number of factors that can be described but not quantitified, they have limited the recommended tests to those concerning strength.

Fire safety

Doors may serve two functions in the event of fire in buildings, firstly as a barrier to limit the spread of fire and secondly to protect escape routes.

To limit the spread of fire it is usual to divide larger buildings into compartments of restricted floor area by means of compartment floors and walls. Where doors are formed in compartment walls the door must, when closed, act as a barrier to fire in the same way as the walls. For this purpose doors must have a notional integrity, which is the period in minutes that they will resist the penetration of fire.

In Approved Document B, giving practical guidance to meeting the requirements of the Building Regulations 1991, Table B1 of Appendix B gives provisions of tests for fire resistance of doors, with minimum requirements in minutes for the integrity of doors. These are usually stated as, for example, FD 20, being a provision of 20 minutes minimum integrity for a fire door.

There should be adequate means of escape from buildings in case of fire, to a place of safety outside, which is capable of being safely and effectively used at all times. To meet this basic requirement it is usual to define escape routes from most buildings along corridors and stairways which are protected by fire barriers and doors from the effects of fire for defined periods. This latter function of a fire door is described as smoke control. The majority of doors along escape routes will need to serve as fire doors to resist spread of fire and to control smoke.

Resistance to the passage of heat

Doors, which do not usually form a major part of the area of external walls, make little contribution to overall heat loss when closed, so considerations of operation, strength, stability and security are of more importance in the construction of a door than resistance to heat transfer. To minimise air infiltration and draughts, weatherstripping should be fitted where it will not appreciably impede ease of operation. As glass offers poor resistance to heat transfer it is sensible to fix double glazing to reduce heat loss.

Resistance to the passage of sound

A door should afford reduction of sound for the sake of privacy and for those functions, such as lecture rooms, where the noise level is of importance. The heavier and more massive a door the more effective a barrier it is in reducing sound transmission. A solid panel door is more effective than a flimsy hollow-core flush door. To be effective as a sound barrier a door should be fitted with air seals all round as a barrier against airborne sound. Fig. 83 shows threshold seals that can be used to improve insulation against airborne sound.

Weather resistance

As a component part of an external wall a door should serve to exclude wind and rain depending on the anticipated conditions of exposure described for windows.

The justification and advantages of weatherstripping around an external door depend on the normal use of the door. While the advantage of weatherstripping in conserving heat and excluding draughts may outweigh the disadvantage of resistance to operating in domestic doors which are only occasionally opened, the disadvantage of weatherstripping that reduces ease of operation in more frequently opened doors, such as a shop door, may suggest dispensing with weatherstripping.

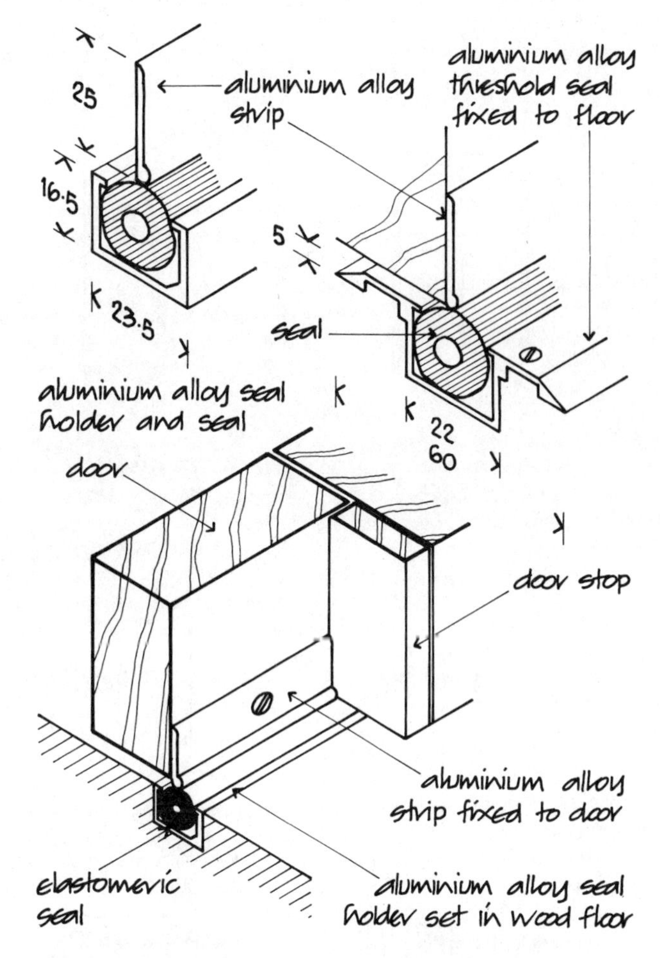

Fig. 83

For doors that may with advantage be weatherstripped, tests similar to those for windows are suggested.

Laboratory tests on doors show that external doors, particularly those opening inwards, are more susceptible to water leakage than windows. It is most difficult to design an inward opening external door which will meet the same standards of water tightness that are expected of windows, without the protection of some form of porch or canopy. For maximum watertightness a door will need effective weatherstripping which will to an extent make opening more difficult, and a high or complex threshold which may obstruct ease of access.

The recommendations in DD 171, therefore, suggest requirements for limited resistance to rain penetration.

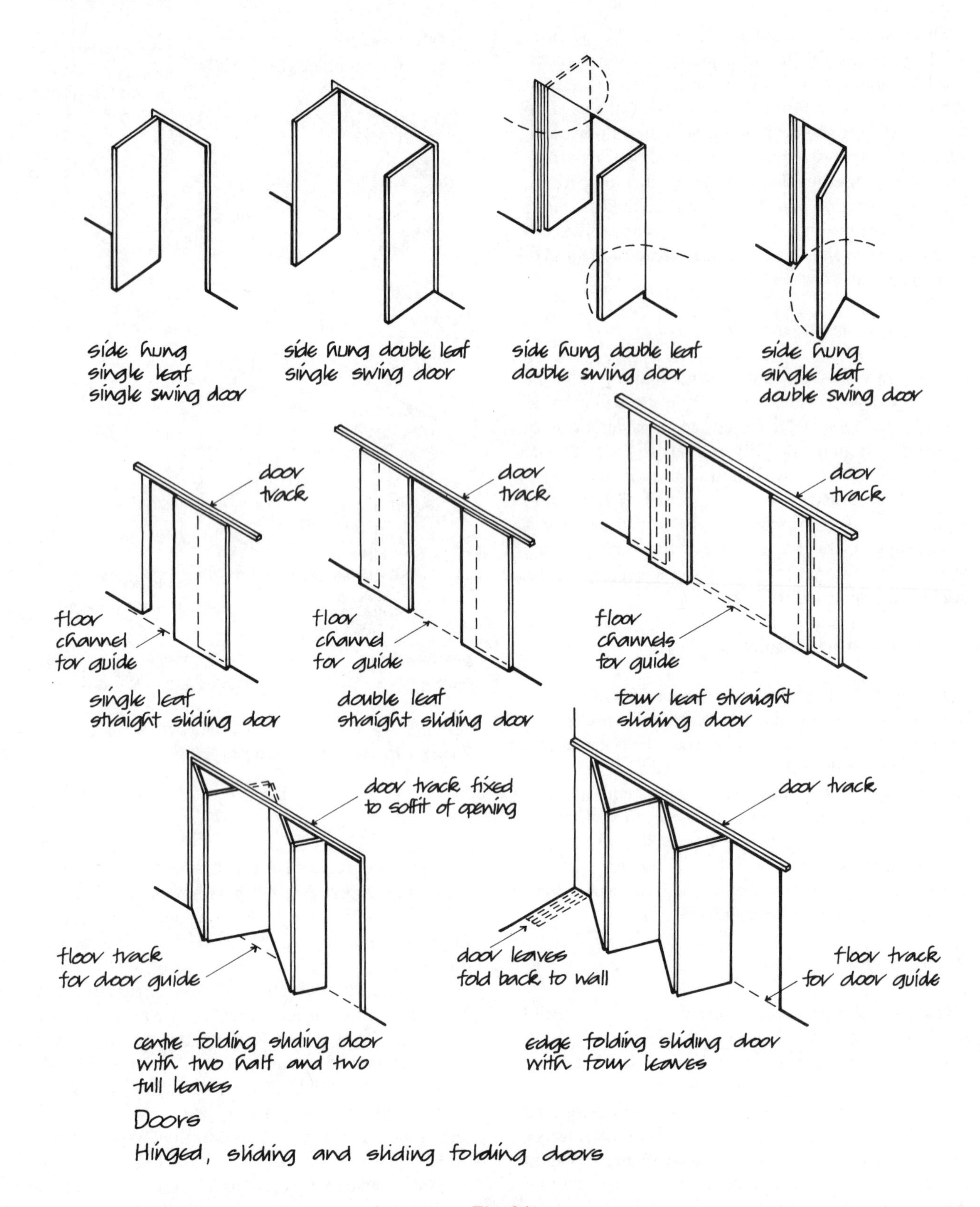
side hung
single leaf
single swing door
side hung double leaf
single swing door
side hung double leaf
double swing door
side hung
single leaf
double swing door
door
track
floor
channel
for guide
single leaf
straight sliding door
door
track
floor
channel
for guide
double leaf
straight sliding door
door
track
floor
channels
for guide
four leaf straight
sliding door
door track fixed
to soffit of opening
floor track
for door guide
centre folding sliding door
with two half and two
full leaves
door track
door leaves
fold back to wall
floor track
for door guide
edge folding sliding door
with four leaves
Doors
Hinged, sliding and sliding folding doors

Fig. 84

Security

An external door, particularly at the rear or sides of buildings, out of sight, is obviously a prime target for forced entry. Glazing and thin panels of wood, brittle fibreglass and beaded plastic panels invite breakage with a view to opening bolts or latches. Solid hinges, locks and loose key bolts to a solidly framed door in a soundly fixed solid frame are the best security against forced entry.

The practical guidance in Approved Document N to the Building Regulations 1991 recommends the use of safety glass to doors and door side panels up to a level of 1500 above finished floor level. It is up to this level that hands, wrists and arms are vulnerable to injuries from broken glass.

DOOR TYPES

The traditional door comprises one leaf which is hinged on one side to a frame or lining, to open in one direction for the convenient entry or exit of people. Because of its comparatively simple construction and operation, a side hung, single leaf, single swing door is used more than any other type of door.

Where a wide door opening is wanted for convenience or for appearance, a side hung, double leaf, single swing door may be used, as illustrated in Fig. 84.

Where there is moderate to heavy traffic through doors, as along access corridors, it is usual to fit double swing doors either on double swing hinges or floor springs, in the form of single or double leaf doors (Fig. 84). As these doors may be pushed open from both sides simultaneously it is usual for a top panel to be glazed to avoid accidents.

Sliding doors are used either where there is limited space that a hinged door would obstruct or for occasional access between rooms. Sliding doors do not afford the same ease and speed of access as a hinged door. Sliding doors may be arranged as single or double leaf or three or more overlapping leaves, as illustrated in Fig. 84.

To convert two smaller rooms into one large room to accommodate various activities, systems of hinged, sliding and sliding folding doors may be used. Centre fold doors open so that the leaves are folded back on the centre of the frame and edge fold into one or other side of the frame as illustrated in Fig. 84.

The up and over door is designed specifically for garages.

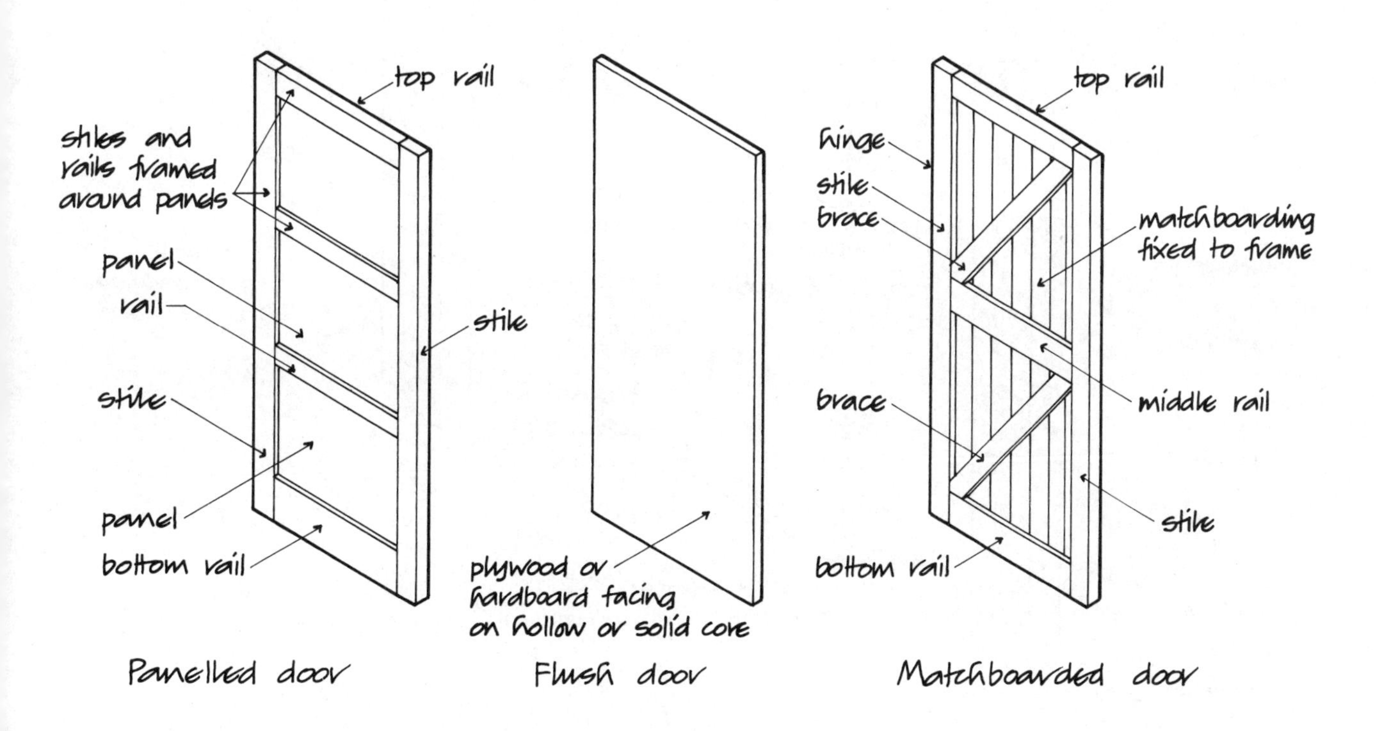

Fig. 85

WOOD DOORS

Wood doors may be classified as (Fig. 85):

Panelled doors
Glazed doors
Flush doors
Matchboarded doors.

The traditional door is formed from solid softwood or hardwood members framed around panels. This traditional construction has been in use for centuries with little modification other than in changes in jointing techniques due to machine assembly and the use of substitute materials for wood.

During the early part of the twentieth century it became fashionable to use flush doors with plain, flat surfaces both sides, devoid of decorative moulding, that matched the then current trend for plain surfaces that was considered 'modern'. This fashion persisted well into the century and it is only in recent years that it is being abandoned in favour of the old, familiar, traditional look of the panelled door. The consequence is that manufacturers now make panelled doors that are indeed framed around panels, and a cheaper substitute in the form of flush doors where the formerly flush facing is press formed to give an often weak simulation of a fully panelled door. These latter doors are by construction flush doors, with a panelled appearance.

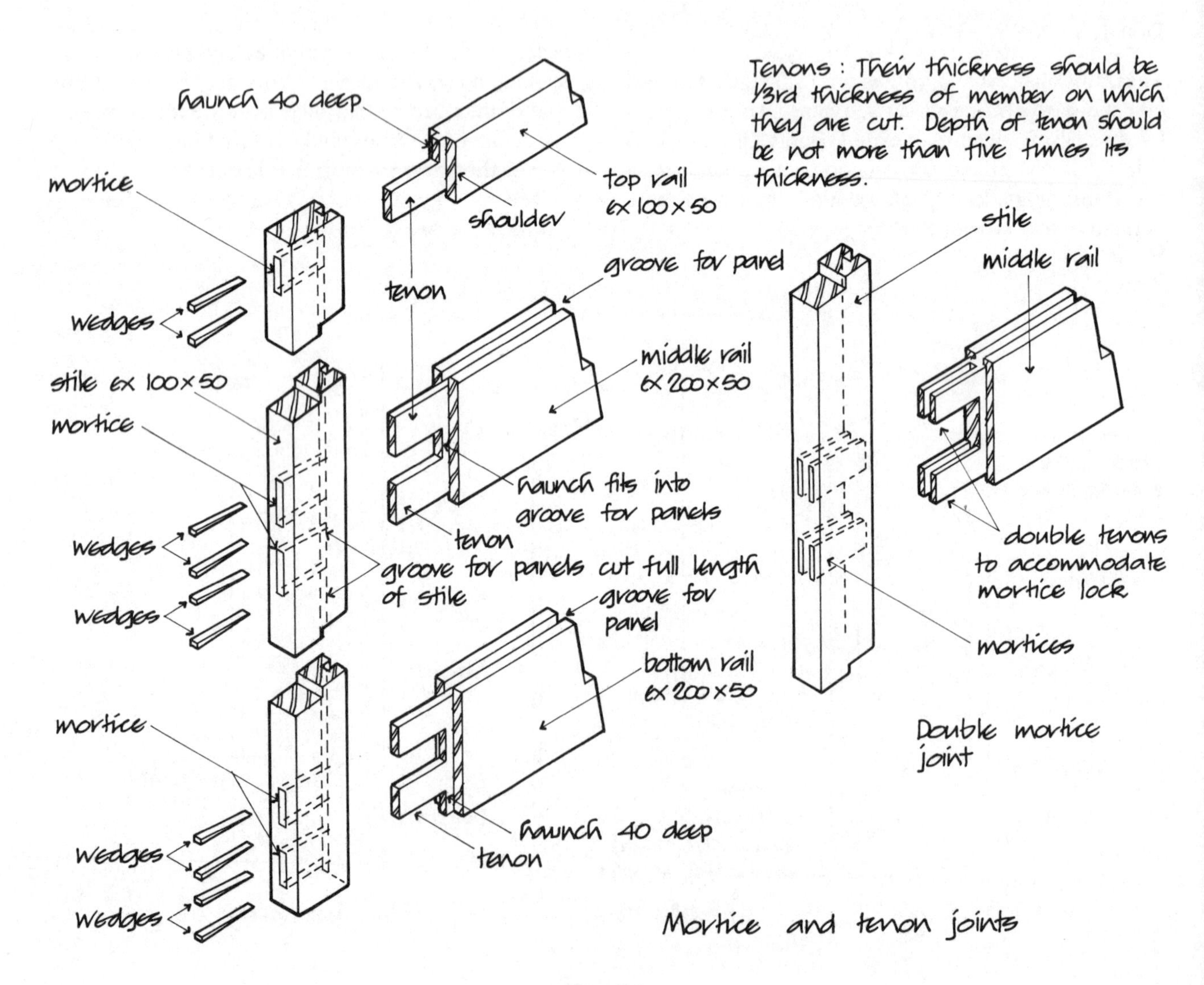

Fig. 86

PANELLED DOORS

Panelled doors are framed with stiles and rails around a panel or panels of wood or plywood. The stiles and rails are cut from timbers of the same thickness and some of the more usual sizes of timber used are: stiles and top rail 100 by 38 or 100 by 50; middle rail, 175 by 38 or 175 by 50; bottom rail 200 by 38 or 200 by 50. Because the door is hinged on one side to open, it tends to sink on the lock stile. The stiles and rails have to be joined to resist the tendency of the door to sink and the two types of joint used are a mortice and tenon joint and a dowelled joint.

Mortice and tenon joint This is the strongest type of joint used to frame members at right angles in joinery work. Fig. 86 shows the stiles and rails of a panelled door before they are put together and glued, wedged and cramped around the panels, which are not shown

Haunched tenon Obviously the tenons cut on the ends of the top rail cannot be as deep as the rail if they are to fit into enclosing mortices, so a tenon about 50 deep is cut. It is possible that the timber of the top rail may twist as it dries. To prevent the top rail moving out of upright a small projecting haunch is cut on top of the tenon, which fits into a groove in the stile.

Two tenons are cut on the ends of the middle and bottom rails. It would be possible to cut one tenon the depth of the rails but the wood around the mortice might bow out and so weaken the joint. Also a tenon as deep as the rail might shrink and become loose in the mortice. To avoid this a tenon should not be deeper than five times its thickness, hence the use of two tenons on the middle and bottom rails. Double tenons are sometimes cut on the ends of the middle rail as illustrated in Fig. 86. The purpose of these double tenons is to provide a space into which a mortice lock can be fitted without damaging the tenons.

It is apparent that the joints between the middle and bottom rails and stiles are stronger than that between the top rail and stile because of their greater depth of contact. For strength it would seem logical to make the top rail as deep as the middle rail. But a panelled door with a top rail deeper than the width of the stiles does not look attractive and by tradition the top rail is made as deep as the width of the stiles. The top and bottom faces of mortices are tapered in towards the centre of the doors so that when the tenons are fitted, small wood wedges can be driven in to make a tight fit.

Glueing, wedging and cramping The word cramp describes the operation of forcing the tenons tightly into mortices. The members of the door are cramped together with metal cramps which bind the members together until the glue in the joints has hardened. Before the tenons are fitted into the mortices both tenon and mortice are coated with glue. When the members of the door have been cramped together small wood wedges are knocked into the mortices top and bottom of each tenon. When the glue has hardened the cramps are released and the projecting ends of tenon and wedges are cut off flush with the edges of the stiles.

Pinned mortice and tenon joint If the timber from which a door is made shrinks, the mortice and tenon joints may in time become loose, and the door will lose shape. To prevent this, panelled doors are sometimes put together with pinned mortice and tenon joints. The mortices and tenons are cut in the usual way and holes are cut through the tenons and the sides of the mortices, as illustrated in Fig. 87. The tenons are fitted to the mortices, and oak pins (dowels) 13 diameter are driven through both mortice and tenon. Because the holes in the tenons and mortices are cut slightly off centre the pins, as they are driven in, draw the tenons into the

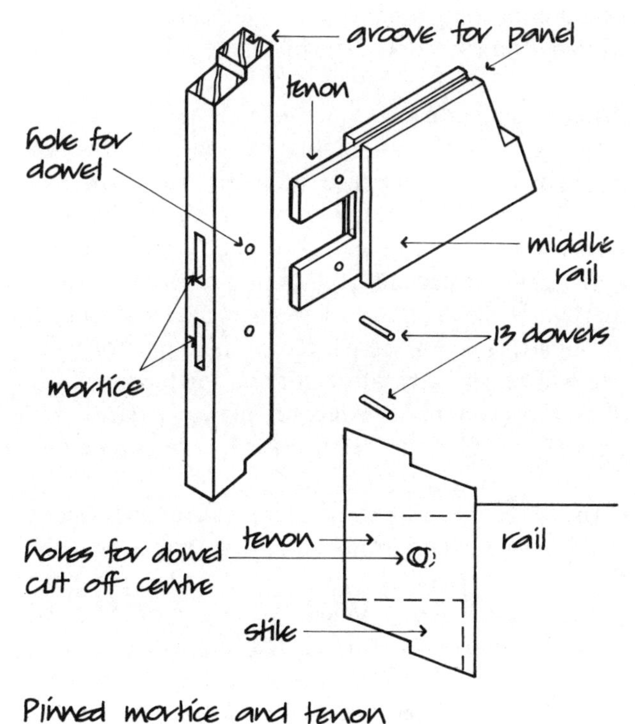

Fig. 87

mortices. Pinned mortice and tenon joints are glued and wedged. This joint should be used for all heavy panelled doors.

Dowelled joints The economic advantage of woodworking machinery cannot be exploited to the full in the cutting, shaping and assembly of mortice and tenon joints because of the number of separate operations involved. It is practice, therefore, to use a jointing system better fitted to woodworking machine operations in the cutting and assembly of mass produced doors. This type of joint is the dowelled joint which in addition to facilitating cutting and assembly by machine also effects a small economy in timber. The rails and stiles of panelled doors are joined by wood dowels that are fitted and glued to holes drilled in the members to be joined, as illustrated in Fig. 88. The dowels or pins, which are of hardwood or the same wood as the door, are 125 × 16 and each dowel is fitted half into each of the members joined. At least two dowels are used for the top rail and three for the middle and bottom rails. The dowels should be spaced not more than 57 apart, measured between centres. Two shallow grooves are cut along the length of each dowel so that when it is driven into the holes in stile and rail, excess glue and air trapped in the holes can escape. Because of improvements in glues this joint will strongly frame the members of a panelled door.

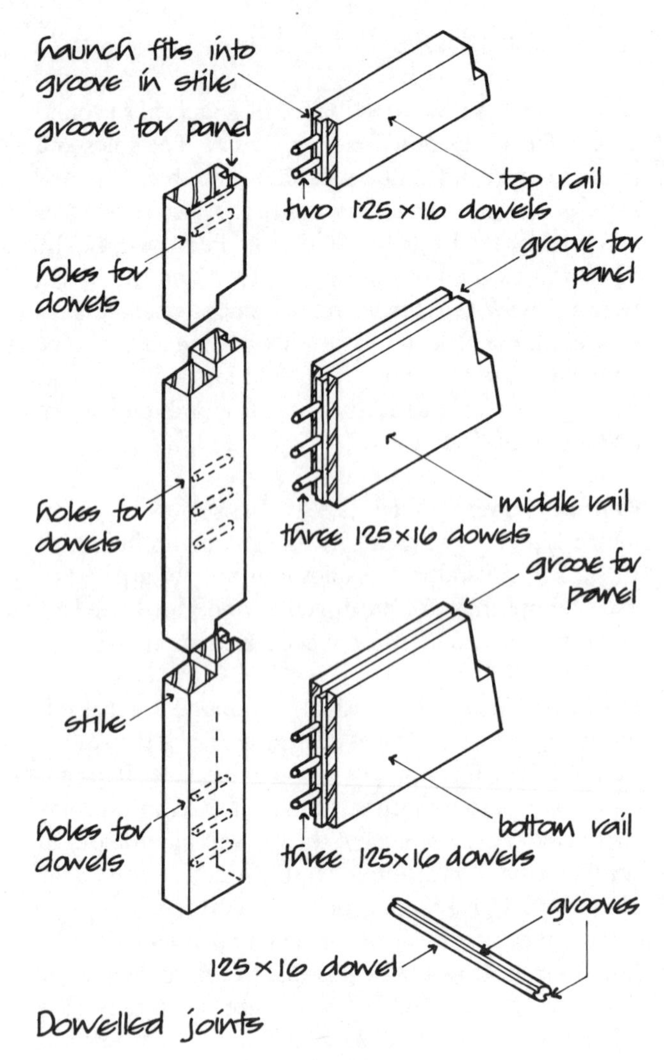

Fig. 88

Panels Timber panels more than about 250 wide are made up from boards 150 wide that are tongued together. The term 'tongued' describes the operation of jointing boards by cutting grooves in their edges into which a thin tongue or feather of wood is cramped and glued, as illustrated in Fig. 89. If the timber from which the panels are made has been adequately seasoned, through-tongued wood panels will not crack or wind and will be perfectly satisfactory during the life of the door. To economise in labour, plywood panels are generally used for most flat panels for doors today.

Plywood is made from three, five, seven or nine plies or thin layers of wood firmly glued together, so that the long grain of one ply is at right angles to the grain of the plies to which it is bonded. The most pronounced shrinkage in wood occurs at right angles to the long grain of the wood and any shrinkage of the centre ply is resisted by the outer plies, hence the odd number of plies used. Plywood does not shrink appreciably and because of the opposed long grains it does not warp or twist. Three-ply wood, 5 or 6.5 mm thick is generally used for door panels.

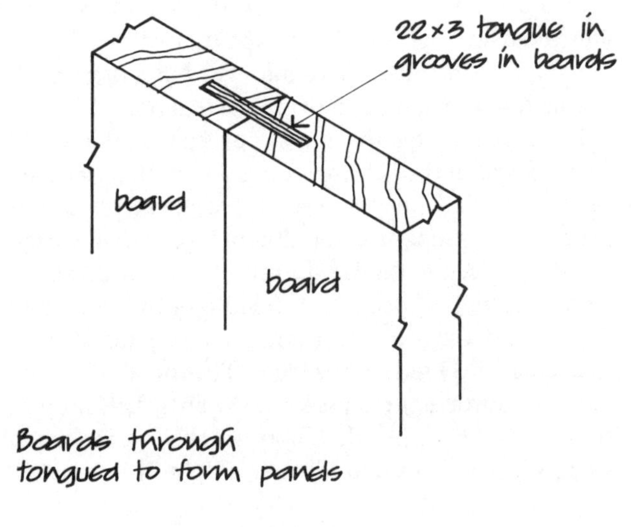

Fig. 89

Fixing panels The usual method of fixing panels is to set them in grooves cut in the edges of the stiles and rails, as illustrated in Fig. 90. If any shrinkage of the members of the door occurs, gaps will not appear around the panels. A panel set in grooves to stiles and rails with square edges may have an unfinished look which can be modified by cutting a moulding on the edges of the members around the panel, as illustrated in Fig. 91. The moulding cut on the edges of the members is described as a 'stuck' moulding.

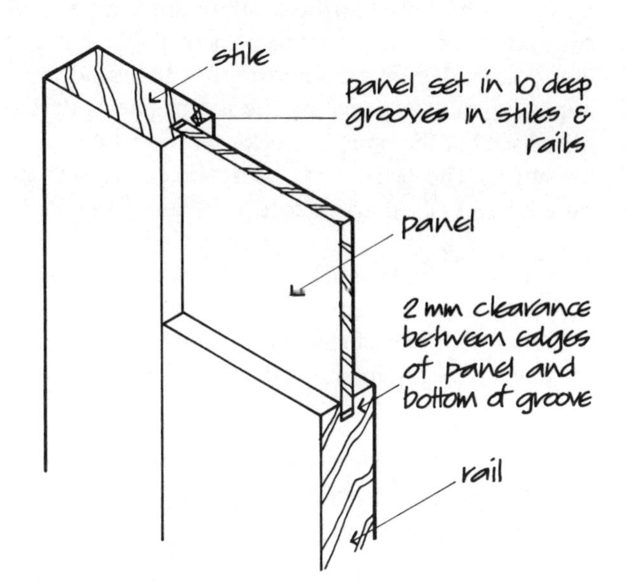

Fig. 90

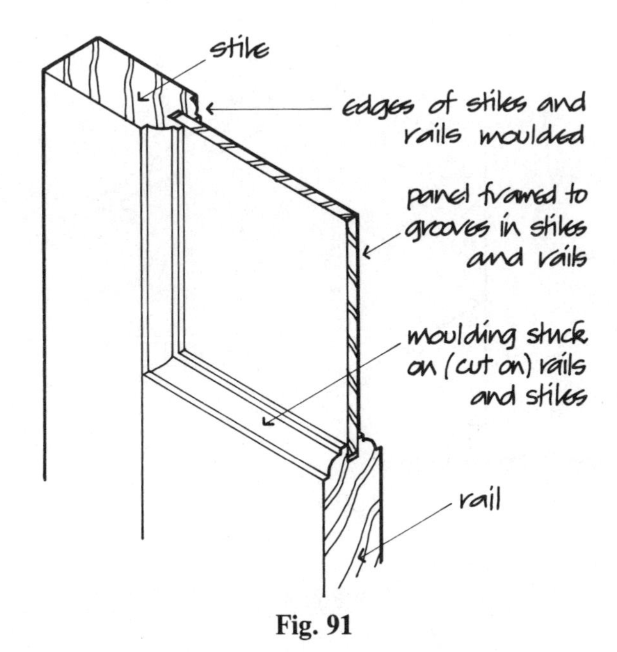

Fig. 91

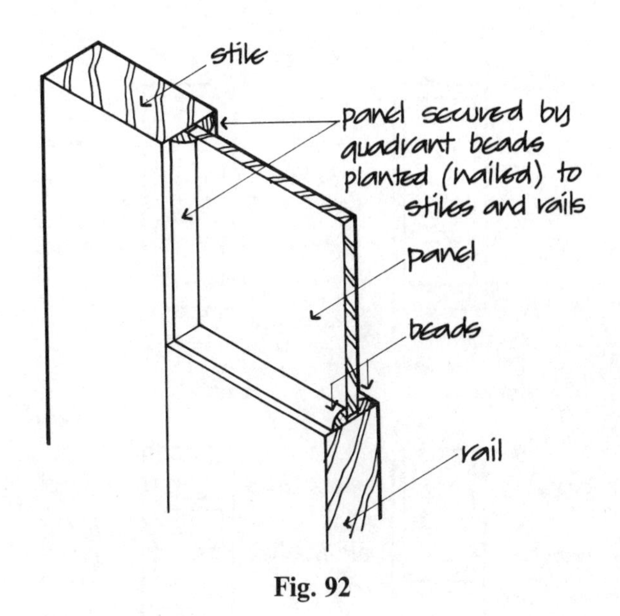

Fig. 92

An inferior method of fixing panels is to plant (nail) timber beads each side of the panel, as illustrated in Fig. 92. When the beads shrink, cracks appear.

The traditional panelled door was constructed with four or six panels, as illustrated in Fig. 93. The centre vertical member separating the panels is termed a muntin and it is either the same size as the stiles or 25 less in width. But fashions change and economic considerations led to the old British Standard panelled door which was arranged without muntins and with one or more horizontal panel, as illustrated in Fig. 94. These mass-produced panel doors are considerably cheaper than similar non-standard doors. The top, bottom and middle rails are through-mortice-and-tenoned or joined with dowelled joints, and intermediate rails are either stub-tenoned or joined with dowels.

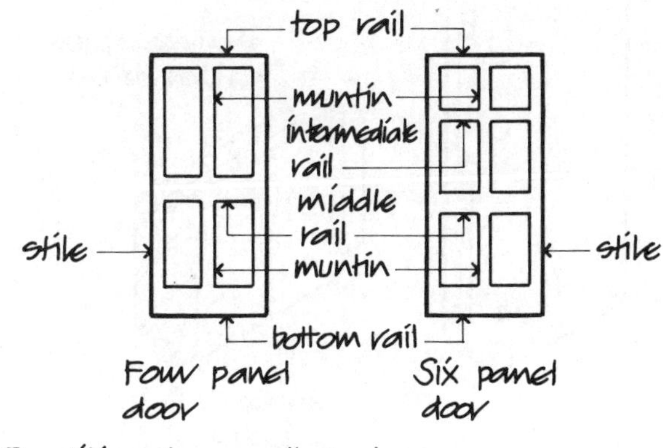

Fig. 93

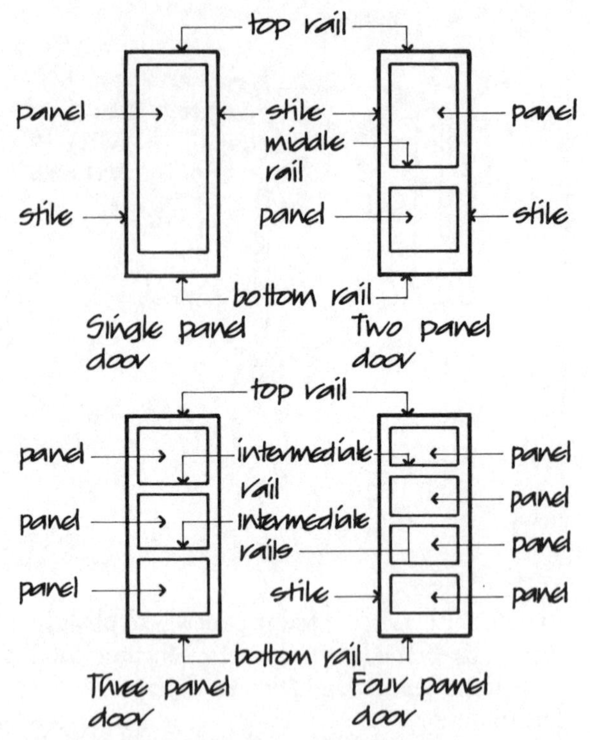

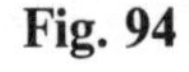
Fig. 94

Stub-tenon describes a tenon fitted to a mortice not cut right through a member, as illustrated in Fig. 95. The tenon is glued and cramped into the mortice.

The panels of the British Standard panel door may be of timber, plywood or hardboard set in grooves in the stiles and rails. There should be a clearance of 2 around each panel when it is set in the grooves, so that shrinkage of the stiles and rails can occur without it buckling the panel.

Doors with raised panels For appearance sake entrance doors and doors to principal rooms in both domestic and public buildings are often made more imposing and attractive by the use of panels that are raised so that the panel is thicker at the centre than at the edges. The usual types of raised panel are bevel raised, bevel raised and fielded, and raised and fielded.

Bevel raised Each panel is cut with four bevel faces on each side, the bevel faces having the same slope so that they rise to a point, if the panel is square, or a ridge, if the panel is rectangular, as illustrated in Fig. 96.

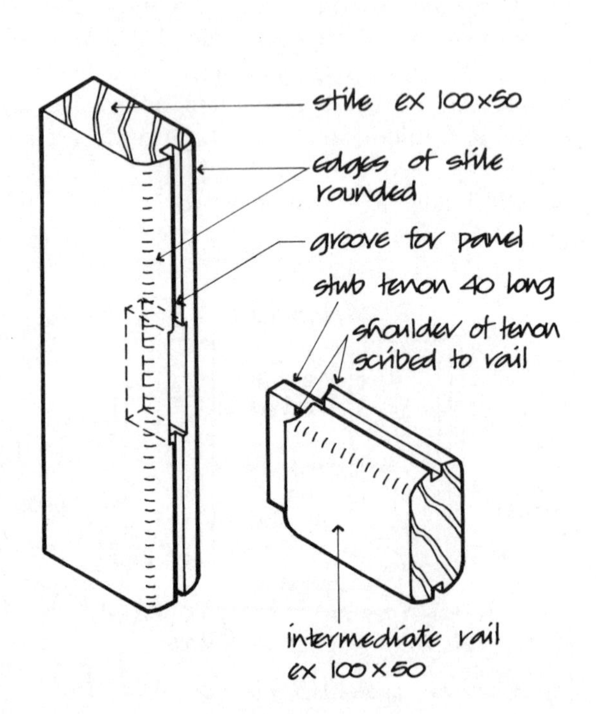

Fig. 95

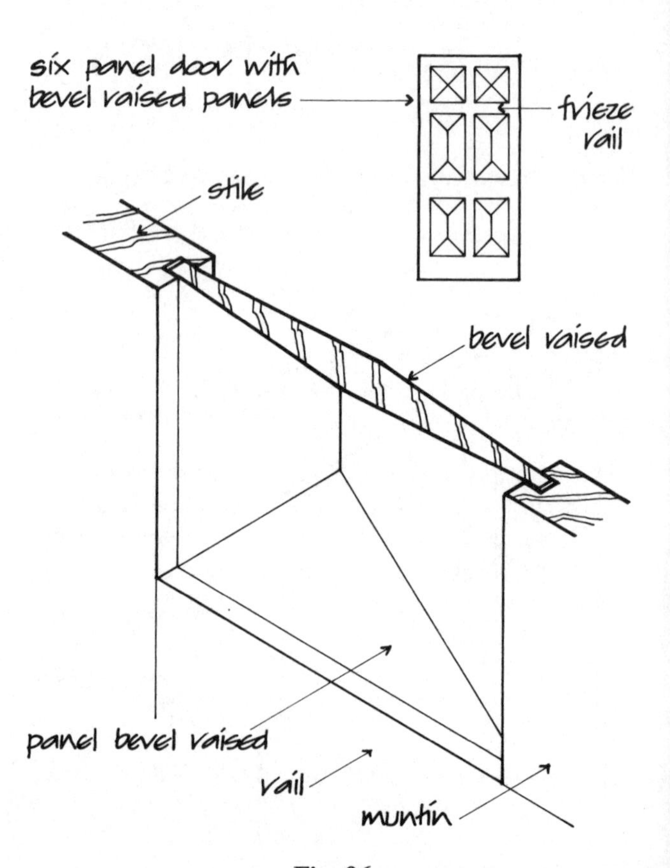

Fig. 96

Bevel raised and fielded The bevels rise to a flat surface at the centre of the panel, with the flat surface, termed the field, emphasised by a slight sinking cut around the field, as illustrated in Fig. 97. The bevels are cut at the same slope and the field is either square or rectangular depending on the shape of the panel. At the field the panel is either as thick as or slightly less thick than the stiles. The proportion of the fielded surface to the whole panel is a matter of taste.

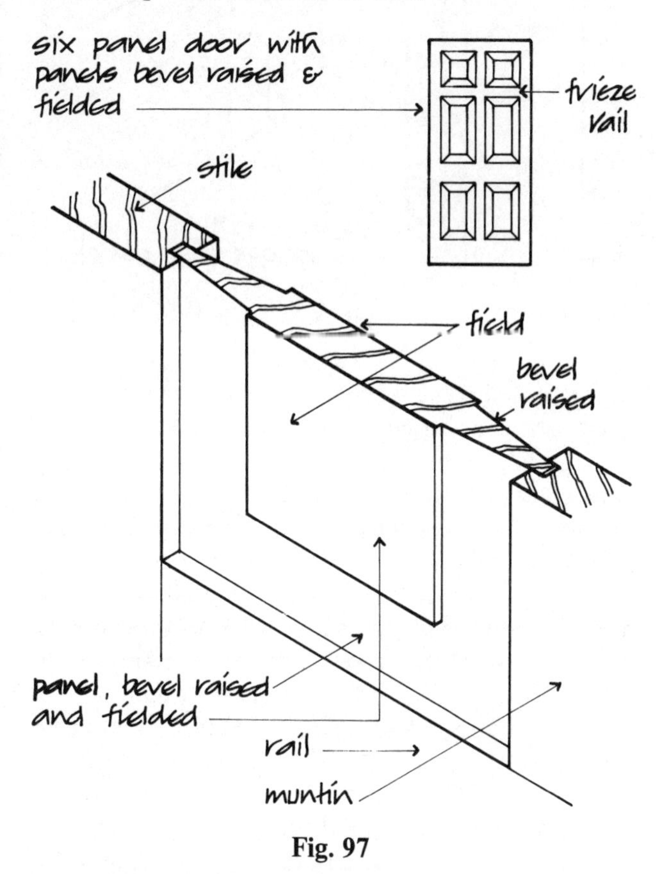

Fig. 97

Raised and fielded panel The panel, which is of uniform thickness around the sides, is raised to a flat field at the centre with a sinking as shown in Fig. 98, the field being either squre or rectangular depending on the shape of the panel.

The panels may be raised on both sides as shown in Fig. 96, or raised on one side only as shown in Fig. 98.

Bolection moulding A timber bolection moulding is planted around the panels of a door purely for decoration. A bolection moulding is cut so that when it is fixed it covers the edges of the stiles and rails around the panel to emphasise the panel, as illustrated in Fig. 99. A bolection moulding may be used around either flat or raised panels on one or both sides of a door.

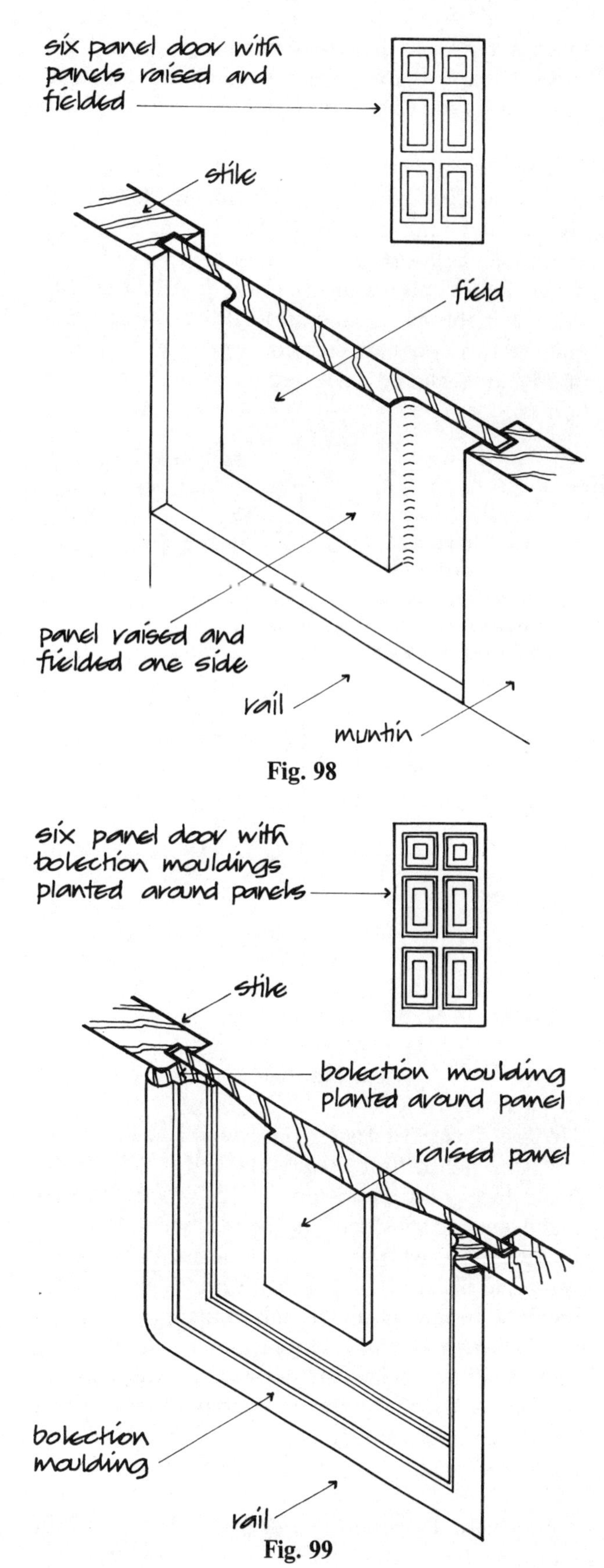

Fig. 98

Fig. 99

Practice is to use bolection mouldings around the panels of large doors as the moulding around small panels looks somewhat bulky and clumsy.

Double margin door A wide panelled door may be made up as if it were two doors with the two doors fixed together and hung as one door. This is done purely for appearance as a wide panelled door made as one door might look clumsy with over-large panels. Fig. 100 shows a double margin door, from which it will be seen that the two doors are secured together with wedges through the middle stiles and metal bars top and bottom.

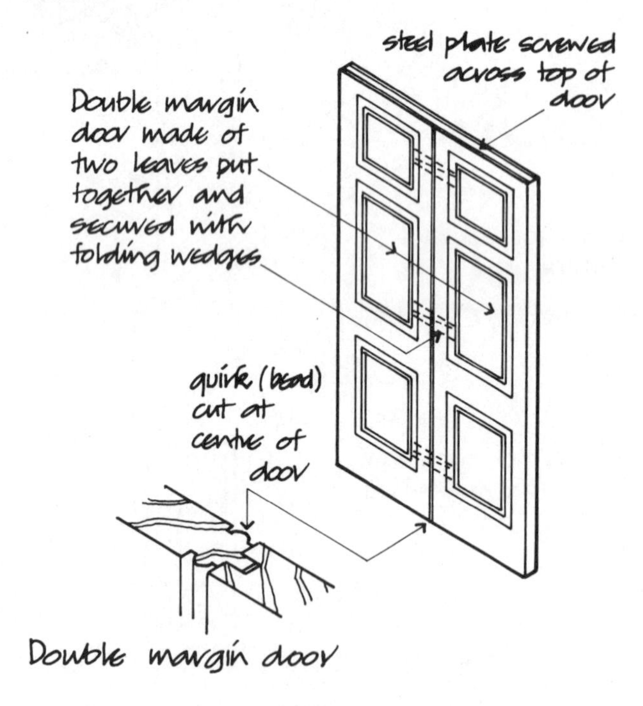

Fig. 100

Hardwood panelled doors Hardwood panelled doors are often polished or French polished to expose the decorative grain and colour of the wood, such as mahogany. To avoid showing the ends of tenons cut on rails it is practice to use stub-tenons secured with foxtail wedges as illustrated in Fig. 101. These foxtail wedges are fitted to saw cuts in the ends of the stub-tenons so that when the tenon is driven into the mortice the wedges are driven into the stub-tenons, which spread and bind inside the mortices. This type of joint, which has to be very accurately cut, makes a sound joint. As an alternative dowelled joints may be used.

Solid panels – flush panels Doors are constructed with panels as thick as the stiles and rails around them for

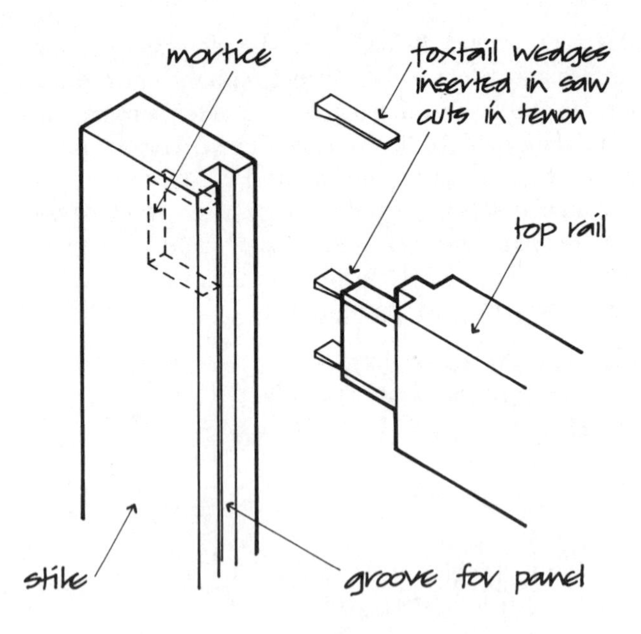

Fig. 101

strength, security or where the door acts as a fire check door. These doors are usually constructed of hardwood, such as oak, that has a better resistance to damage by fire than softwood. The solid panels are tongued to grooves in the stiles and rails and are either cut with a bead on their vertical edges or with a bead all round each panel for appearance sake. Timber shrinks more across than along its long grain and because the long grain in these panels is arranged vertically the shrinkage at the sides of the panels will be more than at top and bottom. For this reason beads are cut on the vertical edges of the panels to mask any shrinkage cracks that might appear, as illustrated in Fig. 102. Where beads are cut top and bottom they are cut on the rails as it is easier to cut a bead along the grain of the rails than across the end long grain of the panels. A panel with beads on its vertical edges only is described as beadbutt and one with beads all round as bead-flush (Fig. 103). As an alternative to horizontal beads cut on the stiles a planted bead can be used, as illustrated in Fig. 103. This is not satisfactory with external doors as water may get behind the bead which may then swell and come away.

Double swing doors Doors are hung to swing both ways to provide ready access to and from parts of buildings used in common by the occupants and users, at points where it is convenient to provide an opening barrier, for example from halls to corridors, to provide some separation of the public and the more private parts of the building. Double swing doors which are

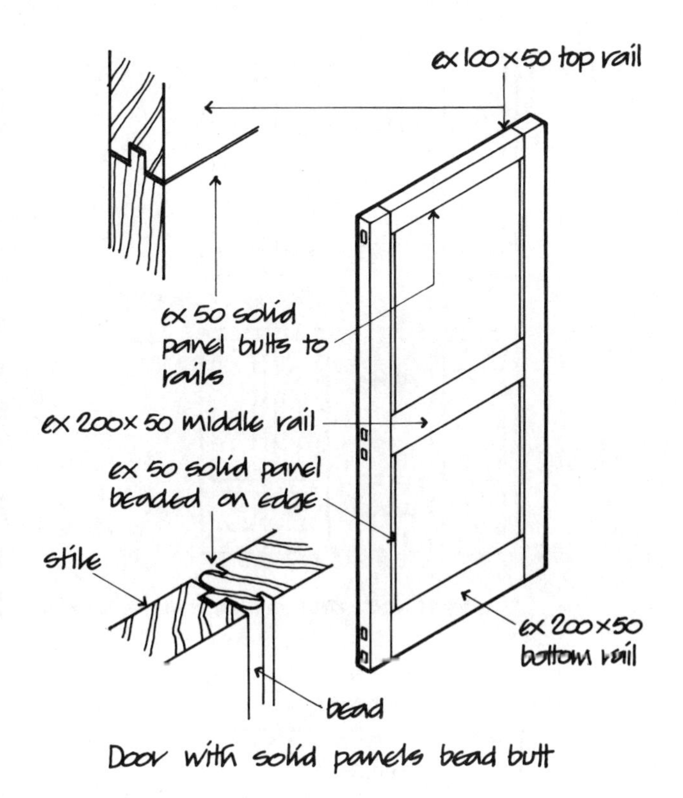

Fig. 102

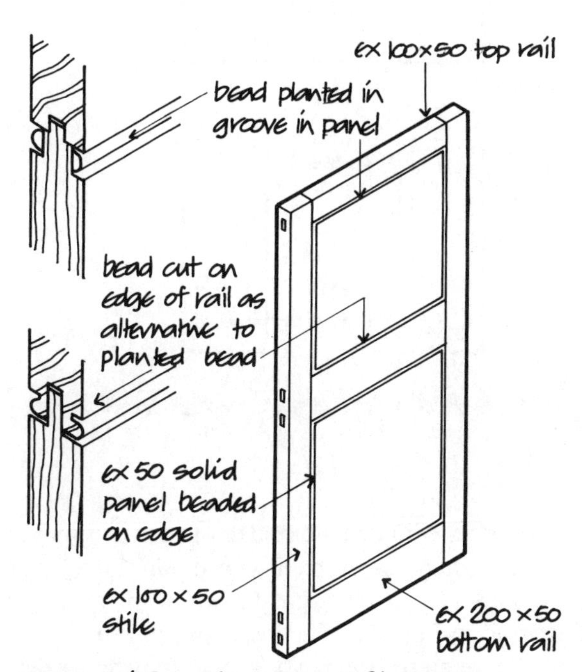

Fig. 103

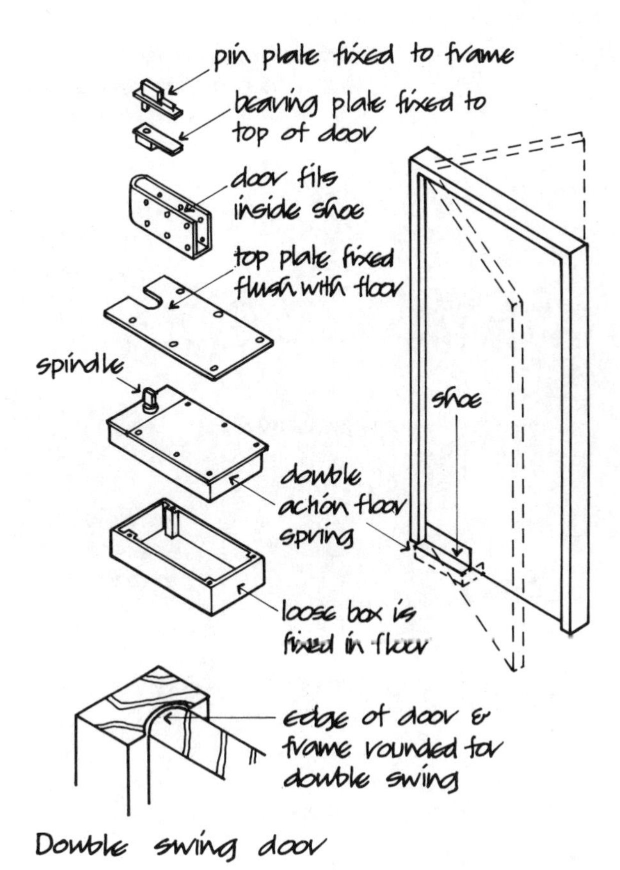

Fig. 104

fixed to provide a visual barrier usually serve neither as thermal, acoustic or security barriers. These doors, which are liable to heavy use, are usually constructed as panelled doors with a glazed panel at eye level to prevent accidents due to simultaneous use from each side. The door leaf is hung either on double-action hinges or pivoted on a double-action floor spring and top pivot, as illustrated in Fig. 104.

Sliding and sliding folding doors Sliding doors are designed mainly for intermittent use to provide either a clear opening or a barrier between adjacent rooms or spaces to accommodate change of use or function, and in narrow spaces to avoid the obstruction caused by a hinged leaf. Sliding folding doors are also designed for intermittent use to provide a larger opening than is practical with sliding doors, and to divide large spaces into smaller by closing, and back to one by opening.

Sliding and sliding folding door leaves may be constructed as flush or panelled doors in either wood or metal. The sliding and sliding and folding door gear consists of a metal track fixed over the opening, in which trolleys run and support the weight of the doors.

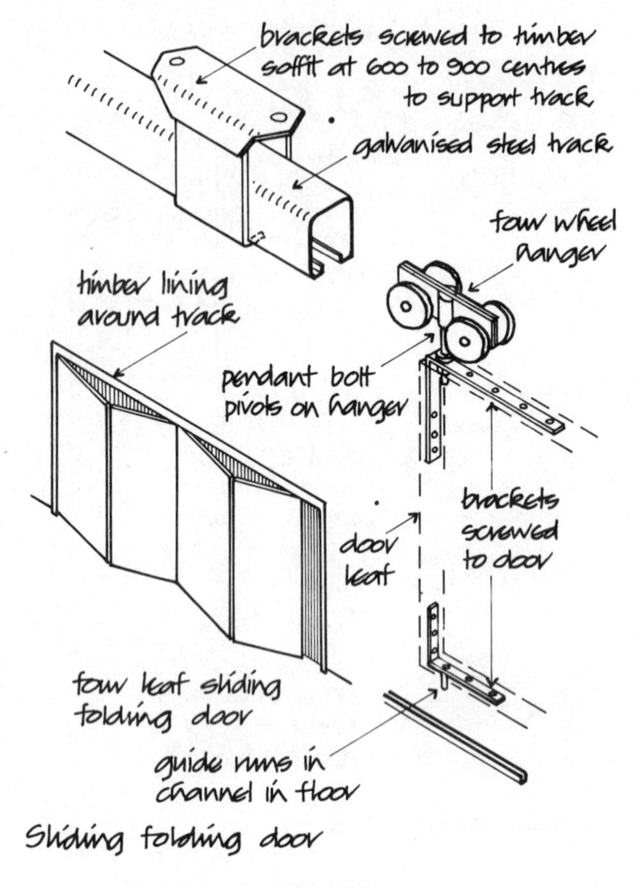

Fig. 105

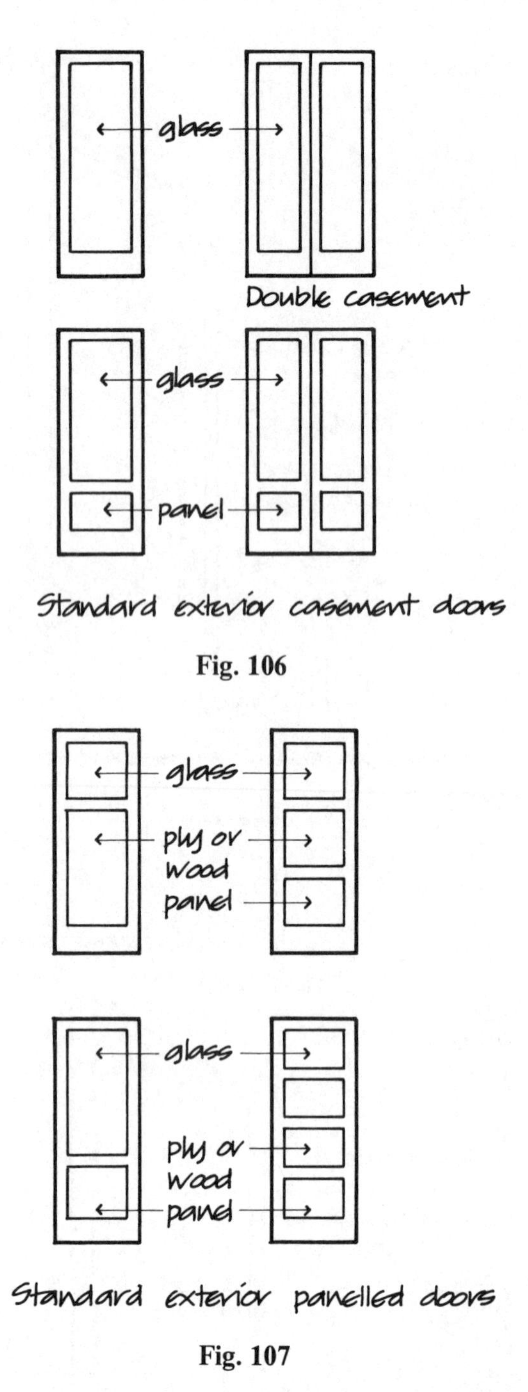

Fig. 106

Fig. 107

A track at floor level takes a runner fixed to the door to act as a guide for the movement of the door. Fig. 105 shows a sliding folding door.

GLAZED DOORS

Doors with one or more glazed panel are used to give some daylight to spaces such as halls that have no windows, and to give some borrowed light from a window through an internal door to an otherwise unlit space, hence the term 'borrowed light'.

The cheapest type of glazed door is a standard panelled door with one or more of the panels glazed, as illustrated in Figs. 106 and 107. The doors illustrated are adaptations of the old British Standard exterior panelled doors which are prepared with rebates for glass panels to be fixed with either bead or putty glazing.

Purpose-made glazed doors The appearance of a British Standard door is not to everyone's taste and the range of sizes is limited. Glazed doors that are purpose-made are often constructed with diminishing or gun stock stiles to provide a greater width for glazing, as illustrated in Fig. 108. The lower panels are of wood, with a large glazed panel above that may be glazed in one square of glass or with several squares to glazing bars as illustrated in Fig. 108. The members of the door are framed with mortice and tenon joints, with the glazing bars through-tenoned to stiles and stub-tenoned to rails as illustrated in Fig. 109.

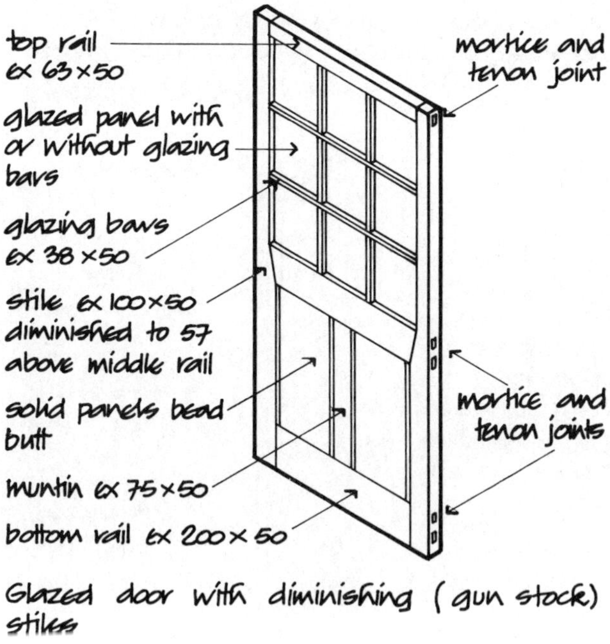

top rail
ex 63×50
glazed panel with
or without glazing
bars
glazing bars
ex 38×50
stile ex 100×50
diminished to 57
above middle rail
solid panels bead
butt
muntin ex 75×50
bottom rail ex 200×50
mortice and
tenon joint
mortice and
tenon joints
Glazed door with diminishing (gun stock)
stiles

Fig. 108

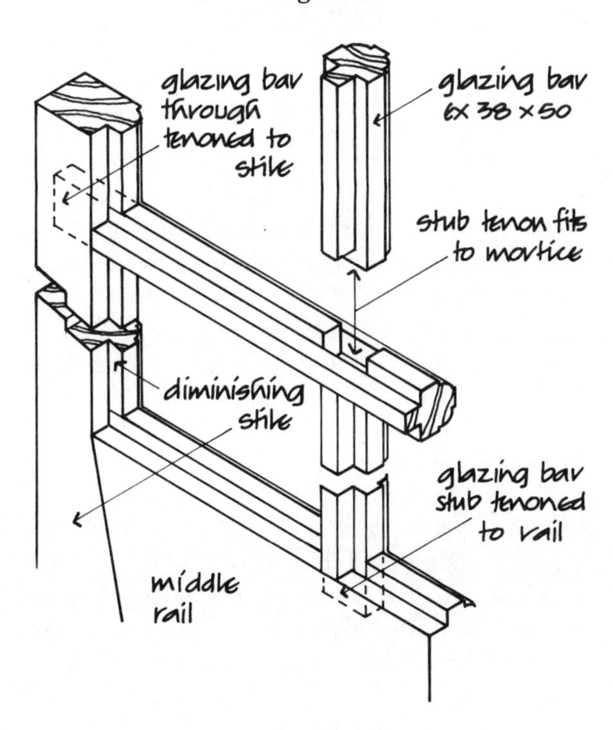

Fig. 109

Fully-glazed door or French casement Many of the windows of houses in France are arranged as casements with the sill of the window at or just above floor level. They are used as windows or as windows and doors giving access to balconies. Fully-glazed doors are often described as French casements or French doors in this country. Fig. 110 is an illustration of a purpose-made French casement or door with glazing bars. The stiles and rails are framed with mortice and tenon joints and the glazing bars are through-tenoned to stiles and stub-tenoned to rails. Standard single and double leaf casement doors are made as illustrated in Fig. 106.

The need for safety glass in doors and door side panels up to a level of 1500 above finished floor level is detailed in Approved Document N to the Building Regulations 1991.

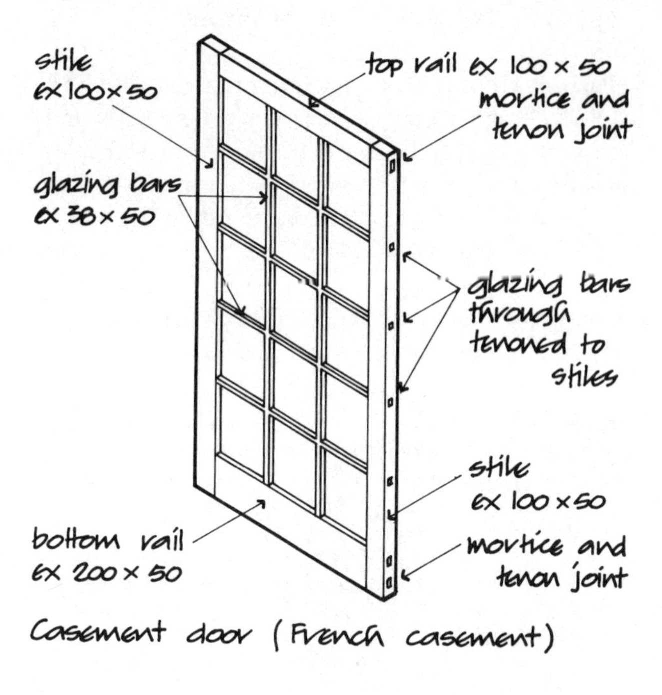

Fig. 110

FLUSH DOORS

A variety of flush doors is manufactured with plain flush faces both sides and fibreboard facings press moulded, often with comparatively shallow sinkings, to resemble the appearance of panelled doors.

The shape stability of these doors depends to an extent on the fixing of the flush or moulded facings to the core; the lighter and thinner the core the more the facings provide stability. With a lightweight cellular or skeleton core the fixing of the facings is generally adequate to maintain the square face shape of the door but may well not be substantial enough to resist torsion, where one free corner of the door will no longer fit into the frame or lining.

With cellular core and skeleton framed doors there may be a tendency for the flush facings to show the pattern of the core or skeleton, particularly where the faces are painted with a gloss paint.

The heavier solid core flush doors with a core of laminated timber, flaxboard, chipboard or compressed fibre strips will tend to maintain shape stability and uniformity of surface facings better than the light cellular and skeleton core doors.

Where the facings are of hardboard, press moulded to simulate door panels, the core is of light section softwood framed as a fixing for the four edges of the door and the fake internal rails and muntins.

Cellular-core flush doors These doors are made with a cellular, fibreboard or paper core in a light softwood frame with lock and hinge blocks covered with plywood or hardboard both sides, as illustrated in Fig. 111. These flimsy, lightweight doors are for light duty such as internal domestic doors. They do not withstand rough usage, and they provide little acoustic privacy, thermal or sound insulation, fire resistance or security. They are mass produced and are cheap.

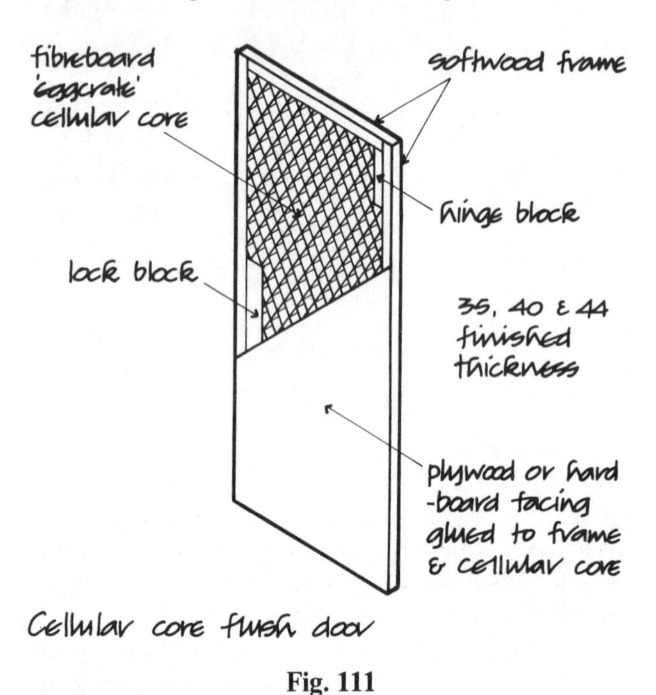

Fig. 111

Skeleton-core flush doors A core of small section timbers is constructed, as illustrated in Fig. 112. The main members of this structural core are the stiles and rails, with intermediate rails as shown as a base for the facing of plywood or hardboard. The framing core members are joined with glued tongued-and-grooved joints. The door illustrated in Fig. 112 has a skeleton core occupying from 30% to 40% of the core of the door. This is a light duty door suitable for internal domestic use. A similar skeleton-core flush door with more substantial intermediate rails in the core, where the core occupies from 50% to 60% of the core, is a medium duty door suitable for use internally in domestic and public buildings and for external use in sheltered positions.

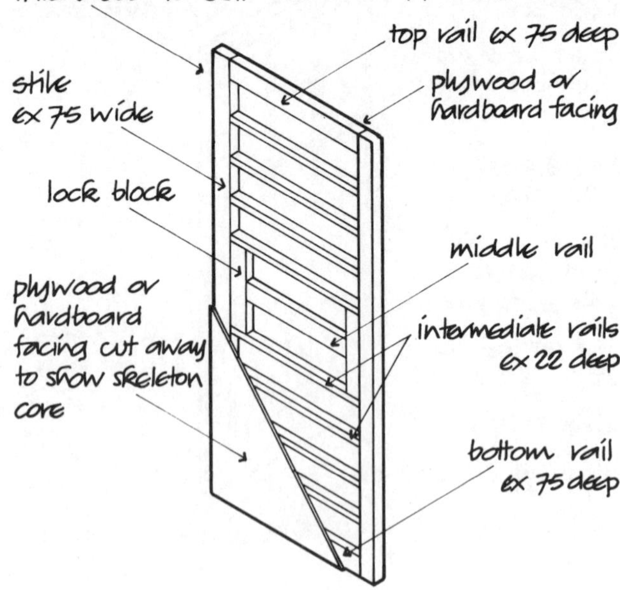

Fig. 112

Solid-core flush doors

A flush door with a solid core of timber, chipboard, flaxboard or compressed fibre board strips is a light or medium duty door. It has better thermal and acoustic properties than cellular-core or skeleton-core flush doors. The solid-core door illustrated in Fig. 113 has a core of timber strips glued together, with plywood facings lipped on both long edges. The chipboard, flaxboard or compressed fibre board strip core doors are made with a solid core enclosed by a light timber frame to which hardboard or plywood is fixed. These solid doors are more expensive than cellular-core or skeleton-core doors.

These doors may be used as fire doors with an integrity rating of 20 or 30 minutes, indicated as FD 20 and FD 30.

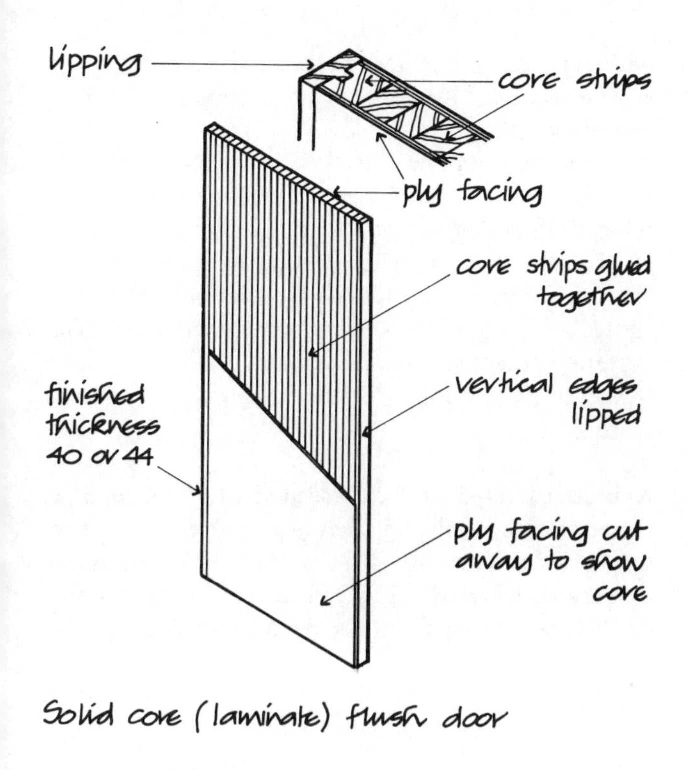

Fig. 113

FIRE DOORS

Fires in buildings start, more often than not, by the heating of some material of the contents to ignition and the beginning of a fire. In the early stages of the growth of a fire the considerable volume of smoke produced rises to fill rooms from the ceiling downwards and spreads through gaps around doors and through unsealed void spaces, making a danger to people in the building. Fires grow and spread through the release of the hot, flammable, gaseous products of combustion to other materials of the building and its contents.

To provide safe means of escape for the occupants of buildings in the early stages of a fire, smoke control fire doors are fixed in enclosures to and along escape routes.

To limit the spread of fire in buildings it is usual to divide all but small buildings into compartments surrounded by floors and walls capable of limiting the spread of fire for a stated period of minutes or hours. Doors in compartment walls and doors to a protected escape route should be capable of resisting the spread of fire to the same extent that the enclosing walls do. These doors, which in the past have been variously described as fire check doors, fire break doors, fire protection doors and fire resisting doors are now simply termed fire doors.

Fire doors serve to protect escape routes and the contents and structure of buildings by limiting the spread of smoke and fire. Fire doors that are fixed for smoke control only should be capable of withstanding smoke at ambient (surrounding) temperatures and limited smoke at medium temperatures by self closing devices and flexible seals. Fire doors that are fixed to protect means of escape routes should withstand smoke at ambient and limited smoke at medium temperatures and have a minimum fire resistance, for integrity only, of 20 minutes.

Fire doors that are fixed as part of a fire compartment and as isolation of special risk areas should have a minimum fire resistance, for integrity only, of a period of minutes or hours appropriate to the periods set out in Approved Document B giving practical guidance to the requirements of the Building Regulations 1991: Fire safety.

To conform to international practice, doors and other non-loadbearing elements are no longer assessed for stability (resistance to collapse) or insulation. The test for integrity is assumed to include performance in regard to stability and insulation. The notation for fire doors is FD followed by the figure in minutes for integrity, as for example FD 20 or FD 30, and doors that serve for smoke control as for example FD 20S.

The performance test for fire doors that serve as barriers to the spread of fire is determined from the integrity of a door assembly or door set in its resistance to penetration by flame and hot gases. The test is carried out on a door assembly which includes all hardware, supports, fixings, door leaf and frame, representative of a door assembly that will be used in practice. Each face of the door assembly is exposed separately to prescribed heating conditions from a furnace, on a temperature-time relationship, to determine the time to failure of integrity. Failure of integrity occurs when flame or hot gases penetrate gaps or cracks in the door assembly and cause flaming of a cotton wool pad on the side of the assembly opposite to the furnace.

A fire door should at once be easy to operate, serve as an effective barrier to the spread of smoke and fire when closed and be fitted with some effective self closing device. For ease of operation there must be clearance gaps around the door leaf. These clearance gaps are effectively sealed when a door leaf closes into and up to the rebate in a door frame. Where a door leaf has distorted in use and when the leaf is distorted by the heat of a fire then the leaf will no longer fit tightly inside the rebate of the frame and smoke and flame can spread through the gaps around the door leaf. As a barrier to the spread of smoke, flexible seals should be fixed to door leafs of frames and as a barrier to the spread of fire heat activated (intumescent) seals should be fitted.

Smoke control door assemblies (FDS) that serve only as a barrier to the spread of smoke without any requirement for fire resistance, such as fire doors along an escape route, may be fitted in rebated frames or hung to open both ways. To provide an effective seal against the spread of smoke through gaps around these doors, flexible seals should be fitted.

Smoke control door assemblies that serve as a barrier to the spread of smoke and fire, such as doors leading to a protected escape route, should be hung in rebated frames and tested for a minimum integrity of 20 minutes against the spread of fire, and should be fitted with heat activated (intumescent) seals and flexible edge seals (Fig. 114) against the likelihood of the door leaf deforming.

Fire door assemblies fixed in compartment walls and to enclosures to special risk areas should be hung in rebated frames and tested for integrity for not less than 30 minutes or such period as detailed in Advisory Document B to the Building Regulations 1991, and should be fitted with the intumescent seals. The currently accepted minimum size of a softwood door frame for a fire door is 70 × 30, exclusive of a planted stop.

A heat activated or intumescent seal is made of a material that swells by foaming and expanding at temperatures between 140° and 300°C. The intumescent seals illustrated in Fig. 114 are cased in aluminium or PVC cover strips that are fixed to the edges of the door or frame.

For a fire door to be effective against the spread of smoke and fire the door leaf should, when not in use, be positively closed to the frame by some self closing

aluminium holder fixed in rebate in frame or door
intumescent material in holder
9
7
7.5mm
21
edge of frame or door
½ hour seal
edge of frame or door
neoprene blade as draught and smoke seal
1 hour seal
Heat from fire makes intumescent material expand out of slots to seal gaps around door

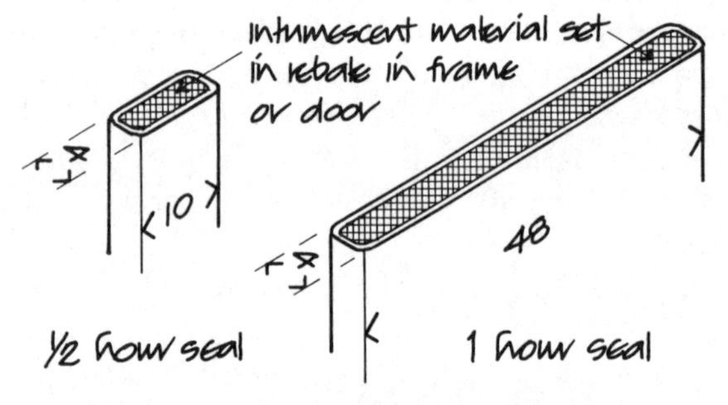

Intumescent Fire Seals
for fixing in rebates of frame or door

Fig. 114

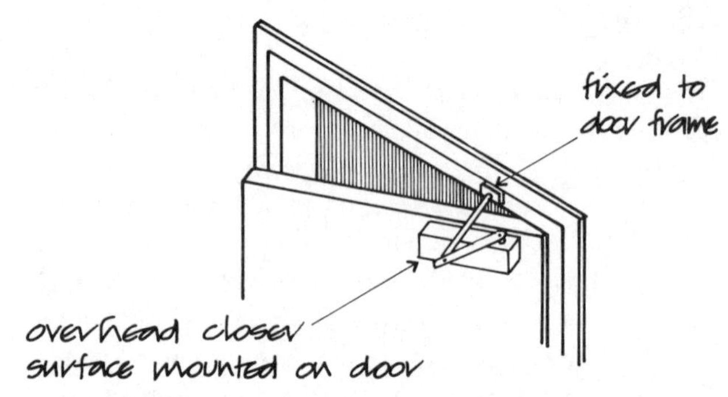

Hydraulic check, surface mounted overhead door closer

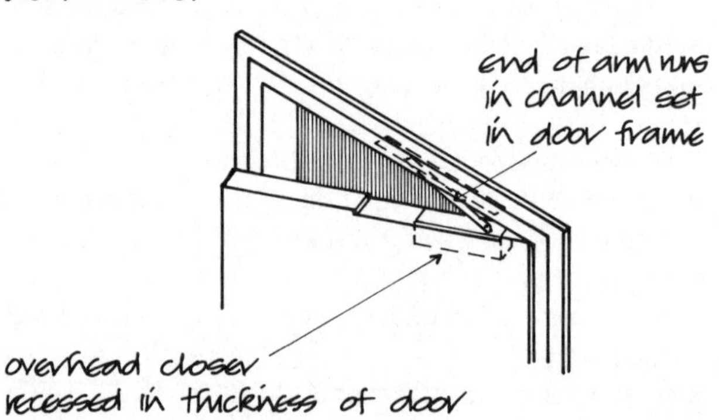

Hydraulic check, concealed fixing overhead door closer

Fig. 115

device. The conventional door closers that are used are the overhead closers illustrated in Fig. 115 and the floor closers in Fig. 104.

The current requirement for fire doors is that complete door assemblies be tested in accordance with BS 476 and certified as meeting the recommendations of performance for integrity and noted, for example, as FD 20, FD 30 as satisfying the requirements for 20 and 30 minutes integrity respectively.

The door leafs illustrated in Fig. 116, which include the earlier British Standard fire check flush door, should match the performance requirements for integrity noted when hung in softwood frames and fitted with self closing devices. The flush steel fire door illustrated in Fig. 117 has lipped edges and a mineral wool core.

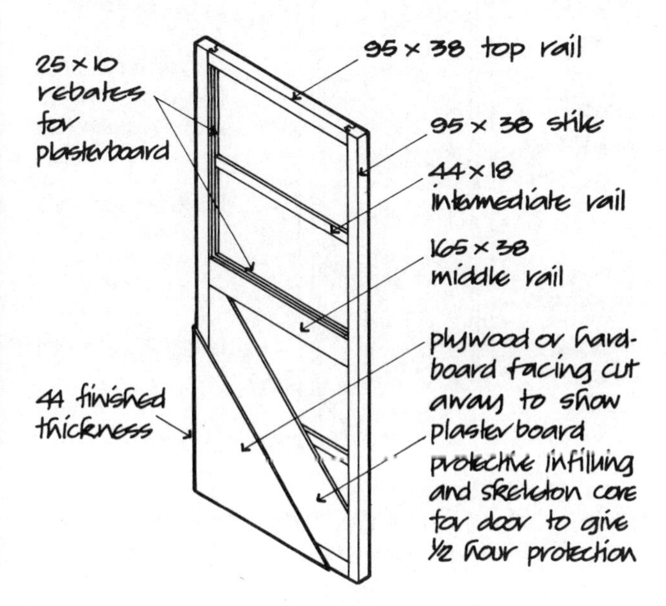

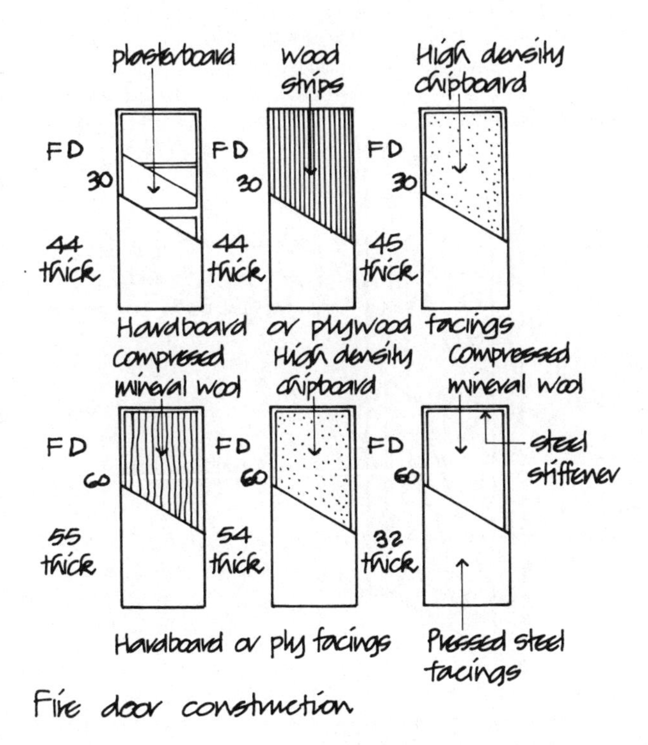

Fig. 116

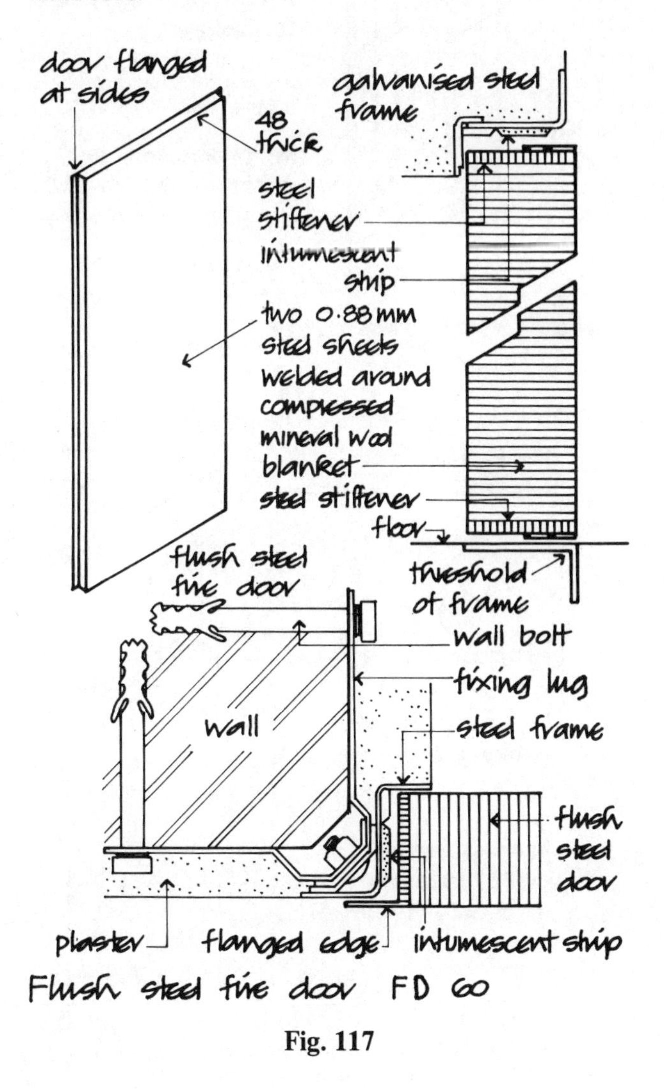

Fig. 117

MATCHBOARDED DOORS

Matchboarded doors are made with a facing of tongued, grooved and V-jointed boards fixed vertically to either ledges and braces or a frame. These doors are used for cellars, sheds and stores where the appearance of the door is not considered important.

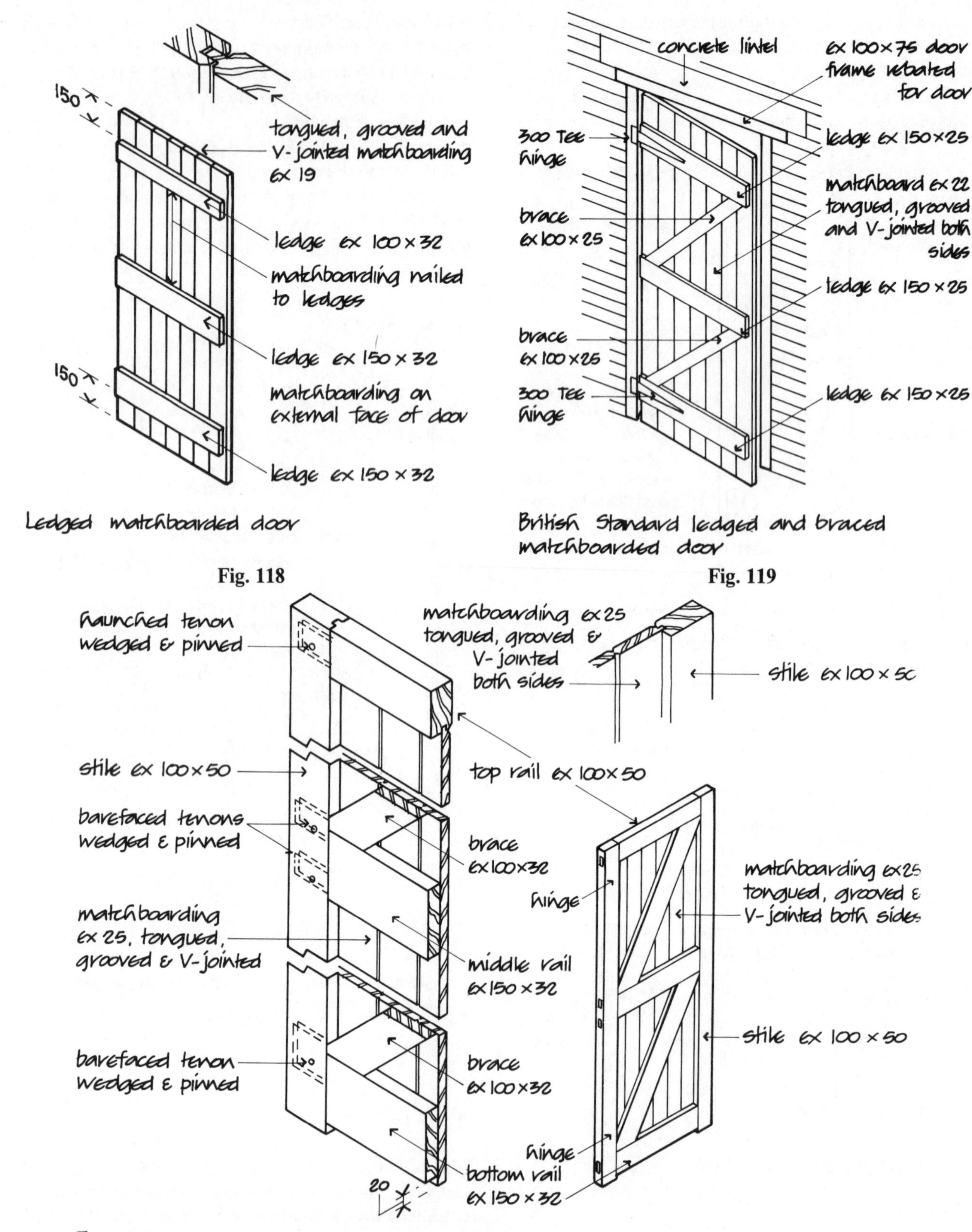

Fig. 118

Fig. 119

Fig. 120

Ledged matchboarded door Matchboarding is nailed to horizontal ledges, as illustrated in Fig. 118.The nailing of the boards to the ledges does not strongly frame the door which is liable to sink and lose shape. This door is used for narrow openings only.

Ledged and braced matchboarded door This type of door is strengthened against sinking with braces between the rails that are fixed at an angle to resist sinking on the lock edge (Fig. 119). The braces are nailed to the boarding.

Framed and braced matchboarded door The matchboarding is fixed to a frame of stiles and rails that are framed with mortice and tenon joints with braces to strengthen the door against sinking, as illustrated in Fig. 120. The boarding runs from the underside of the top rail, to protect the end grain of the boards from rain, down over both middle and bottom rails. To allow for the boards running over them the middle and bottom rails are less thick than the stiles to which they are joined with a barefaced tenon joint (Fig. 121). This joint is used instead of the normal joint with two shoulders so that the tenon is not too thin. These doors are used for large openings to garages, factories and for entrance gates.

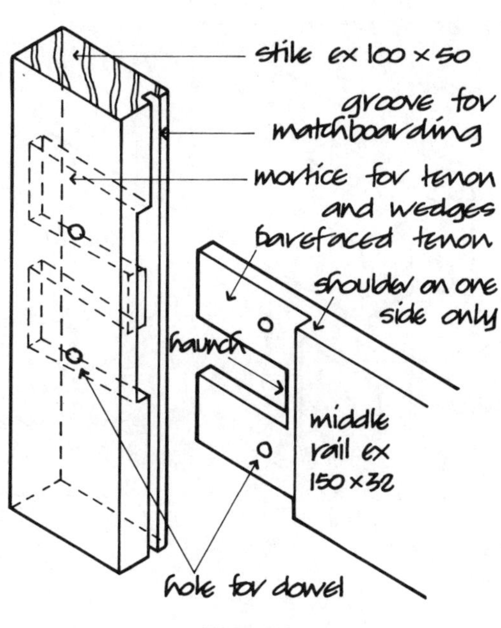

Fig. 121

DOOR FRAMES AND LININGS

A door frame is a surround in a doorway or opening, to which the door is hung and to which it closes, which has sufficient strength in itself to support the weight of the door. A door lining is a surround inside a doorway or opening, as wide as the reveal of the opening, to which the door is hung and closes, which is not in itself strong enough to support the weight of the door without support from the surrounding wall or partition. Door frames and linings may be made of wood, metal or plastic. Fig. 122 illustrates the difference between a wood frame and a lining, from which it will be seen that the frame is a solid rectangular section and the lining a thin section as wide as the wall or partition and plaster which it covers or lines. Door frames are used for external doors and heavy doors and linings for internal doors.

Wood door frames A door frame consists of three or four members which are either rebated 13 deep for the door, as illustrated in Fig. 123, or a wood stop 13 deep is planted (nailed) to the frame. The wood frame illustrated in Fig. 123 consists of two posts and a head member. External doors may also be framed with a fourth member, a threshold or sill, to assist in weather exclusion.

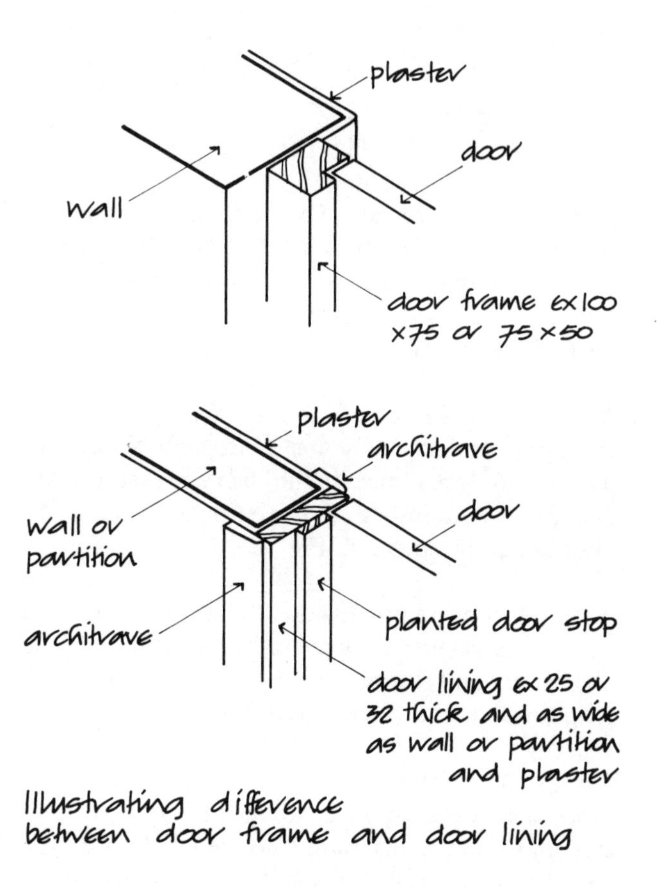

Fig. 122

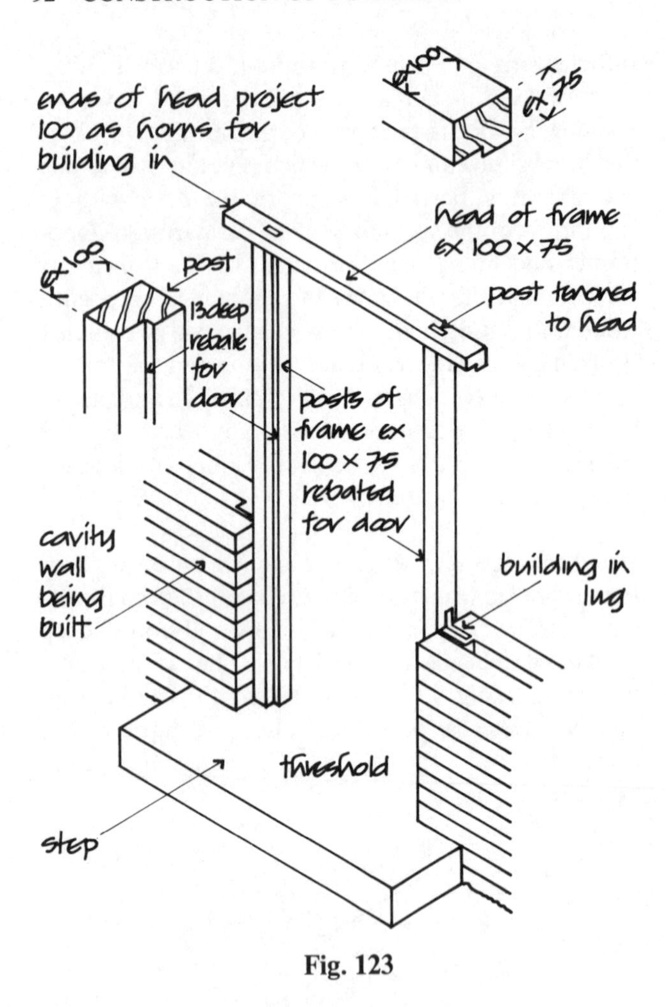

Fig. 123

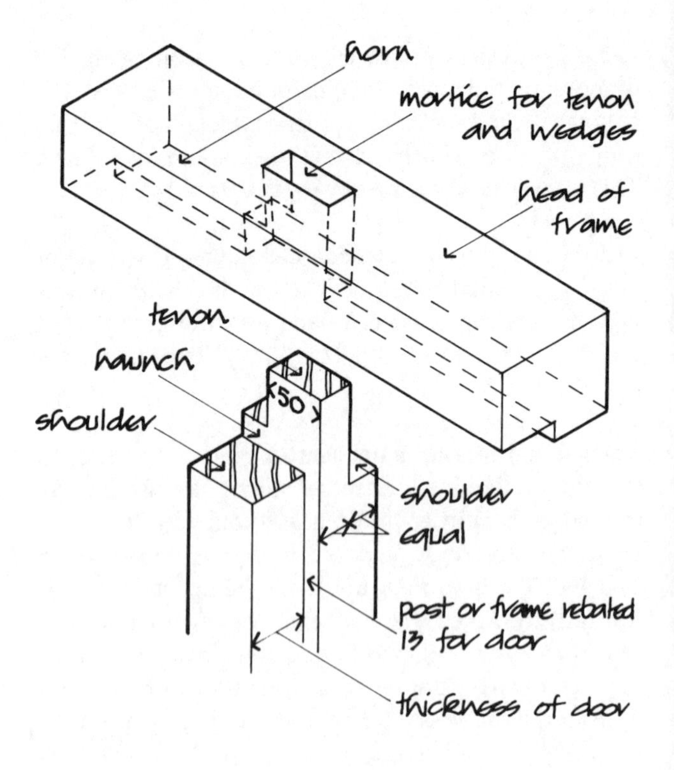

Fig. 124

Mortice and tenon joint The posts of the frame are usually tenoned to mortices in the head as illustrated in Fig. 124. The head is usually cut so that it projects either side of the frame as horns which can either be built into the jambs or cut off.

Slot mortice and tenon This is an alternative joint to the mortice and tenon. Tenons on the ends of the posts are fitted to slot mortices in the head and secured with glue, and wood dowels are driven through the tenon and head, as illustrated in Fig. 125.

Dowels Door frames that do not have a threshold or sill are often secured to the floor by a mild steel dowel, 12 diameter and 50 long, that is driven into the foot of the posts and set in the concrete floor.

Fixing Door frames are usually built-in, which describes the operation of building walls or partitions around the frame. The frame is secured to the wall with L-shaped galvanised steel building-in lugs which are screwed to the back of the frame and built into

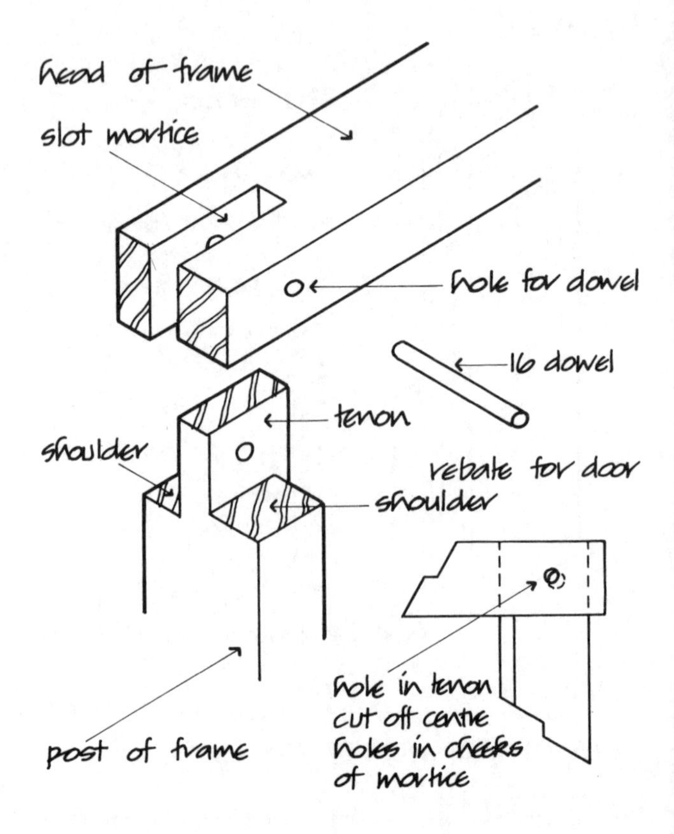

Fig. 125

horizontal joints, as illustrated in Fig. 123. These lugs are usually 75 × 150 × 38 wide. As an alternative a PVC cavity closer and ties, as illustrated in Fig. 126, may be used. Frames that are fixed in after the wall has been built are secured with screws through the posts to wooden plugs or grounds in the reveals of the opening.

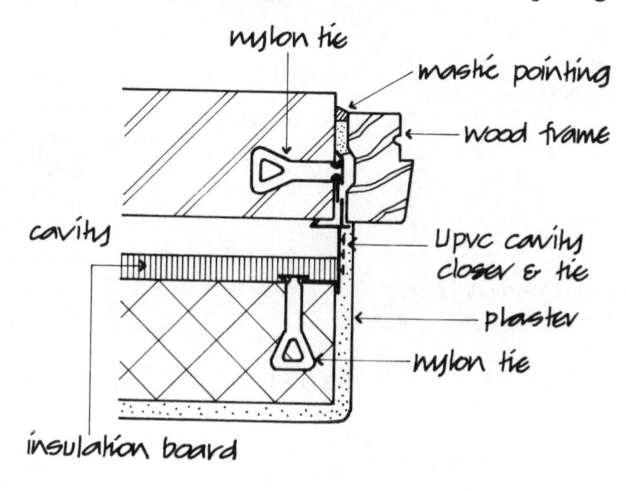

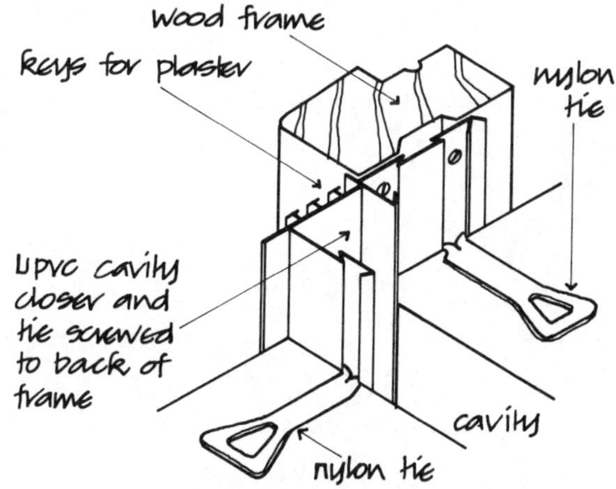

Upvc cavity closer and tie for wood window or door

Fig. 126

Threshold or sill A wood sill to an external door is usually of some hardwood, such as oak, and the sill is joined to the posts of the frame with haunched mortice and tenon joints. The sill is usually wider than the frame and is rebated for the door 13 deep for an outward opening door and grooved for a water bar for an inward opening door and weathered and throated as illustrated in Figs. 127 and 128.

Exclusion of wind and rain The traditional method of excluding wind and rain with outward opening external doors was to rely on the rebates in the door frame, as

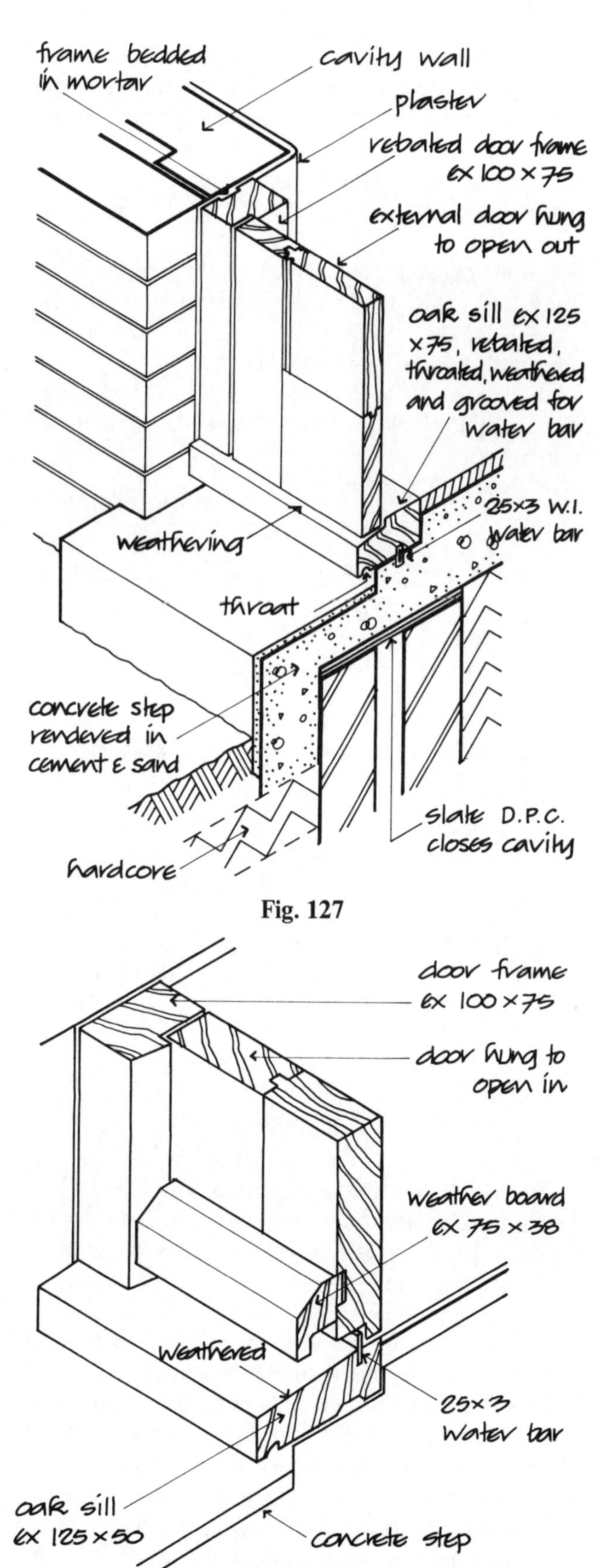

Fig. 127

Fig. 128

illustrated in Fig. 127, often with the addition of a weatherboard to throw water out from the vulnerable bottom edge of the door. This system is not entirely satisfactory, particularly in exposed positions, as there is a deal of air infiltration around the door and the bottom edge of the door is not watertight against driven rain.

Inward opening doors relied on the rebates in the frame and an upstand galvanised steel water bar set into the wood sill or directly into the stone or concrete threshold, as illustrated in Figs. 128 and 129, with weatherboards fixed to the bottom of the door. Again there is appreciable air infiltration and the sill is not watertight.

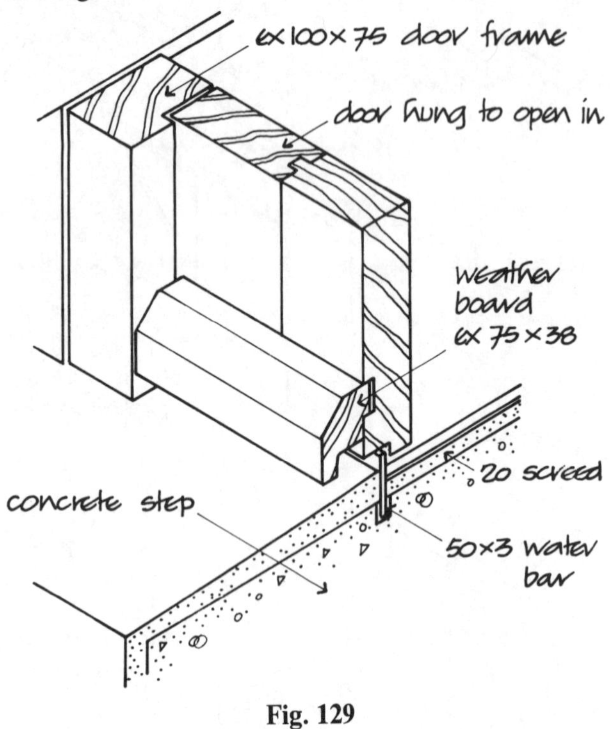

Fig. 129

To reduce air infiltration around the door it is necessary to fix seals on or towards the inside face of the frame. These air seals, which play a small part in excluding rain, are of the flexible type or the compression type shown in Figs. 130 and 131 or the nylon fibre type illustrated for use around windows. The most effective way of excluding rain is by generous drainage channels in the frame and door members, as illustrated in Fig. 132, down which rain runs to the sill where it drains out from the sill channel. The drainage channels which are on the outside of the air seals are designed to relieve the pressure build-up of water that would occur in narrow gaps between the door and the frame. The weatherboard, which is tongued to a groove in the

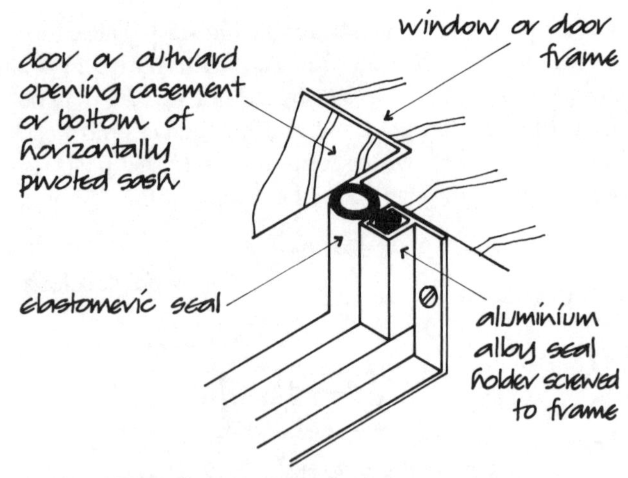

Compression seal for weatherstripping windows and doors

Fig. 130

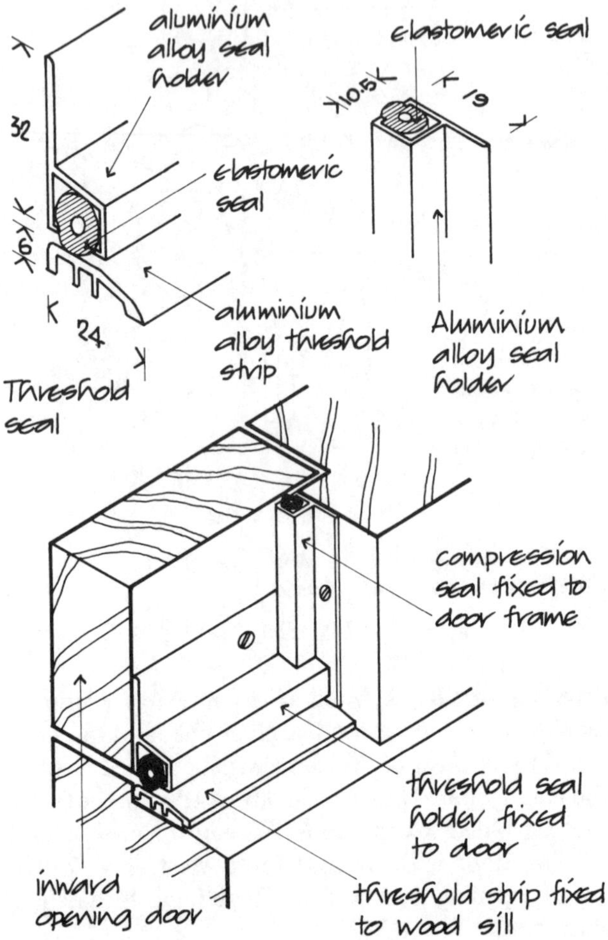

Threshold seal

Threshold and side weather seal to inward opening external door

Fig. 131

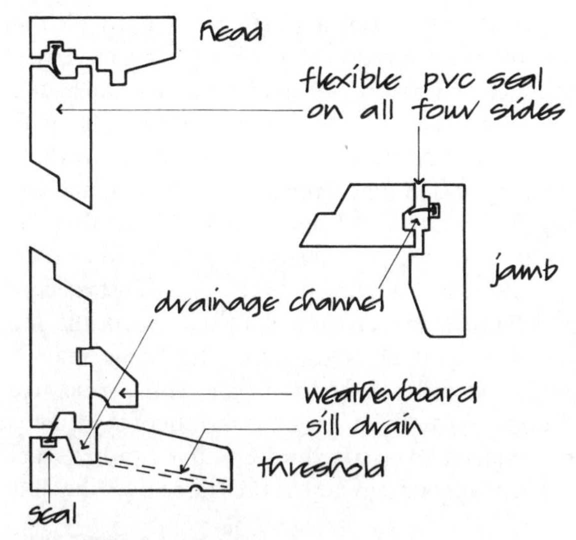

Fig. 132

bottom rail of the door, sheds water away from the vulnerable bottom edge of the door.

Wood door linings Wood door linings (door casings) may be plain with planted (nailed) door stops, rebated for doors or panelled, as illustrated in Fig. 133. Plain linings with planted stops are used for light doors in thin non-load-bearing partitions, rebated linings for more solid doors in thicker brick or block internal walls or partitions, and panelled linings for panelled doors with raised panels in substantial walls or partitions.

Plain linings are as wide as the wall or partition in which they are fixed and the three members of the lining are joined with a tongued-and-grooved joint as illustrated in Fig. 134. The linings are nailed to wood plugs driven into joints in the jambs or to wood pallets built into joints or to fixing blocks. The planted stops are nailed to the lining to accommodate the door thickness. Rebated linings are similarly fixed.

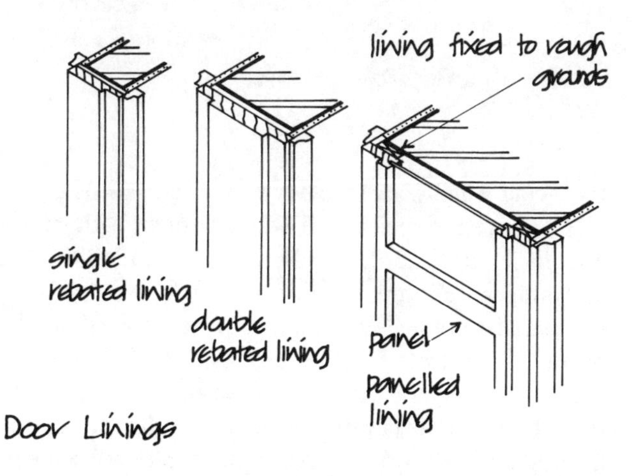

Fig. 133

Plain door linings with planted stops are illustrated in Fig. 135.

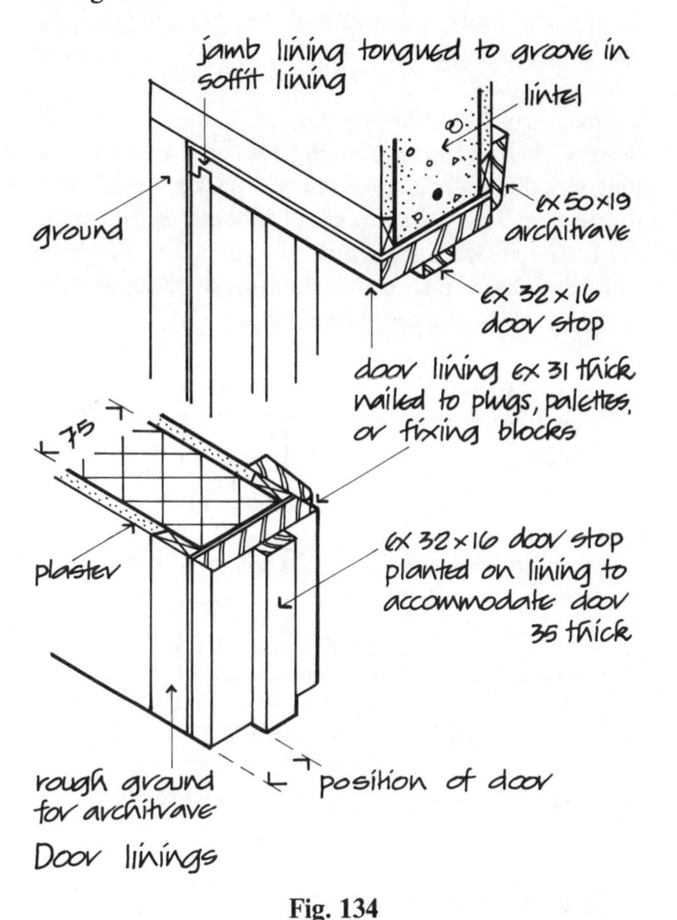

Fig. 134

ex 31
ex 115
75 block partition
with 12 plaster
ex 32×16
planted stop
ex 138
½ brick partition
with 13 plaster
ex 31

Door linings for use with block or brick partitions

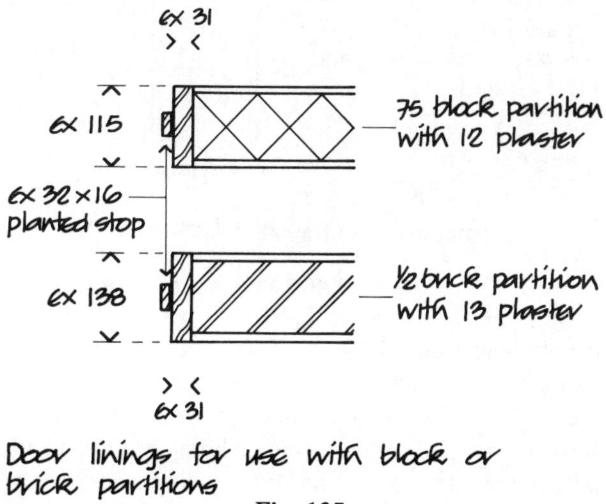

Fig. 135

Wood door frames, linings and door sets Manufacturers offer standard frames and linings for standard size doors. The door frames are cut from sections of 102 × 64 and 89 × 64, rebated for doors and with sills for external doors, as illustrated in Fig. 136. Door linings, or casings as they are sometimes called, are cut from sections 138 × 38, 138 × 32, 115 × 38, and 115 × 32, rebated for doors. The width of these linings may not match the overall thickness of some partitions and finishes.

Door sets (door assemblies) are combinations of doors with door frames or linings and hardware such as hinges and furniture, prepared as a package ready for use on site. This plainly makes economic sense where many similar doors are to be used and packets of doors can be ordered and delivered instead of separately ordering doors, frames and hardware.

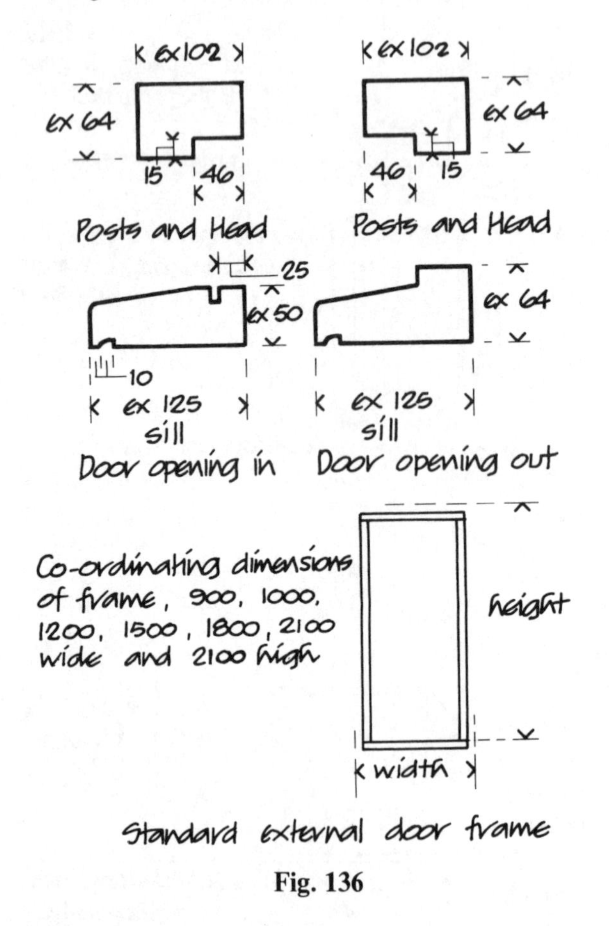

Fig. 136

Storey-height door frames There is often inadequate fixing for a door frame or lining in a thin non-loadbearing partition so that the door, in use, may cause some movement in the frame or lining relative to the partition, to the extent that cracks in finishes around the frame or lining and particularly in the partition over the head of the door may appear. To provide a more secure fixing for doors in thin partitions it is often practice to use storey-height frames that can be fixed at floor and ceiling level. These storey-height or floor-height frames are cut to line the reveal of the door opening and in that sense serve as linings, and are put together with floor-height posts, a head that can be fixed to the ceiling, a transom at the head of the door and also a sill for fixing to the floor, as illustrated in Fig. 137. The frame sections may be rebated for the door or be plain with planted stops. The frame may be of uniform width for the full height with a panel fixed in the space over the door, or the width of the frame may be reduced over the door so that finishes, such as plaster, may be run across the frame over the door.

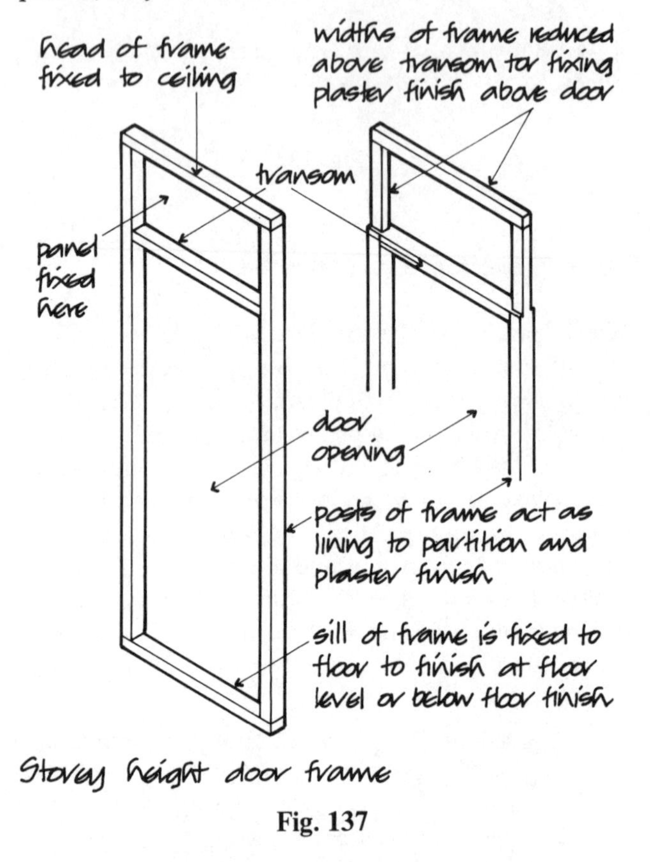

Fig. 137

PVC-clad wood door frames Door sets consisting of PVC covered frames and PVC architraves with standard flush door and hardware are available. The advantages of the solid timber core with a maintenance-free PVC coating are combined in the frame, as illustrated in Fig. 138. The PVC cladding is clipped around the visible parts of the frame and the PVC architrave is clipped into a slot in the frame. The frames are fixed by screwing to plugs in the reveal and the screw heads are covered with a plastic plug.

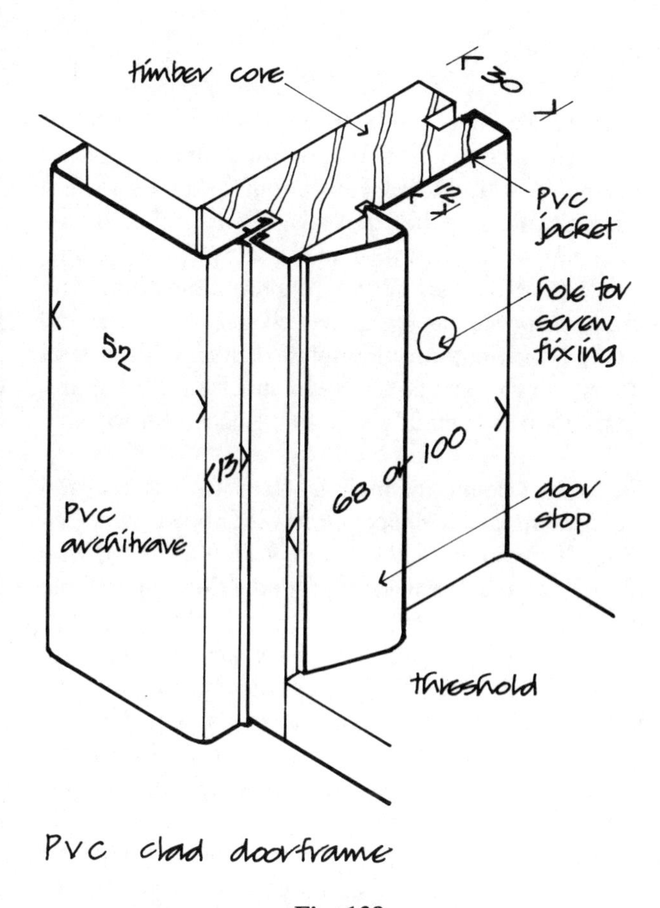

Pvc clad doorframe

Fig. 138

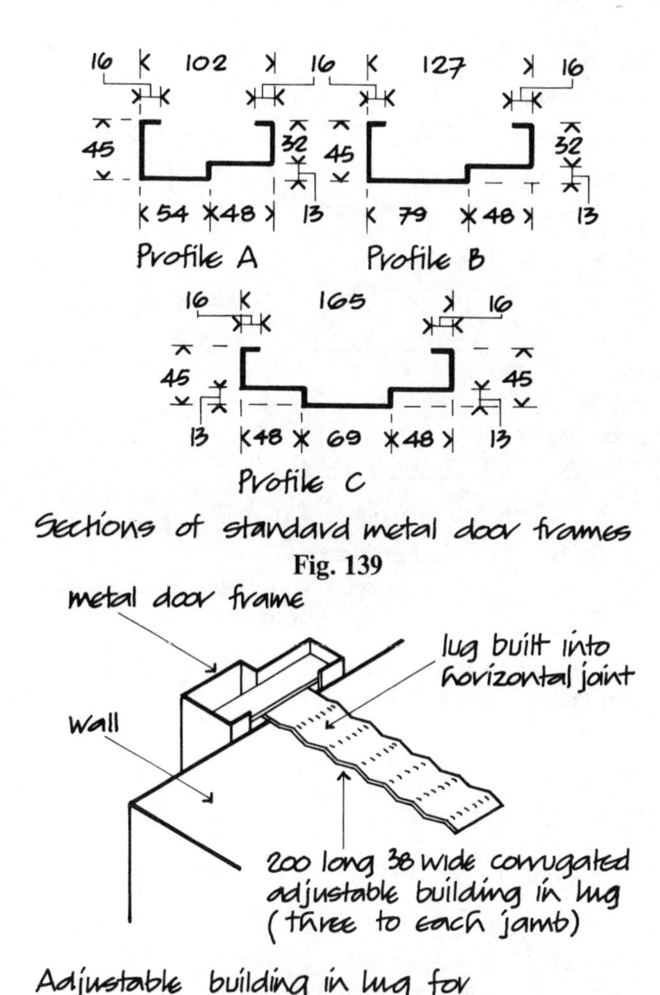

Sections of standard metal door frames

Fig. 139

Adjustable building in lug for metal door frames

Fig. 140

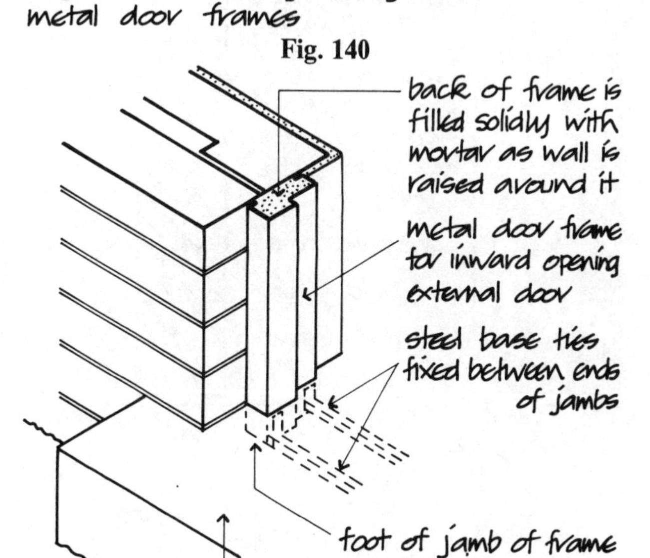

Building in external metal door frame

Fig. 141

Metal door frames for wood doors are manufactured from mild steel strip pressed into one of the three standard profiles. The same profile is used for head and jambs of the frame. The three pressed steel members are welded together at angles. After manufacture the frames are hot-dip galvanised to protect the steel against corrosion. Two loose pin butt hinges are welded to one jamb of the frame and an adjustable lock strike plate to the other. Two rubber buffers are fitted into the rebate of the jambs to which the door closes to cushion the impact sound of the door closing. Fig. 139 is an illustration of standard metal door frame sections. The frames are made to suit standard door sizes.

Metal door frames are built-in and secured with adjustable metal building-in lugs as illustrated in Fig. 140. These frames may be used for either internal or external doors. The frames are bedded and filled with mortar as they are built in and are supplied with metal ties to maintain the spacing of the posts (Fig. 141). For internal doors these frames are wider than the finished thickness of the partitions, so obviating the necessity of an architrave to mask the joint between the frame and the plaster, as illustrated in Fig. 142.

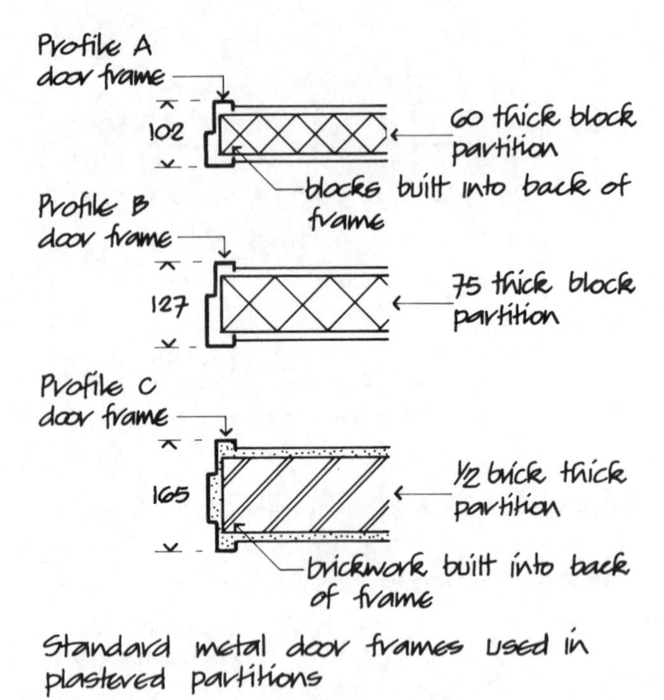

Fig. 142

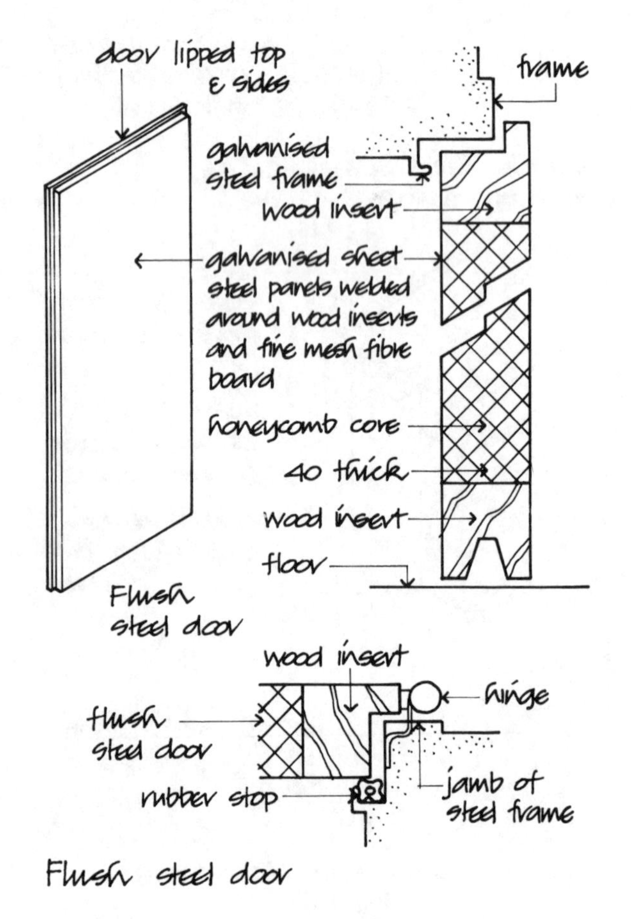

Fig. 143

STEEL DOORS

Glazed steel doors are fabricated from the hot rolled W20 steel sections used for windows. The sections are assembled with welded corner joints. The doors and frames, which are hot dip galvanised after manufacture, may be finished with an organic powder coating. Single glass is either putty or clip-on aluminium bead glazed. Double glazing is bedded in mastic tape and secured with clip-on aluminium beads. Glazed steel doors, which have largely been superseded by aluminium doors, are mainly used for replacement work.

Flush steel doors are manufactured from sheet steel which is pressed to shape, often with lipped edges, hot dip galvanised and either seam welded or joined with plastic, thermal break seals around a fibre board, chip board or foamed insulation core, generally with edge, wood inserts as framing and to facilitate fixing of hardware. The sheet steel facings may be flush faced or pressed to imitate wood panelling or glazed panels. The exposed faces of the doors may be finished ready for painting or with a stoved on organic powder or liquid coating.

These comparatively expensive, robust, heavy duty doors are generally used in this country in commercial and industrial buildings and as fire doors. In North America and parts of Northern Europe they are extensively used in all types of buildings. These doors are generally supplied as door sets complete with frame, door leaf and hardware and fittings. Fig. 143 is an illustration of a flush faced steel door.

ALUMINIUM DOORS

An extensive range of partly-glazed and fully-glazed doors is manufactured from extruded aluminium sections. The slender sections possible with the material in framing the doors provide the maximum area of glass. These glazed doors, commonly advertised as 'patio doors', are made as both single- and multi-leaf doors to hinge, slide or slide and fold to open. Glazed doors serve as a window by virtue of the large area of grass which provides no privacy, and as doors by the facility to open them from floor level. As windows they afford little insulation against loss of heat, unless double glazed, and as doors give poor security because of the extensive use of glass. The particular use of these doors is to provide a large area of clear glass for an unobstructed view out to gardens and to give ready access from inside to outside.

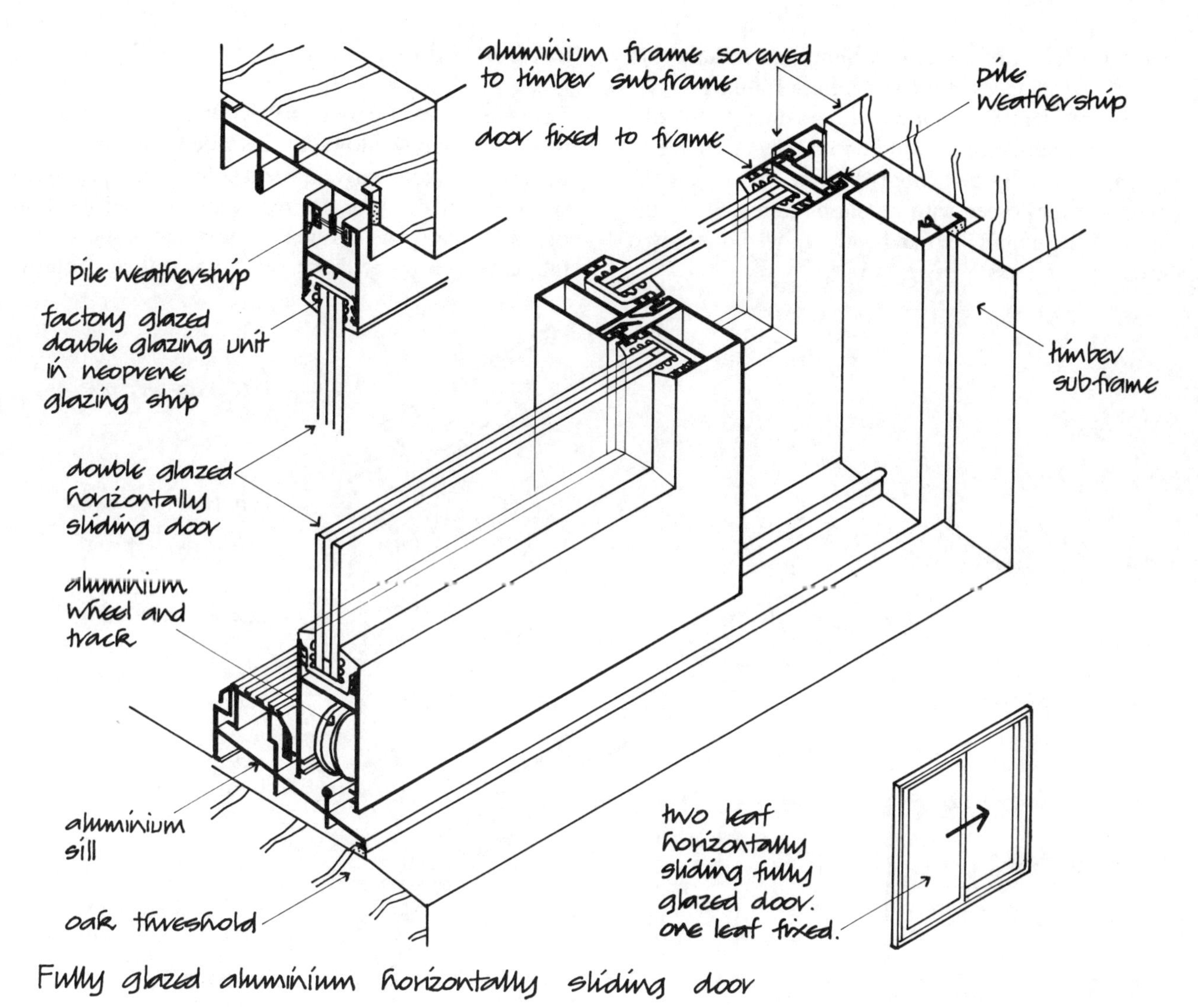

Fig. 144

Fig. 144 is an illustration of an aluminium section glazed door designed to slide open. The doors are double glazed to reduce heat loss and have weather-stripping and drainage channels to exclude wind and rain.

A disadvantage of these doors is that there may be appreciable condensation on the inside faces of the aluminium framing. To minimise condensation on the inside faces of these double glazed doors it is practice to fabricate them as thermal break doors. The main framing sections of the doors, which are joined with corner cleats, are fixed to aluminium facings through plastic sections that act as a thermal break.

Of recent years single leaf aluminium doors have been made to resemble traditional panelled wood doors. These, so called, residential doors are framed from extruded aluminium sections in the same way that windows and fully glazed doors are fabricated with the addition of a middle, horizontal rail to imitate the middle or lock rail of a wood door. The sections are made to take either glazed or solid panels secured with internal pop-in glazing beads. The solid panels which are fabricated from PVC or glass fibre reinforced plastic sheets around an insulating core, may be moulded to imitate traditional wood panels.

An advantage of these doors is that they may be finished in a range of coloured powder or liquid coatings that do not require periodic painting for maintenance. These doors are sufficiently robust for use in domestic buildings and may be fabricated as thermal break construction to minimise condensation on the internal faces of the aluminium framing.

These residential, aluminium doors do not look like the traditional panelled wood doors they are made to replicate and may be a poor security risk unless the panels are reinforced with an aluminium sheet in the core and the panels are fixed with screwed or secured beads.

uPVC DOORS

Following the success of uPVC windows as replacement for wood windows, the extruders and fabricators of uPVC sections have in the last few years produced single leaf uPVC doors for replacement or substitution of traditional panelled wood doors. These doors are fabricated from a frame of comparatively bulky,

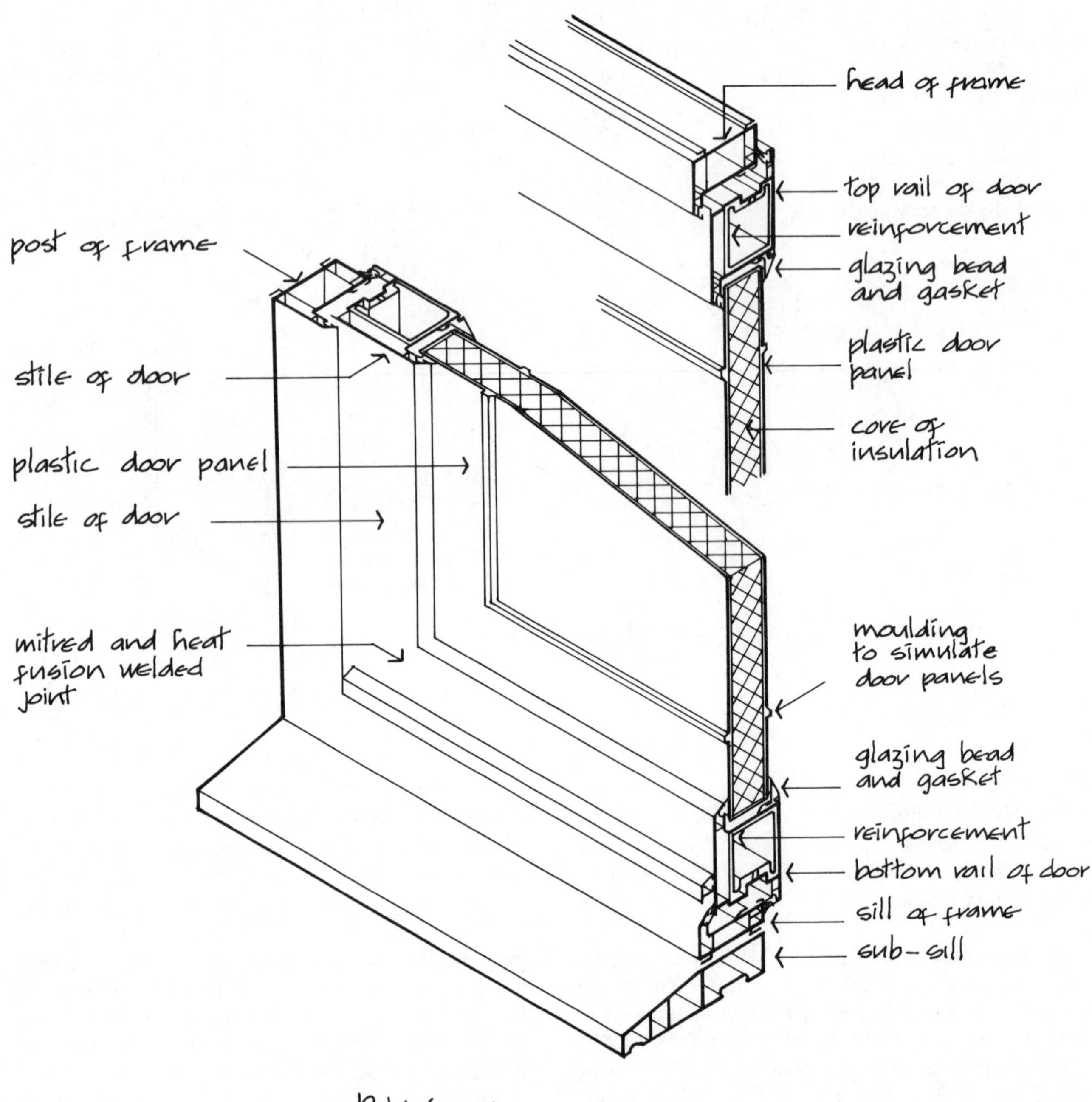

Fig. 145

extruded uPVC hollow sections similar in size to the stiles and rails of wood framed doors. The hollow framing sections are reinforced with galvanised steel or aluminium sections in the main cell of the hollow sections that are mitred and heat fusion welded at corners. A mid rail member is fitted to match the middle or lock rail of a wood door. These door leafs are hung to extruded, hollow section uPVC frames and thresholds as illustrated in Fig. 145.

The uPVC door leafs are framed for glazing with single or double glazing, secured with internal pop-in beads and weathered with wedge and blade gaskets. As an alternative to glazed panels a variety of plastic panels is produced from press moulded acrylic, generally moulded to imitate wood door panels either as full door height panels or as two panels fitted to a middle rail. So that the panels may have plain edges for fitting to the rebate and glazing beads of the hollow uPVC framing, they are moulded to represent the stiles, rails and panels of wood doors. In consequence these doors do not look like the panelled wood doors they are fabricated to replicate.

The hollow panels may have a core of some insulating material and a foil or thin sheet of aluminium as a barrier to breaking and entering by fracturing the panel.

The majority of these doors are made as white or off white impact modified uPVC to minimise the considerable thermal expansion that this material suffers due to solar radiation. Coloured and wood grain finishes are also supplied.

The advantage of uPVC doors is that they require no maintenance during their useful life, other than occasional washing. The disadvantage of these doors is that they may jam shut due to thermal expansion, knocks and indentations cannot be disguised by painting, and they are not as robust to heavy use as a traditional wood framed door.

Garage doors Pressed metal doors are suited for use as garage doors because they are lightweight and have adequate stiffness and shape stability for a balanced 'up-and-over' opening action. The doors are manufactured from pressed steel or aluminium sheet which is profiled to give the thin sheet material some stiffness. The sheet is welded or screwed to a light frame to give the door sufficient rigidity. Steel doors are hot-dip galvanised and primed for painting or coated with PVC, and aluminium doors are anodised. Fig. 146 is an illustration of a steel up-and-over garage door.

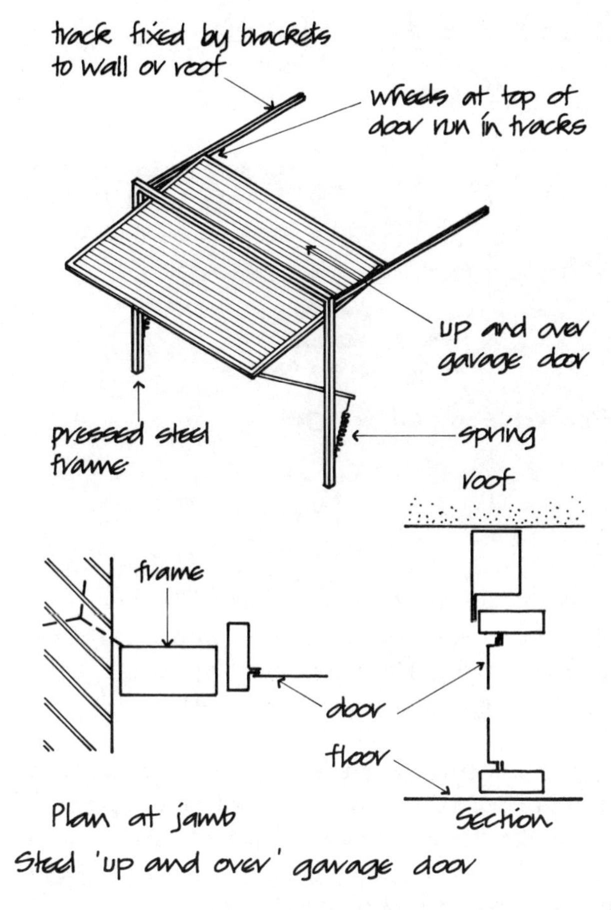

Fig. 146

HARDWARE FOR DOORS

Hardware is the general term for the hinges, locks, bolts, latches, and handles for a door. Ironmongery was a term used when most of these were made of iron or steel. The term 'door furniture' is sometimes used to describe locks, handles and levers for doors.

Hinges

Pressed steel butt hinges The cheapest and most commonly used hinges are pressed steel butt hinges. They are made from steel strip which is cut and pressed around a pin, as illustrated in Fig. 147. They are used for hanging doors, casements and ventlights. The pin of the standard butt hinge is fixed inside the knuckle. Loose pin butt hinges are made, the advantage being that the door can be taken off by taking out the loose pin instead of unscrewing the hinge.

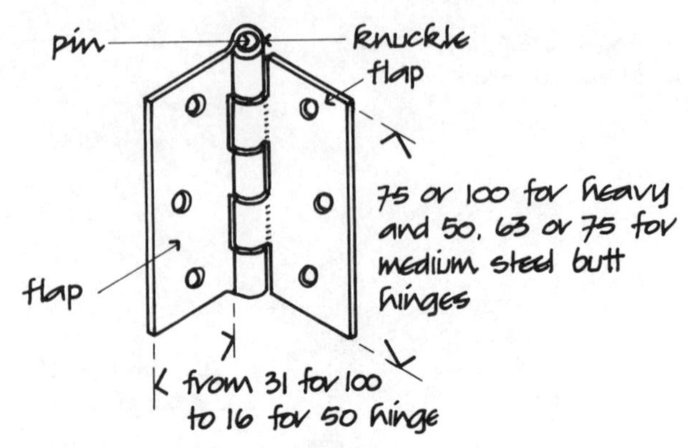

Fig. 147

Double pressed steel butt hinges are made of two strips of steel each folded back on itself around the pin, as illustrated in Fig. 148. They are stronger than ordinary steel butt hinges and are used for heavy doors.

Cast iron butt hinges are heavier and more expensive than steel butts of the same size and have a longer useful life as the bearing surfaces of the knuckle are more resistant to wear. They are used for heavy doors in frequent use, such as entrance doors.

Brass butt hinges are more expensive than steel and cast iron butt hinges and are used for polished doors and cupboards as they do not rust and have a pleasing appearance.

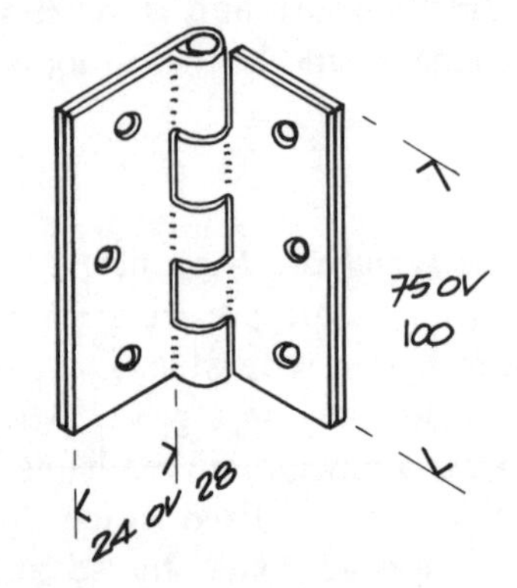

Fig. 148

Steel skew butt hinges (rising butts) The bearing surfaces of the knuckle are cut on the skew, so that as the hinge opens one butt rises, as illustrated in Fig. 149. They are used for hanging doors so that as the door opens it rises over and so reduces wear on carpets. Because of the action of the rising butt, doors tend to be self-closing and this type of hinge is also used for this effect.

Steel tee hinges consist of a rectangular steel flap and a long tail with knuckles around the pin, as illustrated in Fig. 150. The flap is fixed to the frame and the tail to the door. These hinges are used mainly for matchboarded doors as they assist in supporting the door by being fixed along the ledges or rails.

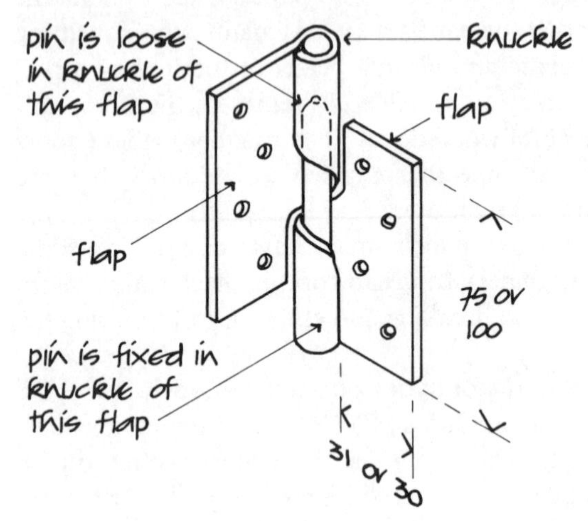

Fig. 149

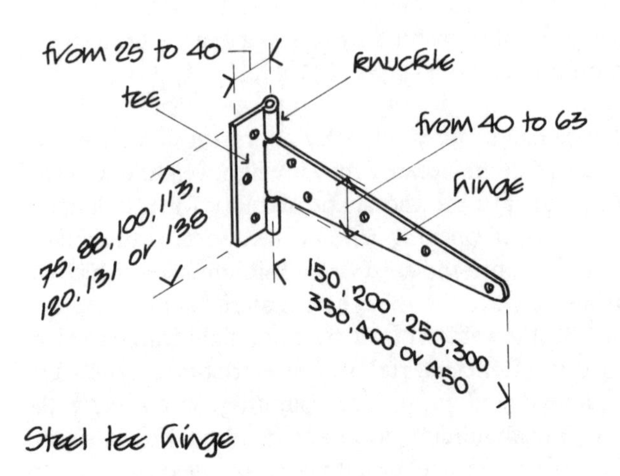

Fig. 150

Hook and band hinges consist of a rectangular steel plate in which a pin is fixed and a steel band folded around the pin (Fig. 151). They are made of heavier steel than tee hinges and are used for hanging heavy doors such as those to garages and workshops.

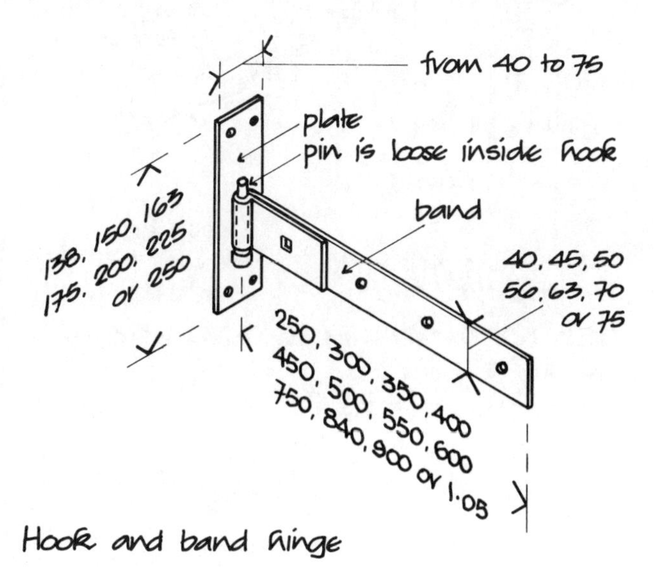

Fig. 151

Latches and locks

The word latch is used to describe any wood or metal device which is attached to a door or window to keep it closed and which can be opened by the movement of the latch operated by a handle, lever or bar. A lock is any device of wood or metal attached to a door which can be used to keep it closed by application of a loose key.

Mortice lock The mechanism most used today is the mortice lock which combines the operation of both a latch and a lock through a latch operated by a handle or knob and a lock operated by a loose key, as illustrated in Fig. 152. These locks are set into a mortice in the door and are latched and locked to a striking plate in the frame.

Where a mortice lock is fixed to a solid lock rail of a door it is usually a horizontal two-bolt lock as shown in Fig. 152, and where it is fitted to the stile of a door, for example a glazed panel door, it is an upright lock as illustrated in Fig. 153. The security of the lock itself depends on a sound fixing of the striking plate to a solid frame and the loose key being taken out when the lock bolt is shot.

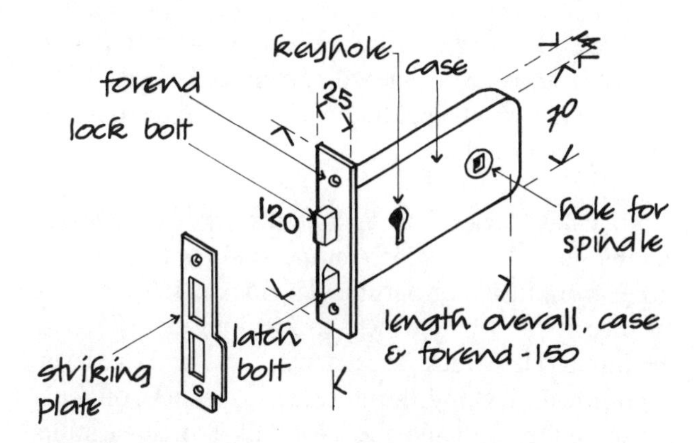

Fig. 152

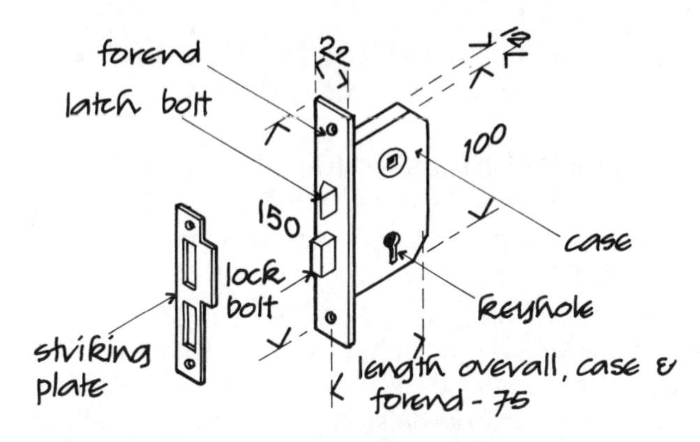

Fig. 153

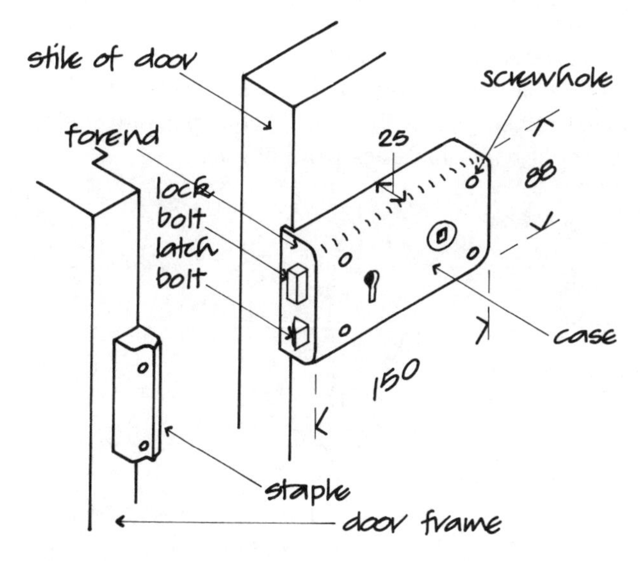

Fig. 154

Rim lock Locks which are made to be screwed to one face of a door are described as rim locks (Fig. 154). They are not much used today as they tend to spoil the appearance of a door.

Mortice dead lock This operates as a lock only, there being one lock bolt which is operated by a loose key either from inside or outside. The lock case is set in a mortice in the door and locks to a striking plate fixed to the frame. These locks are made as three and five lever operation for locking, the more levers the more difficult it is to tamper with and open. Fig. 155 is an illustration of a mortice dead lock. These locks are used in addition to mortice locks or cylinder night latches for additional security to doors.

The rack bolt illustrated in Fig. 156 is used with windows and doors for additional security.

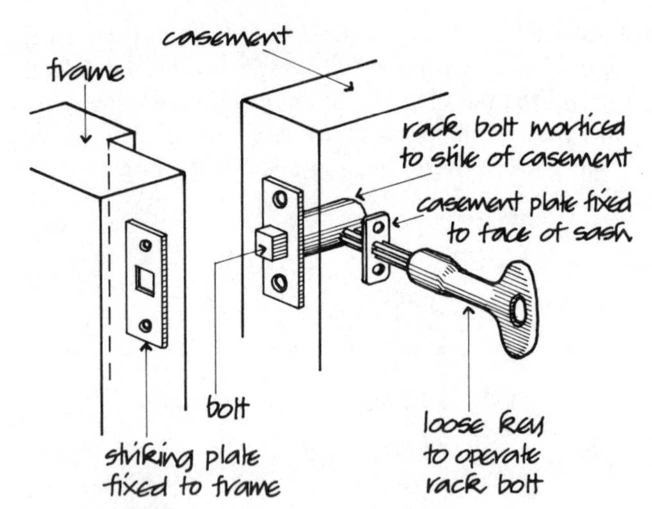

Fig. 156

Cylinder night latch (springlatch) This is designed to act as a latch from inside and a lock from outside for convenience in use on front doors. It is made as a rim latch for fixing to the inside face of doors (Fig. 157). There is one bolt, shaped as a latch for convenience in closing, which is opened by a knob or lever from the inside. For security the latch bolt can only be opened with a loose key from outside. This type of latch offers poor security as it is fairly easy to push back the latch from outside by means of a piece of thin plastic or metal inserted between the door and frame.

A more secure type of night latch is designed as a mortice lock which is opened as a latch from inside by means of a lever and from outside by a loose key. The lock has a double throw action which by two turns of the key from outside locks the latch in position so that it cannot be pushed back from outside.

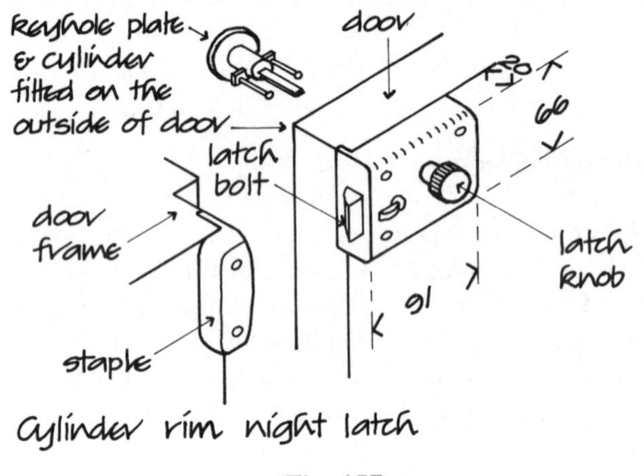

Fig. 157

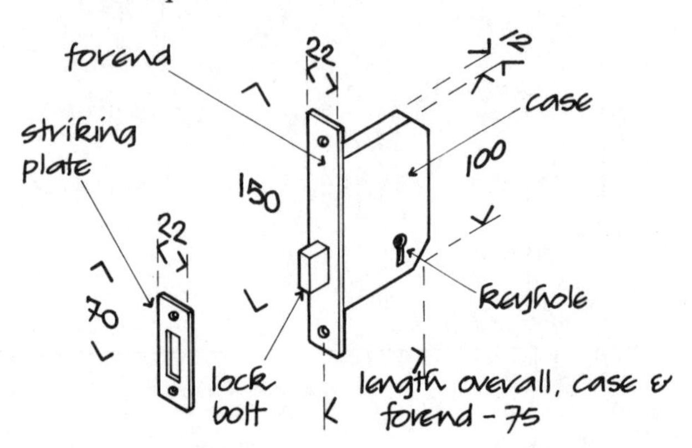

Fig. 155

CHAPTER THREE

FIRES, STOVES AND CHIMNEYS

FUELS

During the last forty years there has been a move from the use of solid fuel burning open fires and stoves to heating by off-peak electricity, then to oil-fired central heating and more recently to gas-fired central heating. These changes were made initially for the convenience of heating by electricity, oil and gas and because of legislation to create smokeless zones to control fog.

The most convenient method of heating is by electricity which can be controlled by the touch of a switch, requires only slender cables to distribute energy and needs very little attention from the consumer. It is extravagant to convert natural fuels such as coal and oil to electricity and this convenient energy source is now so costly that it is much less used for heating. Oil, which was once so cheap, is now an expensive fuel and the limited sources of this fuel should be conserved. In the middle of this century very many heating appliances were changed or converted to burn oil. Following the steep increase in the cost of oil there was a move to gas as a fuel, prompted by the availability of natural gas which was then cheap. The cost of natural gas is increasing steeply.

The disadvantage of coal and its derivatives such as smokeless fuel and coke as a fuel are that they are laborious to deliver, bulky to store and dirty in use and produce ash which is tedious and messy to dispose of. The advantages of coal are that it is plentiful, provides a cheerful flame or glow when it burns and provides a traditional focal point in a room. Smokeless fuels and coke are lighter, cleaner and produce less ash than household coal but nonetheless require space for storage and effort in use, unlike the convenience fuels electricity, oil and gas.

The solid fuels available are bituminous coal (house coal), anthracite, smokeless fuels, coke, wood and peat.

Bituminous coal or house coal is a natural coal that ignites easily and burns with a bright flame. It is the traditional solid fuel in the open fire, much enjoyed for its bright flame and cheerful glow and equally hated for the need for frequent fuelling, cleaning of ash and the smoke pollution in towns that was a prime cause of fog. In Smoke Control Areas, only authorised fuels may be used and house coal is generally prohibited.

Anthracite is a dense natural fuel that burns slowly and is a natural smokeless fuel. The limited supplies of this fuel are used in stoves and boilers.

Manufactured solid fuels are produced by processing coal to produce a smokeless fuel, for use in Smoke Control Areas. Smokeless fuel is less dense and cleaner to handle than natural coal, burns with a glow rather than a flame and produces a fine ash.

Coke is the by-product of the conversion of coal to town gas (see Volume 5). It is light in weight, clean to handle, burns with a glow and is smokeless, but produces hard clinker and ash which is messy to dispose of.

Wood There are very limited supplies of wood for use as fuel for heating in this country. The large volume of wood required for heating necessitates large open fires or stoves and considerable storage space.

Peat, which is compressed, decayed vegetation, is cut and used as a fuel mainly in Ireland and the west of England.

FUNCTIONAL REQUIREMENTS

The functional requirements of fires and chimneys are:

Structure – strength and stability
Resistance to weather
Fire safety
Ventilation

SOLID FUEL BURNING APPLIANCES

Solid fuels are burned in open fires, room heaters or stoves and boilers for heating and hot water.

Open fires The traditional open fire consists of a grate inset in a fireplace recess formed in a brick chimney breast, as illustrated in Fig. 158. As its name implies, an open fire is clearly visible and this is its chief attraction. The disadvantage of an open fire is that much of the air drawn into the fire and up the chimney by convection is not necessary for combustion or burning the fuel and this excess air wastefully takes a large proportion of the heat from the fire up the chimney. The air drawn into the fire is replaced by air drawn into the room which causes draughts of cool air which are uncomfortable and wasteful of heat. An open fire inset in a recess in a chimney built into an external wall will lose some heat to the outside.

The freestanding open fire in Fig. 158 has some advantage in that the fire is contained inside the metal surround and hood from which heat will be transferred

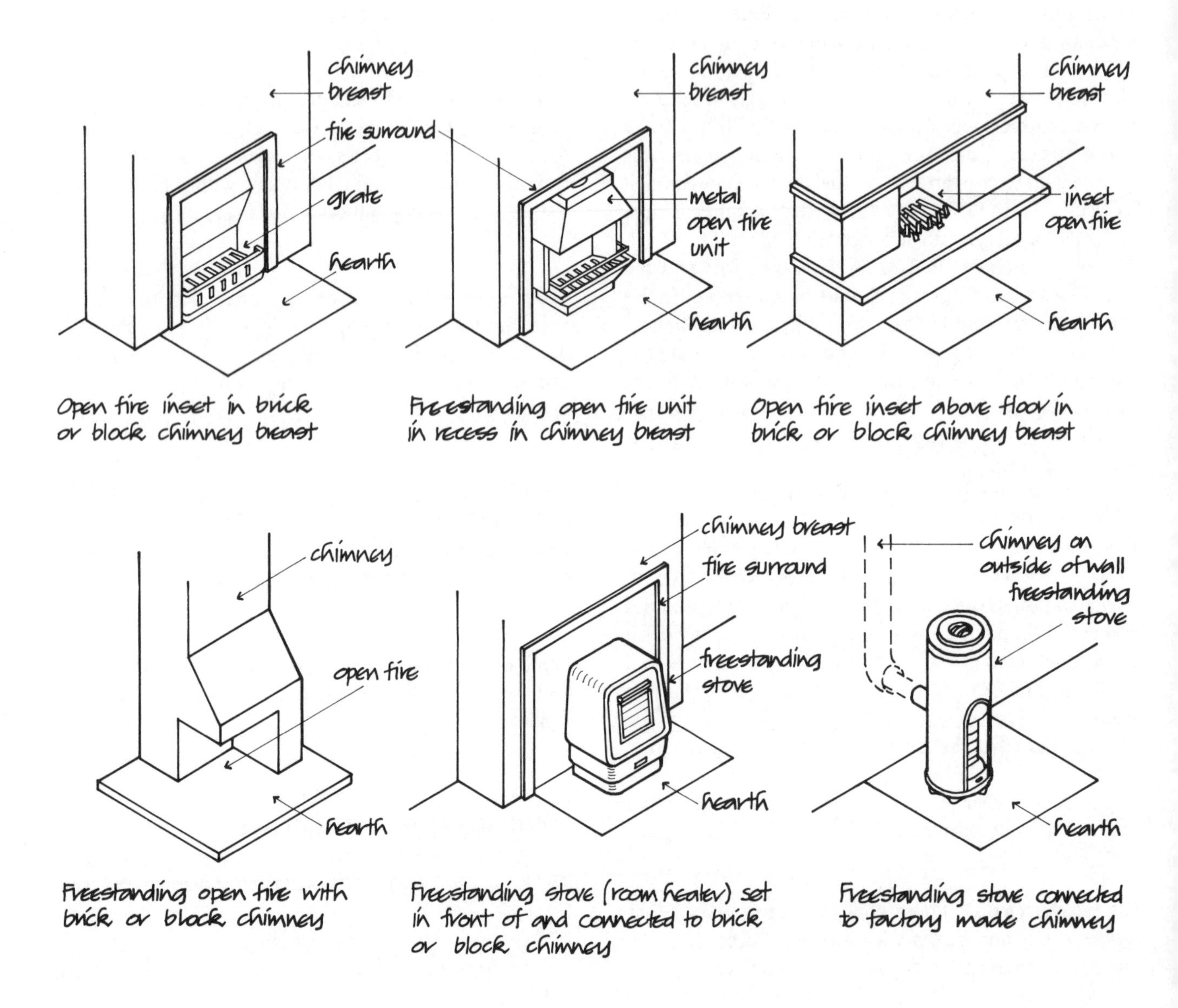

Fig. 158

to the room by the air around the fire. This type of freestanding open fire which is contained inside its metal case can be inset inside a fire recess or be fixed completely freestanding and connected to a metal chimney. It is designed principally for appearance rather than efficiency.

An open fire does not have to be built at floor level. An open fire inset above the floor is illustrated in Fig. 158. The purpose of raising the fire is to make the comparatively small opening of a coal-burning fire more visible; there is no advantage in efficiency.

An open fire does not have to be built into a wall or partition. It can be built as a freestanding structure or fitting where it can be visible from one or more sides. There is some advantage in a freestanding fire in that it will heat the room all round the fire and chimney and be a feature of a room. The solid construction freestanding open fire illustrated in Fig. 158 is visible from two sides.

Room heater is a term recently used to describe an enclosed adaptation of the open fire designed for the more efficient use of solid fuel. It is a modern version of the traditional stove. The advantages of the room heater are that air intake and combustion can be controlled to appreciably reduce the intake of excess air and so reduce the wasteful flow of heated air passing up the chimney, and the whole of the surface area of a freestanding room heater is used to heat the room. To give a view of the fire these room heaters have a glazed panel or door which can be opened to give a clearer view of the fire. Room heaters can be fixed inside a fire recess in the chimney, be freestanding in front of the chimney as illustrated in Fig. 158 or be completely freestanding and connected to a metal chimney.

A down draught room heater is the most efficient type, in which air is drawn on to and through the fuel for combustion to minimise excess air being wastefully drawn in, as illustrated in Fig. 159. For maximum efficiency a down draught room heater is combined with a back boiler to take advantage of the heated combustion gases passing from the fire up the chimney. These back boilers are used to heat water and radiators.

Stoves A stove is an enclosed solid fuel burning appliance of cast iron, steel or brick in which fuel is burned on a hearth and from which a flue pipe is run either directly to outside or to a solid brick or block chimney. The traditional cylindrical European stove, illustrated in Fig. 158, is used for burning wood or coal or both. Like the room heater its advantage is that there is control of the air intake and reduction of flow of heated air up the chimney, and heat is given out from all the exposed surfaces of the stove.

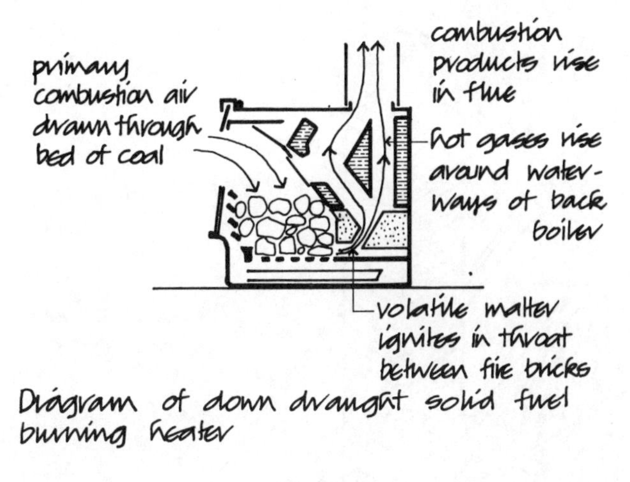

Fig. 159

Boilers A boiler is an enclosed appliance in which fuel is burned specifically to heat water for a hot water supply or for water heating or both. The fuel burned on a hearth heats water run inside cast iron sections around the hearth, the heated water flowing either by gravity or pump to hot water cylinders and radiators and back to the boiler for reheating.

CHIMNEYS AND FLUES

Structure – strength and stability The practical guidance in Approved Document A to the Building Regulations 1991 sets out the least sectional area on plan of chimneys that are built into walls as described in Volume 1.

Ventilation

Air supply Approved Document J gives practical guidance to meeting the requirements of Part J of Schedule 1 to the Building Regulations 1991 for an adequate supply of air to heat producing appliances for combustion and the efficient working of flue-pipes or chimneys.

Section 2 of the guidance, which applies to solid fuel burning appliances with a rated output up to 45 kW, requires a ventilation opening direct to external air of at least 50% of the appliance throat opening area, for open appliances, and at least 550 mm^2 per kW of rated

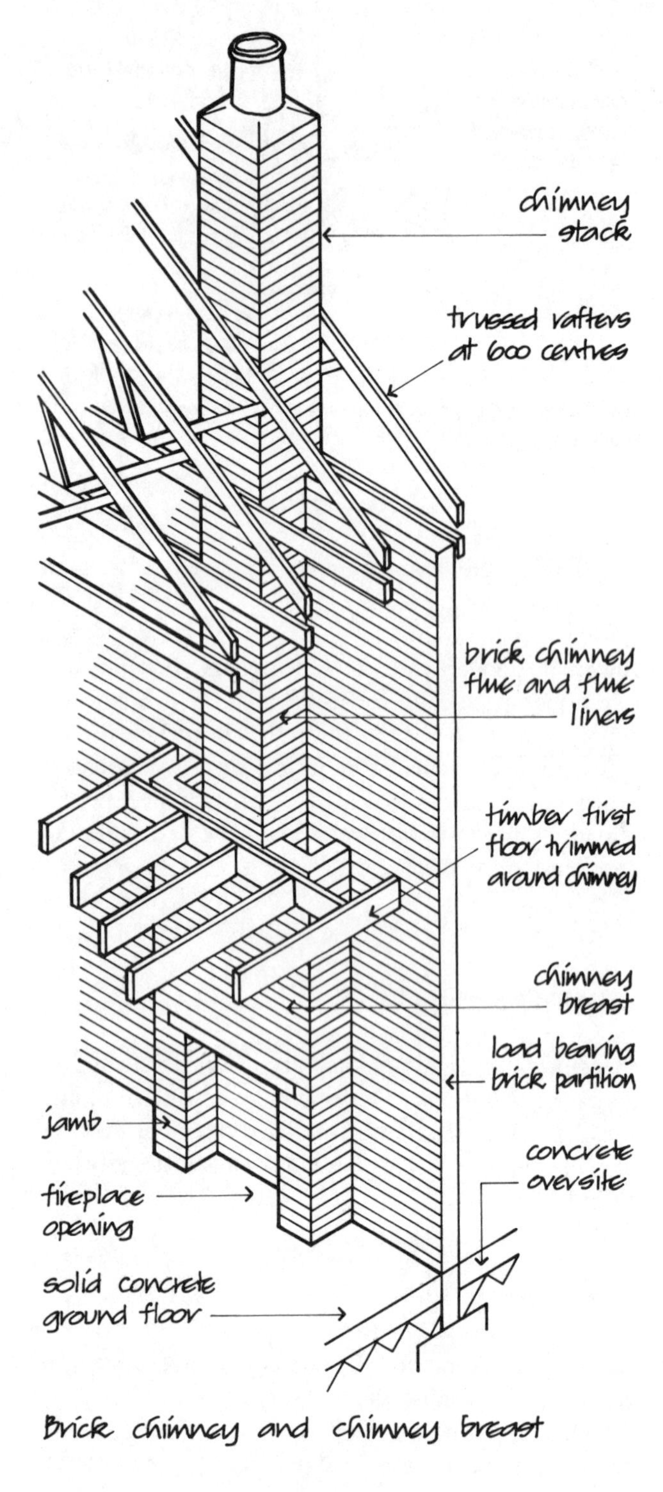

Fig. 160

output above 5 kW for other solid fuel appliances. Where a flue draught stabiliser is used for these appliances, the area of ventilation should be increased by 300 mm² for each kW of rated output.

The requirements for an air supply for combustion, like the requirements for room ventilation, are dictated by the trend over recent years to air sealed windows and doors to contain heat and avoid draughts. The requirement for an adequate air supply for combustion must of necessity suppose a draught of cold outside air entering a room in which an open fire is burning vigorously, unless the fire or appliance is fitted with a separate air intake.

A flue is a shaft, usually vertical, to induce an adequate flow of combustion air to a fire and to remove the products of combustion to the outside air. The material which encloses the flue, brick, block, stone or metal, is termed a chimney. A chimney may take the form of a pipe run to the outside or be constructed of solid brick, block or stone either freestanding or as part of the construction of a partition or wall. The conventional chimney for open fires and stoves is constructed of block, brick or stone and consists of a fireplace recess contained between jambs over which a chimney breast gathers into a chimney that is carried above roof level as a chimney stack (Fig. 160). Because a fireplace and the enclosing jambs are wider than the chimney and flue above, the chimney breast is constructed to accommodate the gathering in of the fireplace opening to the smaller flue.

Some of the heat of combustion of fires and stoves will be transferred to the chimney, which will heat the structure and air surrounding it. To take the maximum advantage of this heat the best position for a chimney is as a freestanding structure in the centre of a room or building where it is surrounded by inside air. As an alternative the chimney may be built as part of an internal partition. Where buildings are constructed with a common separating or party wall it is convenient to construct chimneys back to back on each side of the separating wall. Chimneys constructed as part of an external wall suffer the disadvantage that some of the heat will be transferred to the cold outside air. This loss can be minimised by continuing the cavity of an external wall and cavity insulation behind the chimney. The four positions for a chimney are illustrated in Fig. 161.

A chimney constructed as part of an external wall may project either into or outside the building, as illustrated in Fig. 162, whichever is the most convenient. The gain in floor space inside with an externally projecting chimney is balanced by a slight increase in external surface area from which heat may be lost. The externally projecting chimney will be constructed as a breast the width of the fireplace recess and jambs

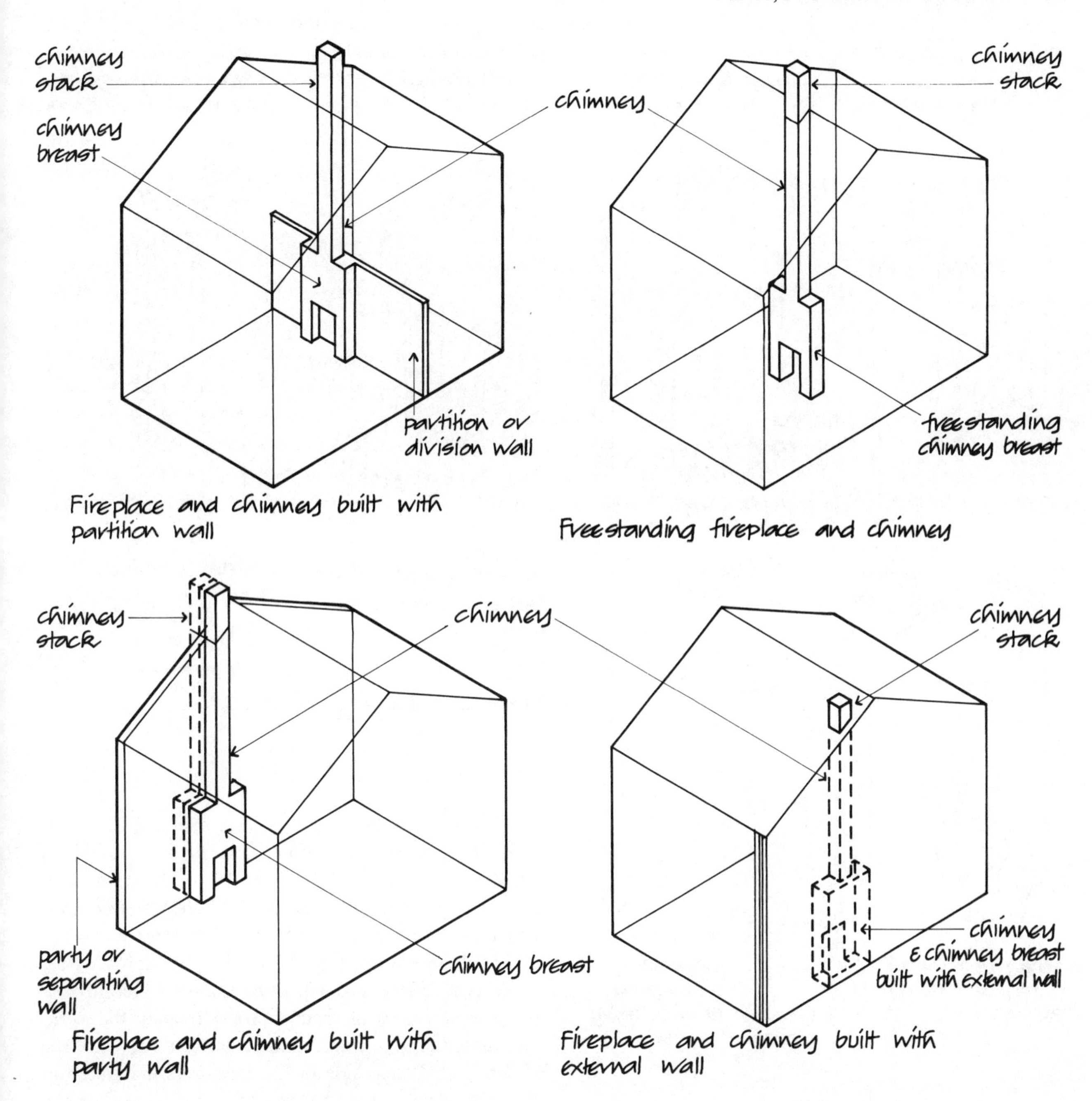

Fig. 161

behind the fire, reducing to a chimney above as illustrated in Fig. 163. The cavity of the external wall and any cavity insulation should be continued behind and around the fireplace and chimney to exclude rain and conserve heat, as illustrated in Fig. 162.

It is not necessary for a fireplace to be constructed on the same side of a partition as the projecting chimney formed as part of the partition. A fireplace or recess may be constructed either on the projecting or the opposite side of the chimney as shown in Fig. 163.

Fireplace or appliance recess The fireplace or appliance recess is an opening in the chimney into which an open fire, room heater or stove is fitted between the jambs of the recess. The traditional recess was usually 575 wide and 625 high to accommodate a domestic solid fuel open fire. It is practice today to form a larger opening into which either an open fire or a room heater can be fitted. The dimensions of this opening are 800 wide by 1000 high, as illustrated in Fig. 164.

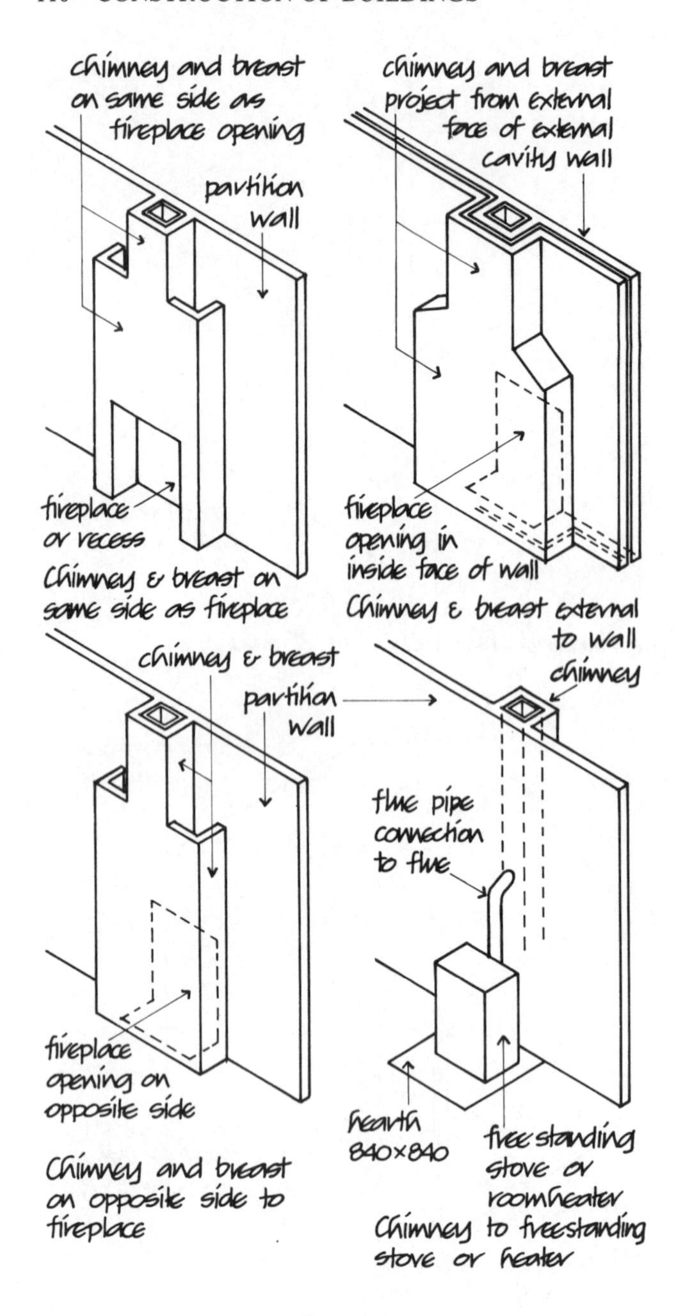

Fig. 162

The jambs each side of the recess serve to support the chimney and also to contain the fire and separate it from combustible materials, such as wood.

Fire safety

Approved Document J gives practical guidance to meeting the requirements of the Building Regulations 1991 for the non-combustible material to be used in the construction of fireplace recesses, hearths and flues.

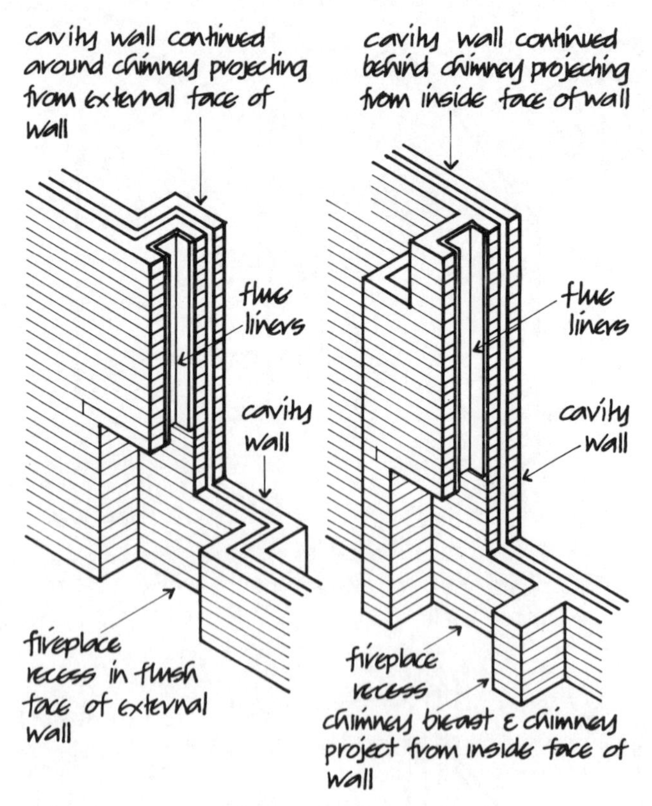

Fig. 163

The minimum dimension of the jambs to fireplaces is 200. The nearest brick dimension to this, i.e. 215, is chosen, as illustrated in Fig. 164. The minimum thickness of solid material at the back of fireplaces, which is shown in Fig. 165, should extend up to the underside of the opening.

Every fireplace recess must have a solid incombustible hearth, usually of concrete, formed inside the recess and extending a minimum dimension beyond the chimney and each side of the recess, as illustrated in Fig. 166. Hearths for appliances not in a recess must be of solid incombustible materials of the minimum dimensions shown in Fig. 166.

Where a hearth is formed in a raised timber ground floor it is constructed on a fender wall built off the concrete oversite on hardcore inside the fender walls, as illustrated in Fig. 164.

To support the chimney breast over the fire recess it is practice today to build in a precast reinforced concrete raft lintel as illustrated in Fig. 167. This lintel which is holed for the flue is built into the jambs of the recess and supports the flue liners and chimney. For appearance, the chimney over the fireplace opening is

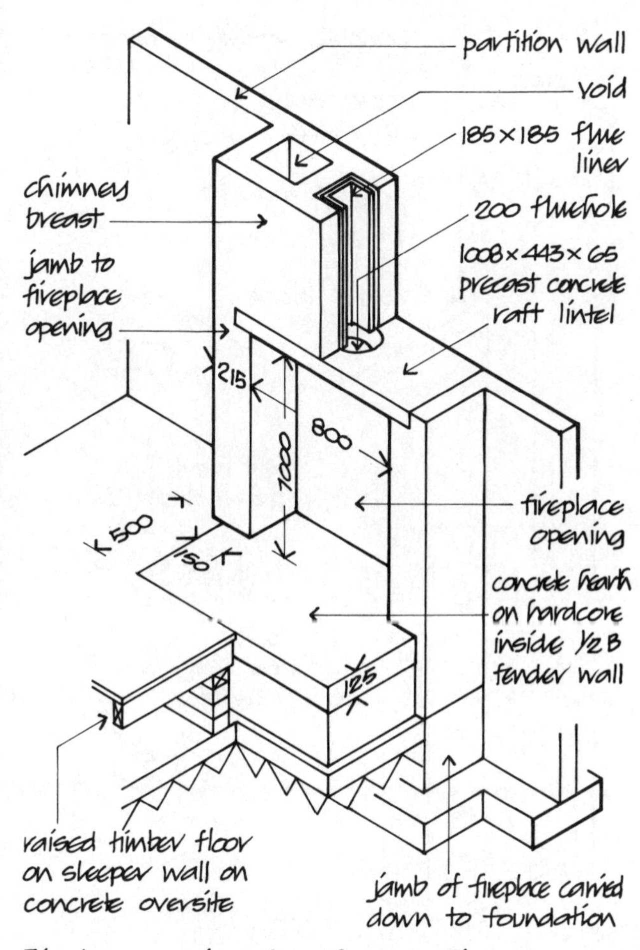

Fig. 164

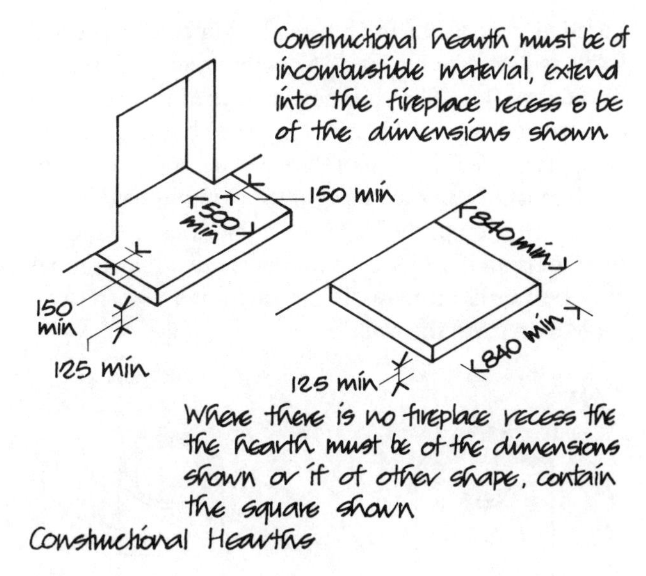

Fig. 166

The back of a fireplace recess must be a solid wall not less than 200 thick or a cavity wall with each leaf not less than 100 thick

In an external wall with no combustible cladding the back may be not less than 100

With fireplace recesses back to back other than in separating walls the back of the recess may be not less than 100

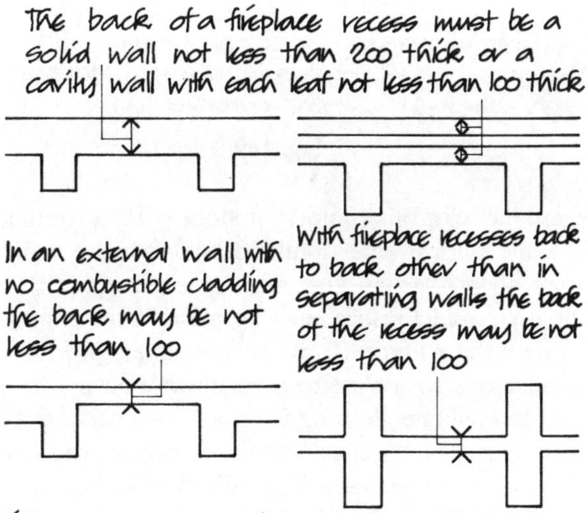

Thickness at back of fireplace recess exclusive of fireback to extend up to underside of fireplace opening

Fig. 165

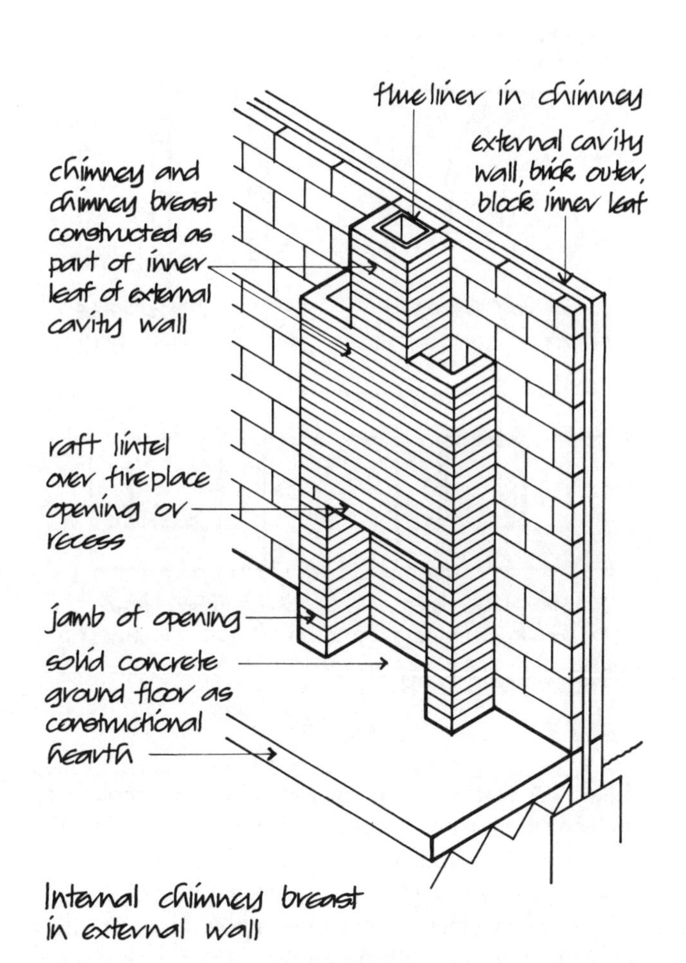

Fig. 167

raised as wide as the fire recess and jambs up to ceiling level, as illustrated in Fig. 167, with voids each side of the central flue. The chimney illustrated in Fig. 167 is constructed in brickwork which is bonded to the concrete block inner leaf of the external cavity wall. The small units of brick are used in the chimney as it would be wasteful to cut the larger units of block around the fire recess and the flue. As an alternative the chimney may be raised directly off the raft lintel as a narrow projection from the wall.

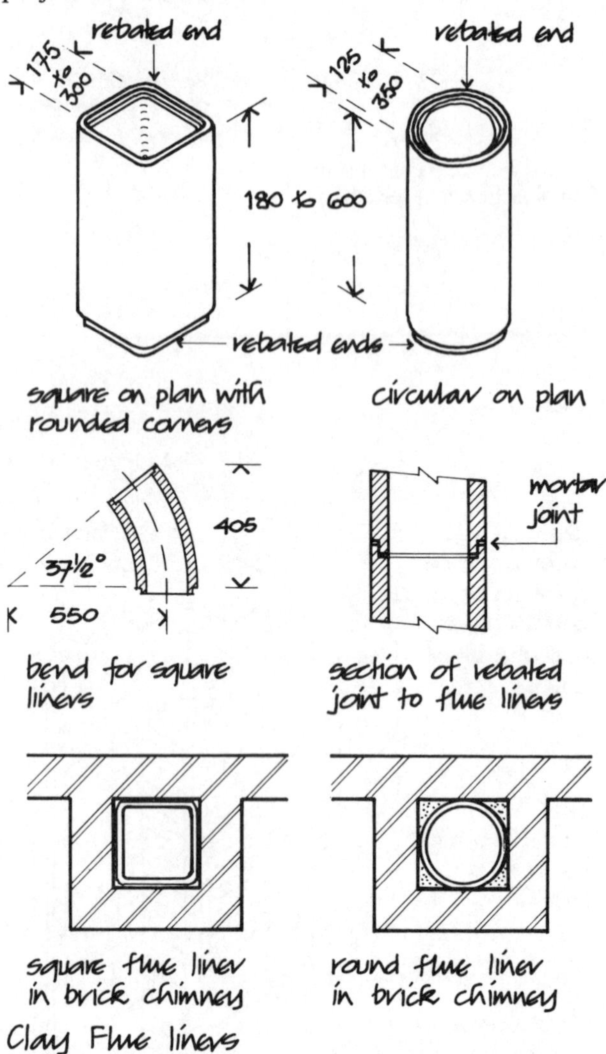

Fig. 168

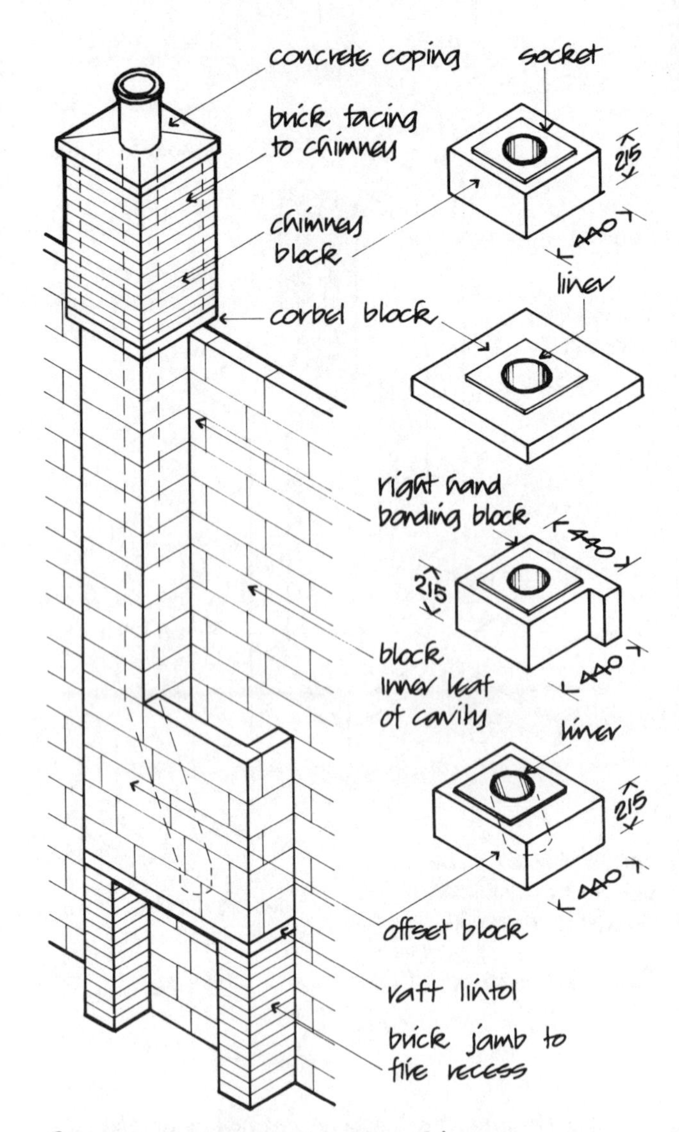

Fig. 169

Flues A flue is a shaft or pipe above a fireplace to induce combustion air to flow and carry away the products of combustion. For maximum efficiency the flue should be straight and vertical without offsets. As the heated products of combustion pass up the flue they cool and tend to condense on the surface of the flue in the form of small droplets. This condensate will combine with brick, block or stonework surrounding the flue to form water-soluble crystals which expand as they absorb water and may cause damage to the chimney and finishes such as plaster and paint. To protect the chimney from possible damage from the condensate, to encourage a free flow of air up the flue and to facilitate cleaning the flue, flue liners are built into flues. The practical guidance in Approved Document J to the Building Regulations 1991 sets out the minimum flue size shown in Table 11.

Flue liners are made of burnt clay or concrete. Clay flue liners are round or square with rounded corners in section and have rebated ends (Fig. 168). These liners

Table 11. Size of flues

Installation	Minimum flue size
Fireplace recess with an opening up to 500 mm × 550 mm	200 mm diameter or square section of equivalent area
Inglenook recess appliances	a free area of 15% of the area of the recess opening
Open fire	200 mm diameter or square section of equivalent area
Closed appliance up to 20 kW rated output burning bituminous coal	150 mm diameter or square section of equivalent area
Closed appliance up to 20 kW rated output	125 mm diameter or square section of equivalent area
Closed appliance above 20 kW and up to 30 kW rated output	150 mm diameter or square section of equivalent area
Closed appliance above 30 kW and up to 45 kW rated output	175 mm diameter or square section of equivalent area

Taken from Approved Document J
The Building Regulations 1991, HMSO

are built in as the chimney is raised and the liners are surrounded with mortar and set in place with the liner socket uppermost so that condensate cannot run down through the joint into the surrounding chimney. Bends are made of the same cross sections for use where flues offset.

Concrete flue liners are made of high alumina cement and an aggregate of fired diatomaceous brick cast in round sections with square ends.

Clay drain pipes with socket and spigot ends may be used instead of flue liners. The pipes are set in place as the chimney is raised with the spigot ends of the pipes uppermost. Because of the appreciable projection of the socket ends of these drain pipes a chimney built with them is larger than one with purpose-made flue liners.

A range of purpose-made precast concrete flue blocks is made for building into blockwork walls and partitions. The blocks are made of lightweight concrete around a flue of dense aggregate and high alumina cement. A range of blocks is made to bond to blockwork, for offsets in flues and as a capping, as illustrated in Fig. 169.

Where there are two or more fireplaces one above the other it is necessary to form offsets in the lower flues so that the flues may run up at the side or sides of the fireplaces above and then offset above the highest

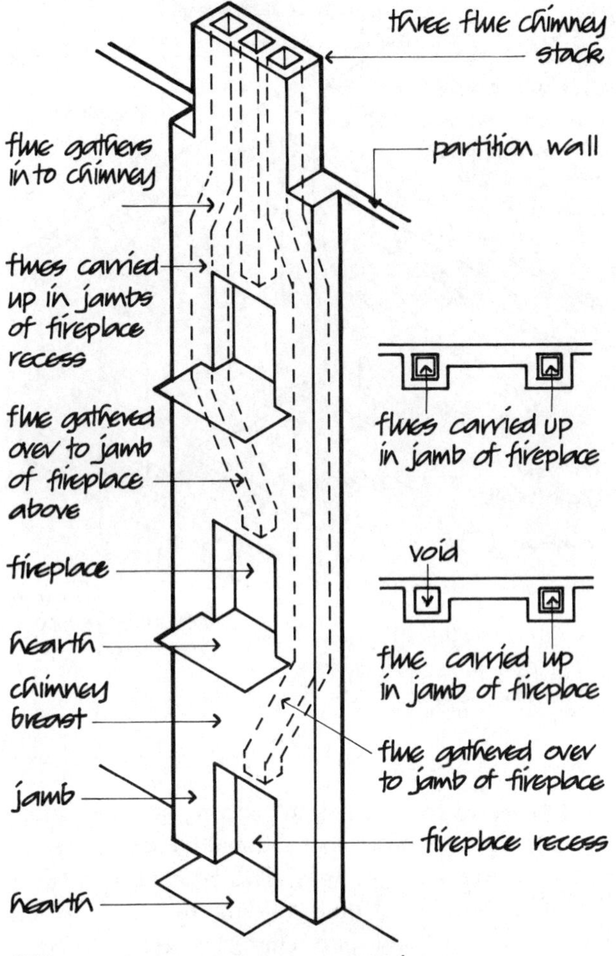

Fig. 170

fireplaces to gather together as a chimney stack above roof level, as illustrated in Fig. 170. It will be seen that the flues from the lower fireplaces run inside the jambs either side of the top fireplace. For the offsets in flues, bends or offset flue blocks are used.

The practical guidance in AD J requires that no part of a flue in a chimney should make an angle of more than 30° with the vertical. An angle of 60° to the horizontal is commonly used for offsets to maintain good updraught to minimise collection of soot and for ease of sweeping.

Every flue in a chimney built of brick, block or stone must be surrounded by and separated from other flues in the same chimney with solid material not less than 100 thick.

Proximity of combustible material to chimneys The practical guidance in AD J to the Building Regulations

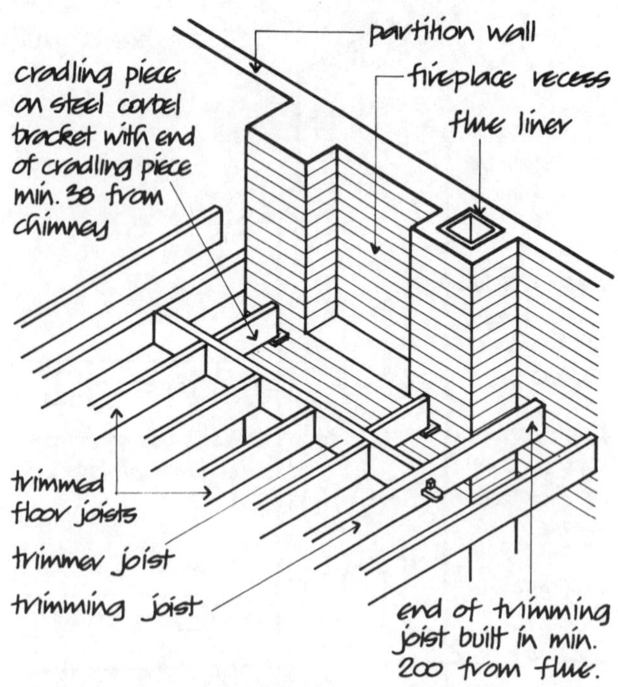

Trimming upper floor for hearth

Fig. 171

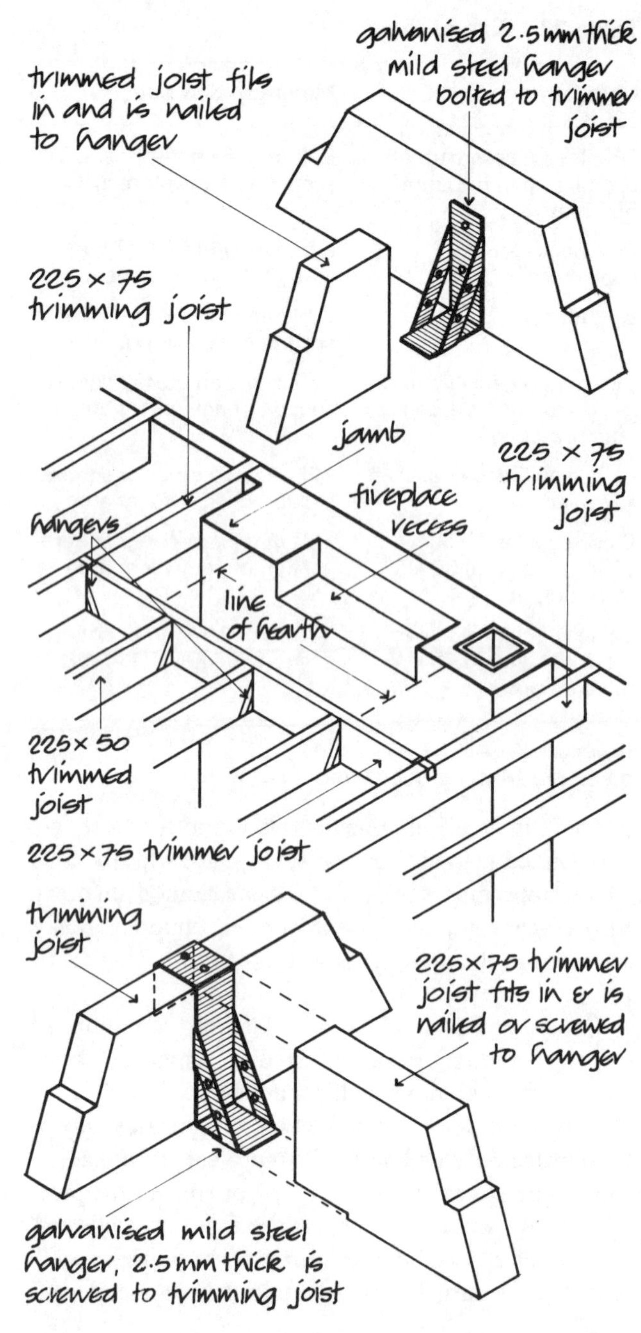

Trimming floor for hearth with metal hanger fixings

Fig. 172

1991 prohibits the building in or fixing of combustible material, that is timber in traditional building, close to flues and fireplace openings. This precaution against fire affects structural timber floors and roofs, wood plugs fixed or built into chimneys and structural timbers close to the face of chimneys. No timber may be placed in a chimney nearer to a flue or the inner surface of a fire recess than 200 and no combustible material other than a floor board, skirting board, dado rail, picture rail, mantel shelf or architrave shall be nearer than 40 to the outer surface of the chimney or fireplace recess. The application of this regulation, where the upper timber floor joists around a fireplace are at least 200 from the flue and the end of a cradling piece for the hearth is at least 40 from the chimney, is illustrated in Fig. 171. Metal fixings in contact with combustible materials should be at least 50 from a flue.

Timber upper floors around hearths Timber upper floors have to be trimmed (cut) around hearths as illustrated in Figs. 171 and 172. The opening in the floor for the hearth is formed by a trimmer joist which supports the trimmed joists and is supported by trimming joists built into the wall each side of the chimney. The two cradling pieces that contain the hearth are nailed to the trimmer and supported on steel corbel plates so that their ends are at least 38 from the chimney face and the built in ends of corbels at least 50 from a flue. The concrete hearth is then cast in the fireplace opening and between the cradling pieces and trimmer on boards fixed to battens nailed to the trimmer and chimney.

To make a strong joint between the trimmer, trimming and trimmed joists either a shaped timber joint or metal hangers are used.

A tusk tenon joint between trimmer and trimming joists and housed joints between trimmed and trimmer joists are used as illustrated in Fig. 173. The tenons on the ends of the trimmer fit through mortices in the centre of the depth of the trimming joists, where this cut will least weaken the joist, and the tenon is wedged in position to make a rigid, strong joint.

The dovetail half depth housed joints illustrated in Fig. 174 provide a secure fixing for the trimmed joists without greatly reducing the strength of the trimmer by cutting housings one third of its thickness. The square housed joint shown in Fig. 174 provides a less secure anchor than the dovetail housed joint. These joints can be accurately and quickly cut and assembled by a skilled carpenter.

As an alternative, purpose-made galvanised steel joist hangers are used, as illustrated in Fig. 172. These hangers are hung and screwed to trimming and trimmer joists respectively and screwed to trimmer and trimmed joists. There is some small saving in labour in the use of these hangers as compared to the cutting of timber joints. Similarly the upper timber floor shown in Fig. 160 has to be trimmed around the chimney with either timber joints or steel hangers.

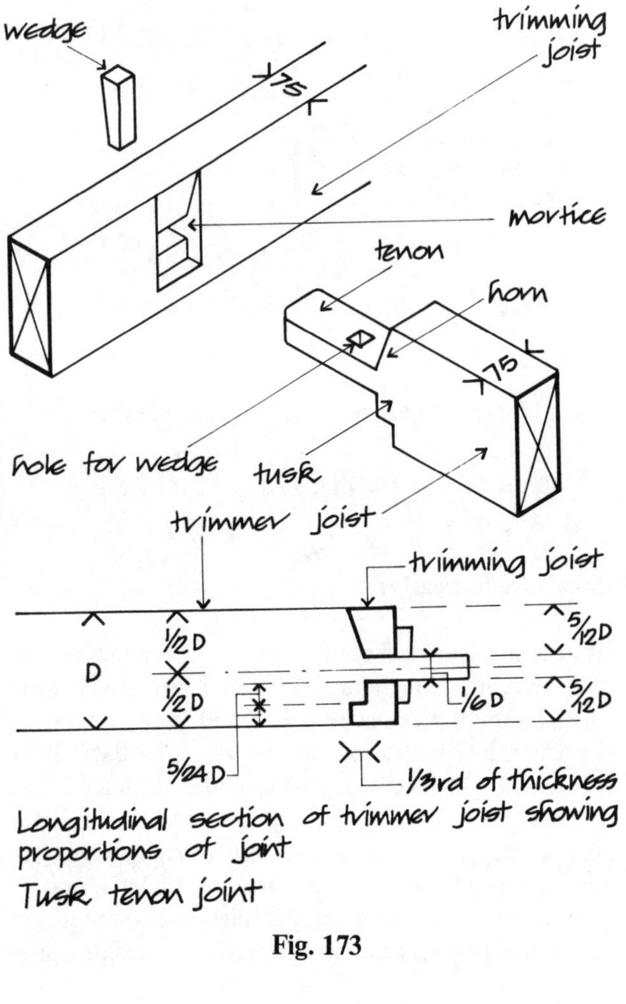

Fig. 173

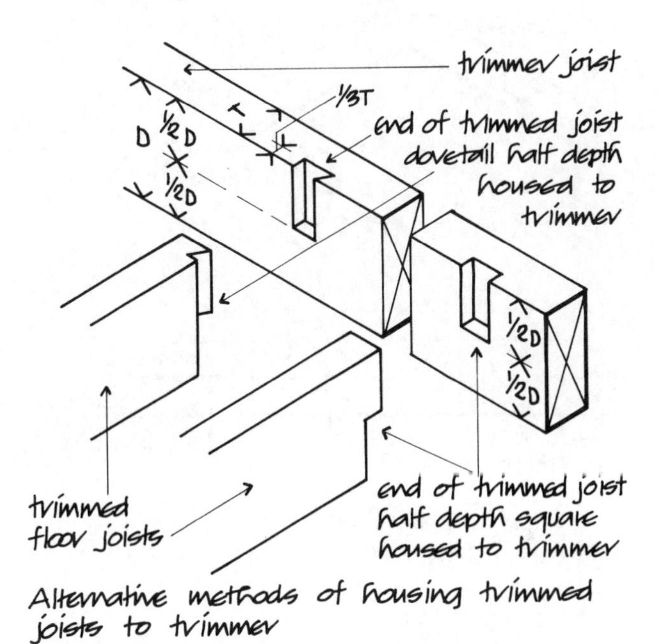

Fig. 174

Timber roofs around chimneys Where the outside dimensions of a chimney stack, such as the one flue stack shown in Fig. 160, are less than the spacing of the roof trusses, there is no need to trim either the roof or the ceiling timbers around the chimney. Where a chimney stack is wider overall than the spacing of the trusses or rafters it is necessary to trim the timbers around the stack using either timber joints or hangers.

Chimney stacks Chimney stacks are raised above roofs to encourage the products of combustion to rise from the flue by avoiding downdraught. The practical guidance in AD J to the Building Regulations 1991 recommends minimum heights for chimney stacks, excluding chimney pots or terminals, as illustrated in Fig. 175.

The practical guidance in Approved Document A to the Building Regulations 1991 sets the max. height of masonry chimneys as 4.5 times the least width of chimneys measured from the highest point of intersection of roof and chimney to the top of chimney pots or terminals, where the density of the masonry is greater than 1500kg/m^3. All stacks should be finished with a chimney pot or terminal of the same section as the flue and at least 150 above the top of the stack. Typical

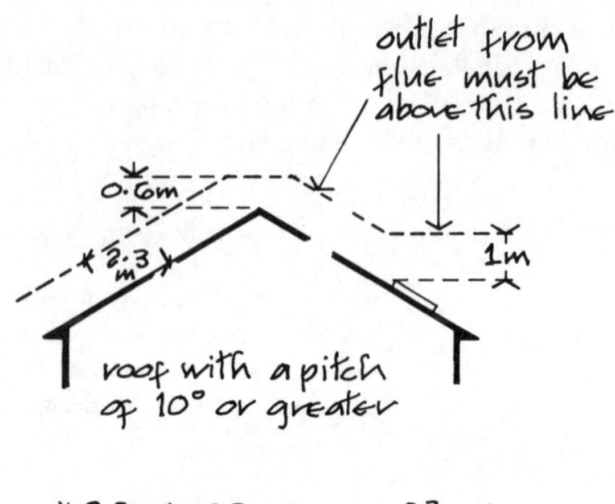

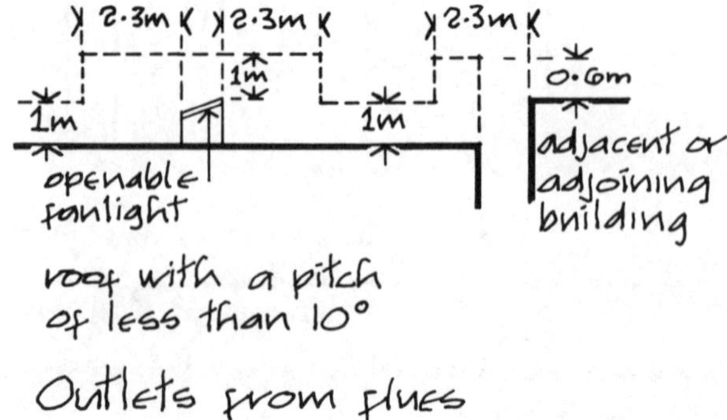

Fig. 175

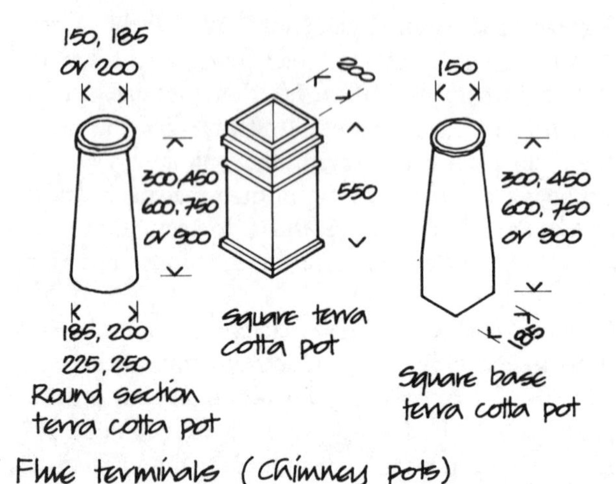

Fig. 176

chimney pots and terminals are shown in Fig. 176. These pots are bedded in mortar on the chimney and finished with fine concrete flaunched up around the pot to shed water (Fig. 179).

A chimney stack is exposed to wind and rain and may become so saturated with rain that the chimney below the roof becomes damp. To prevent this, particularly in exposed positions, it is practice to build in a d.p.c. as illustrated in Fig. 177. The d.p.c. may be of any one of the non-ferrous sheet metals. The horizontal d.p.c. shown gives protection to all but a small part of the lower part of the stack and is the d.p.c. most used. The stepped d.p.c. gives complete protection but involves a deal of wasteful cutting of bricks, particularly with steeply pitched roofs.

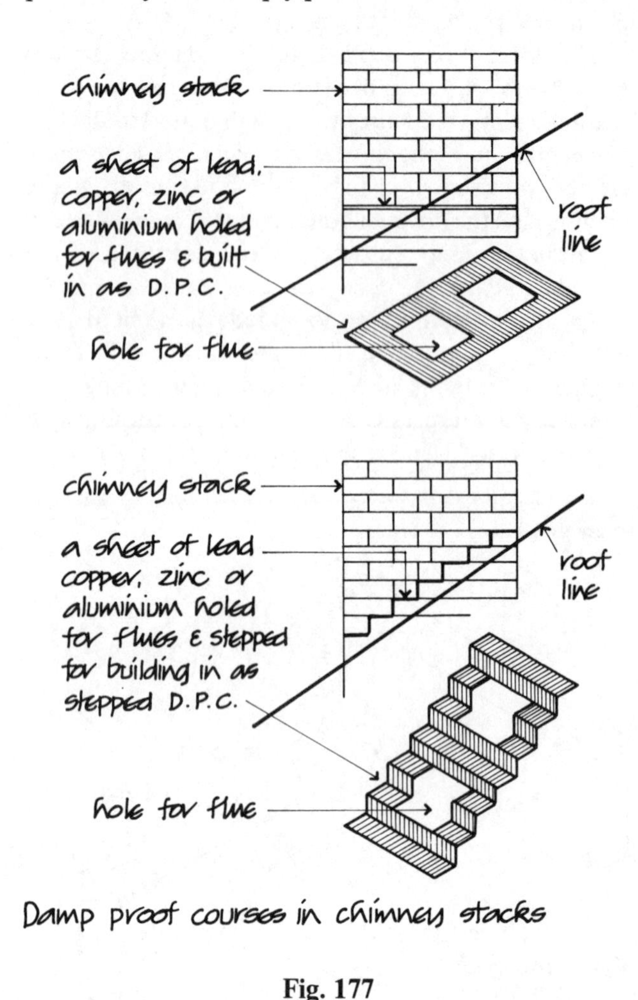

Fig. 177

Resistance to weather

Weathering around chimney stacks At the junction of roof coverings and chimney stacks it is necessary to use some form of flexible material and construction to prevent rain penetrating and to accommodate slight movements between the roof and the stack. The cheapest way of doing this is to run a fillet of mortar around the junction of the roof and the stack. Owing to the drying shrinkage of the mortar and relative movement of the stack and roof, the fillet cracks and lets in water. The only satisfactory method of excluding rain is by the use of some flexible, impermeable material such as non-ferrous sheet metal around the stack. Sheet lead, which is easy to work, is the material most suited to the purpose.

Pitched roofs Where a chimney stack rises through one slope of a roof, weathering around the stack is effected by the use of a back gutter and flashing, stepped flashings and a front apron as illustrated in Fig. 179. The back gutter is shaped to the back of the stack, the timber gutter and dressed up under the tiles and covered with a separate flashing that is turned and wedged into a horizontal brick joint. The stepped flashings to the sides of the stack are cut, shaped and turned and wedged into brick joints in the form of steps and then dressed down over the rolls of the single lap tiles to exclude rain. The front apron flashing is shaped to the stack, turned and wedged into a horizontal brick joint and dressed down over the tiles as illustrated in Fig. 179.

Fig. 178 illustrates the junction of a three-flue stack with the ridge of a plain tiled roof. Because the tiles are

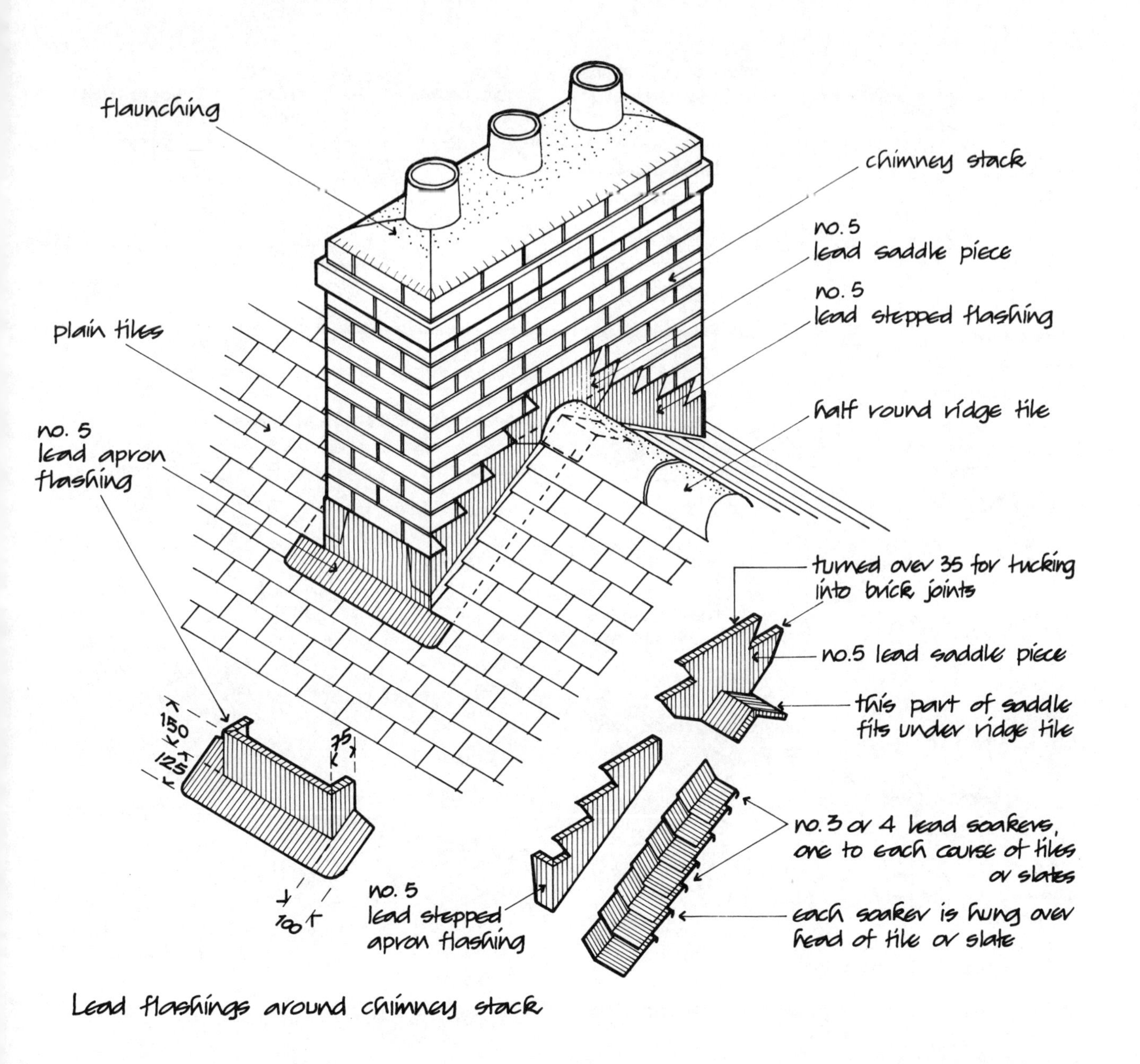

Fig. 178

flat it is not satisfactory to dress the side stepped flashings down on to the back of the tiles as rain would find its way between the flashing and the tiles and into the roof. A system of soakers and stepped flashings is used, as illustrated. One soaker is hung over the back of each tile and turned up against the stack, and the upstand of the soakers is then covered with a stepped flashing. At the junction of the ridge and the stack saddle pieces are turned and wedged into horizontal brick joints. Front apron flashings are fitted. The setting out of the stepped flashings, the size of the soakers and the arrangement of soakers and flashing is illustrated in Fig. 180.

Flashings of copper, aluminium or zinc may be used around chimney stacks, as illustrated in Fig. 181. These sheet metals cannot be shaped as easily as lead and have to be welded, soldered or welted to form flashings. Copper is much less used than other metals as the oxide formed on it can run down and cause unsightly stains.

Flat roofs At the junction of flat roofs and chimney stacks the asphalt, felt or sheet metal covering is dressed

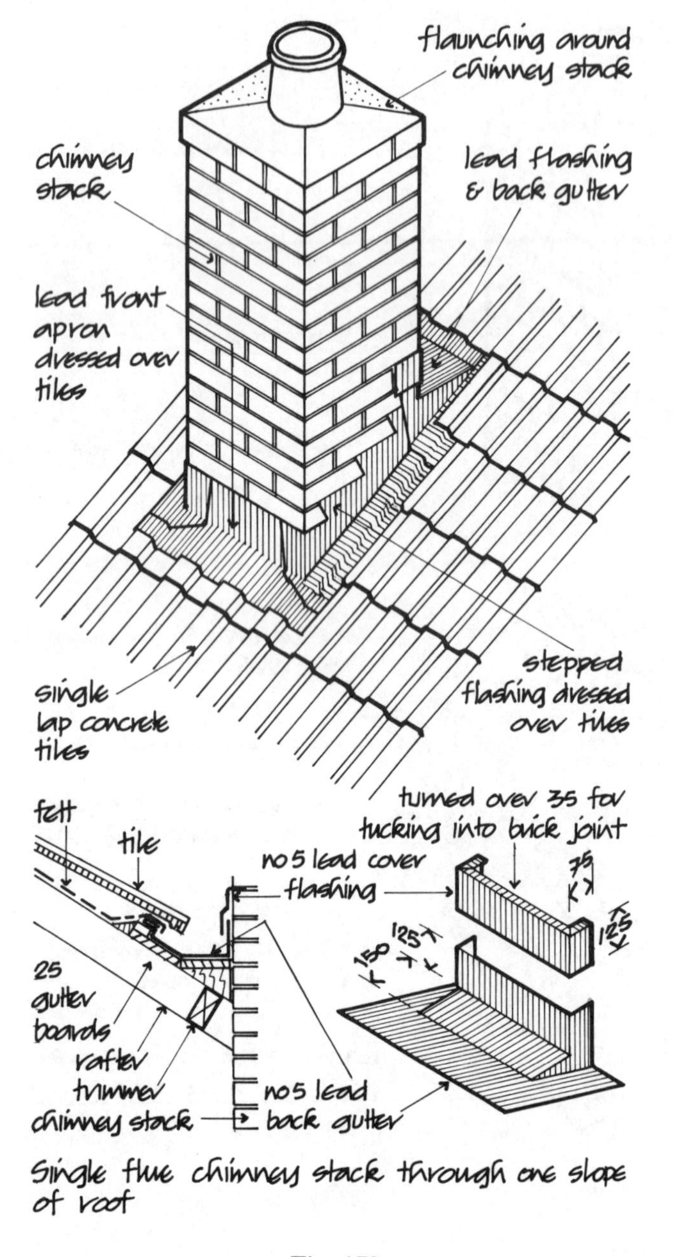

Fig. 179

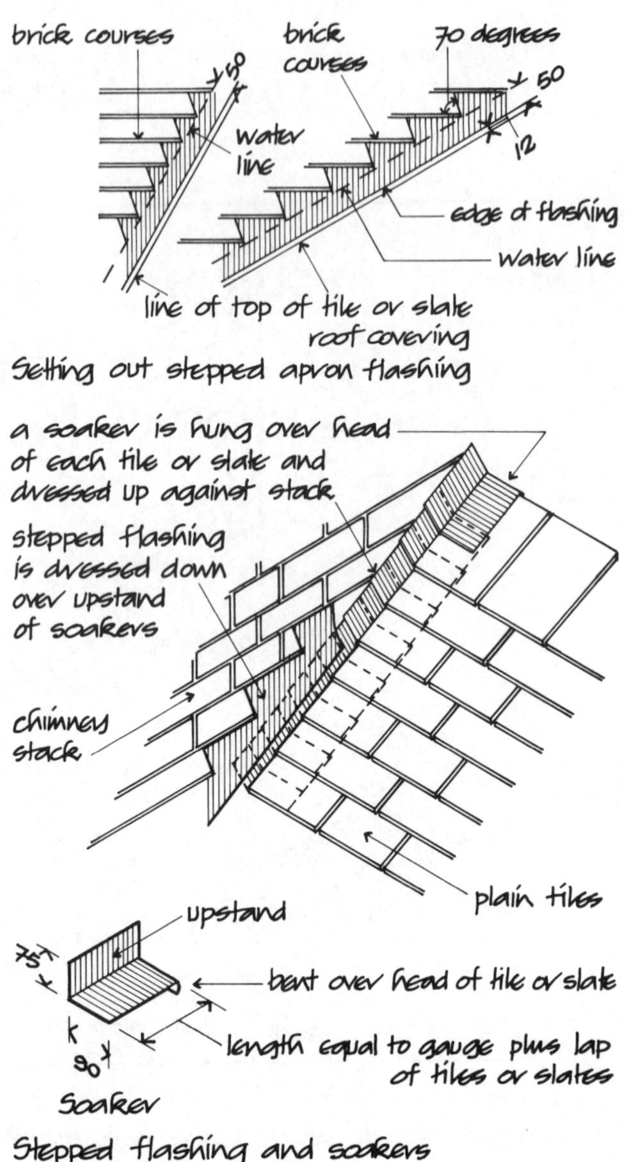

Fig. 180

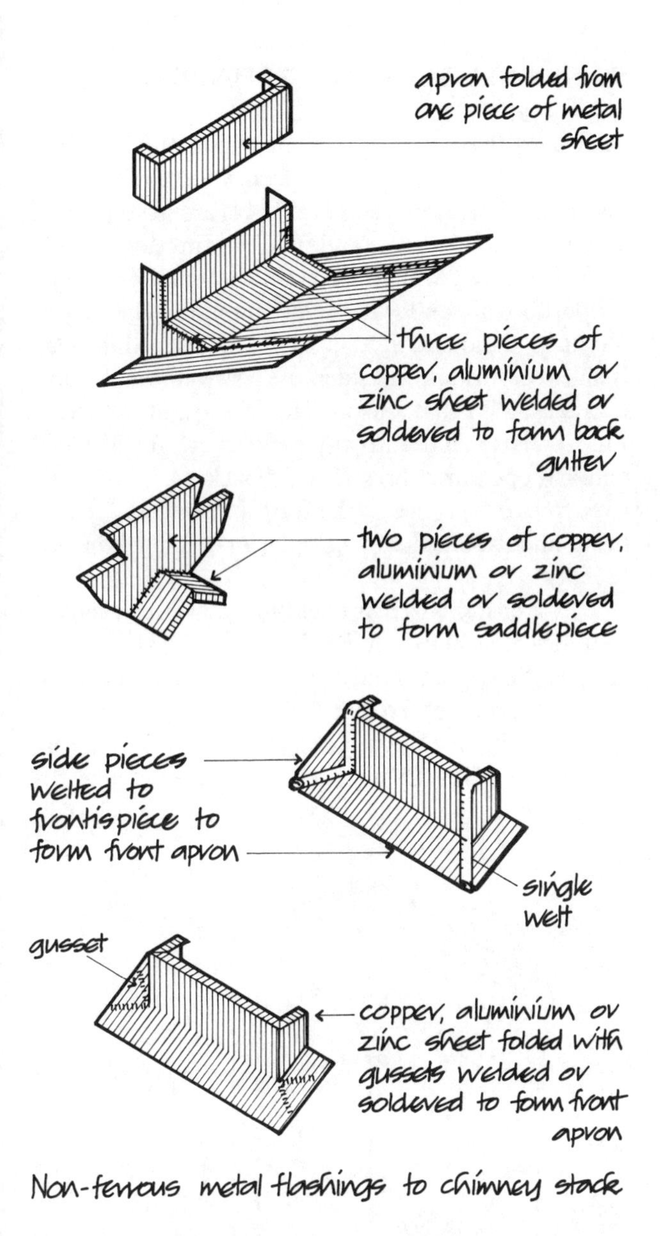

Fig. 181

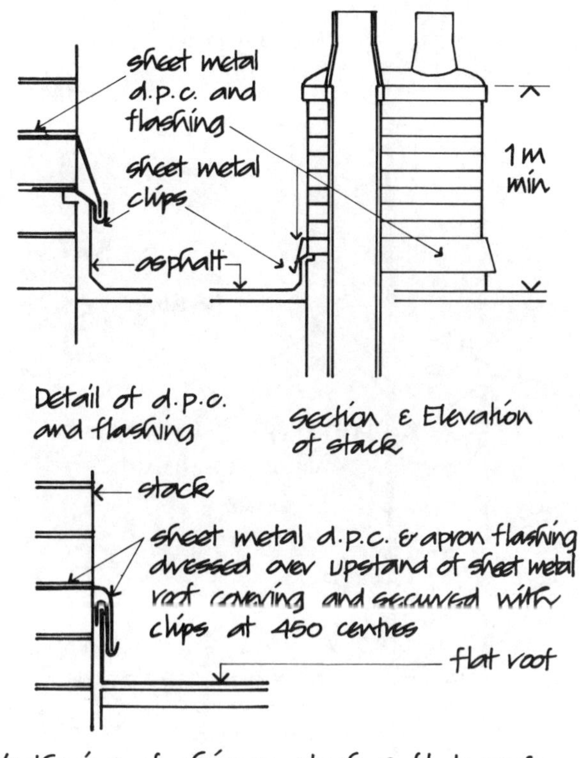

Fig. 182

up around the stack with an upstand that is weathered with a flashing, and a d.p.c. is built into the stack, as illustrated in Fig. 182.

Factory-made chimneys Factory-made sectional chimney sections and fittings are used instead of solid block or brick chimneys in new buildings and where an open fire or stove is to be fitted into an existing building without chimneys. The advantages of these chimneys are their comparatively small cross section, ease of installation, high thermal insulation which conserves heat, and smooth faces to encourage draught and facilitate cleaning. These flue sections require some support at roof and intermediate floor levels and can be fixed anywhere in rooms, either exposed or cased in.

These factory-made chimneys are made in sections that have socket and spigot ends that lock or are clamped together. The sections have stainless steel inner linings and either stainless steel or galvanised steel outer linings around a core of mineral insulation that conserves heat and prevents condensation in the flue. Because of the insulation and construction of these flue sections structural timbers may be as close as 50 to the outside of the flue, which avoids the need for trimming timbers and facilitates supporting the chimney at floor and roof level. Fig. 183 is an illustration of factory-made chimney sections and fittings.

For use with open fires or room heaters these chimneys are run from a fire chest, which is a precast concrete surround into which the open fire or room heater is fitted and from which the factory-made chimney is run, as illustrated in Fig. 183.

Where the chimney runs through timber floors and roofs, timber bearers are fixed to which the chimney is secured and firestop plates and terminal are fitted as illustrated in Fig. 183.

SOLID FUEL BURNING APPLIANCES

Inset open fires An open fire consists of a cast iron grate and fret, set on the hearth inside a fireback, set inside the fireplace recess. Fig. 184 is an illustration of a typical fireback, grate and fret. The fireback is made from fire clays, which are clays that contain a high proportion of sand with some alumina. The clay is moulded, dried and fired in a kiln; the burned fire clay is able to withstand considerable heat without damage and is used for that reason. The fireback illustrated in Fig. 184 is a standard two-piece fireback for 350, 400 and 450 open inset fires. The fireback may be in one, two, four or six pieces, the four and six piece backs being made to facilitate replacing an existing damaged fireback.

The fireback is set in position on a level concrete back hearth as illustrated in Fig. 185, and vermiculite concrete fill is cast behind it as insulation and flaunched up to the raft lintel. To reduce the width of the fireback to that of the flue a two-piece precast concrete throat gather unit (Fig. 185) is used. The space between the sides of the fireback and the jambs of the recess is built

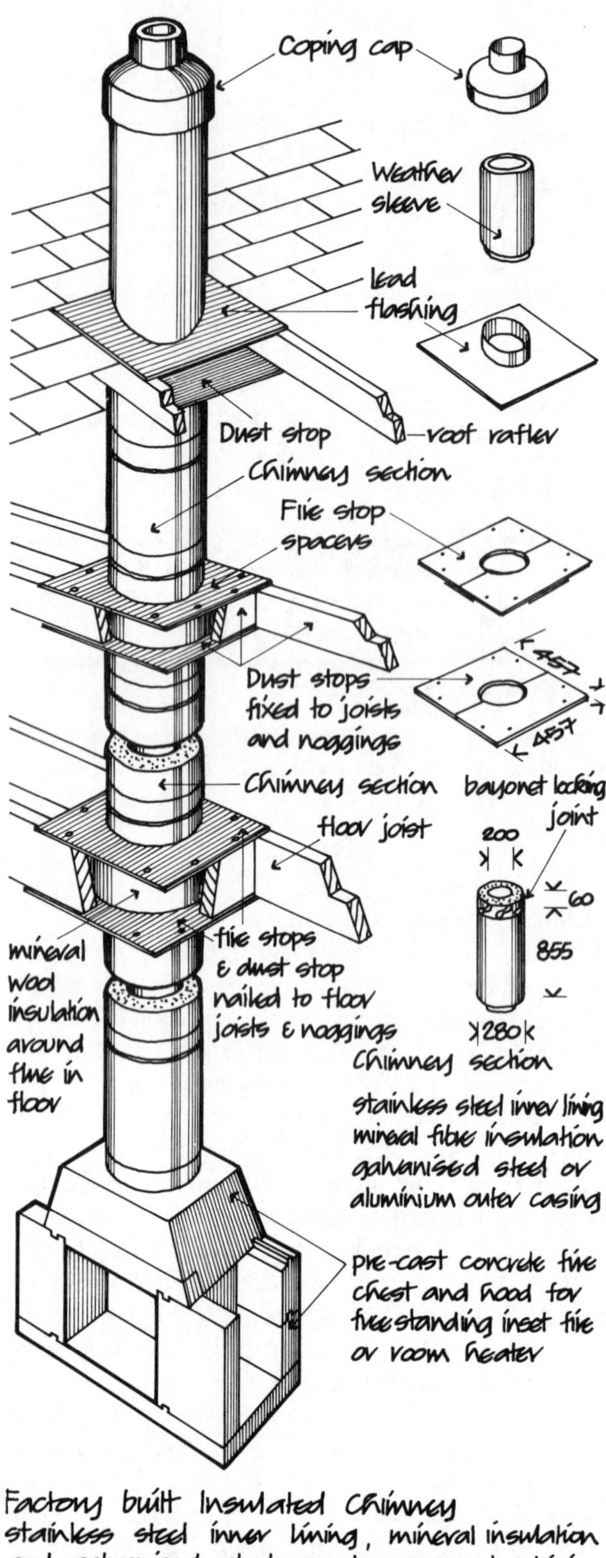

Factory built insulated chimney
stainless steel inner lining, mineral insulation and galvanised steel or aluminium outer lining with bayonet locking joints

Fig. 183

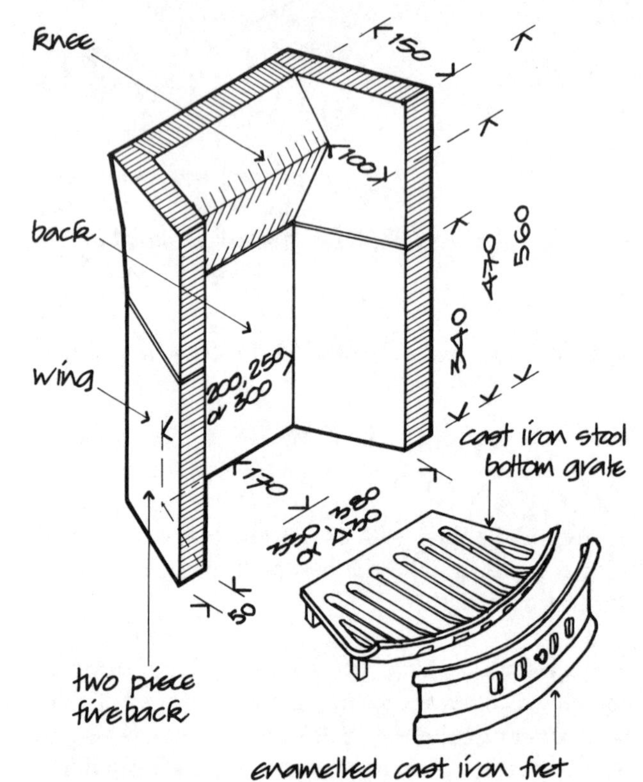

Fireback, grate and fret for open fire

Fig. 184

up with brickwork filling on which the throat unit is bedded, and the removable front insert is fitted in place. Rough brick filling is then formed between the throat unit and the recess opening.

For appearance a fire surround is usually fixed in front of the fireplace opening, as illustrated in Fig. 186. The tiled slab surround illustrated is precast with a core of concrete and tile finish. This surround is bedded in position and secured to the chimney through cast-in lugs screwed to lead plugs. The loose grate and fret are placed in the fireplace.

Sunk hearth open fire An open fire draws a considerable volume of air from a room, which is wasteful of heat and causes draughts. The sunk hearth open fire is designed to reduce and control the volume of air drawn into the fire through a duct under the floor to the hearth, as illustrated in Fig. 186. The grate is level with the hearth and a deep ash pan is set in a pit below the hearth to which is connected a duct with a draught control. The duct is run under the floor to a balancing chamber from which two ducts run to opposite outside walls. The back boiler is fitted to improve the efficiency of the fire by using heat that would otherwise be wasted for warming hot water.

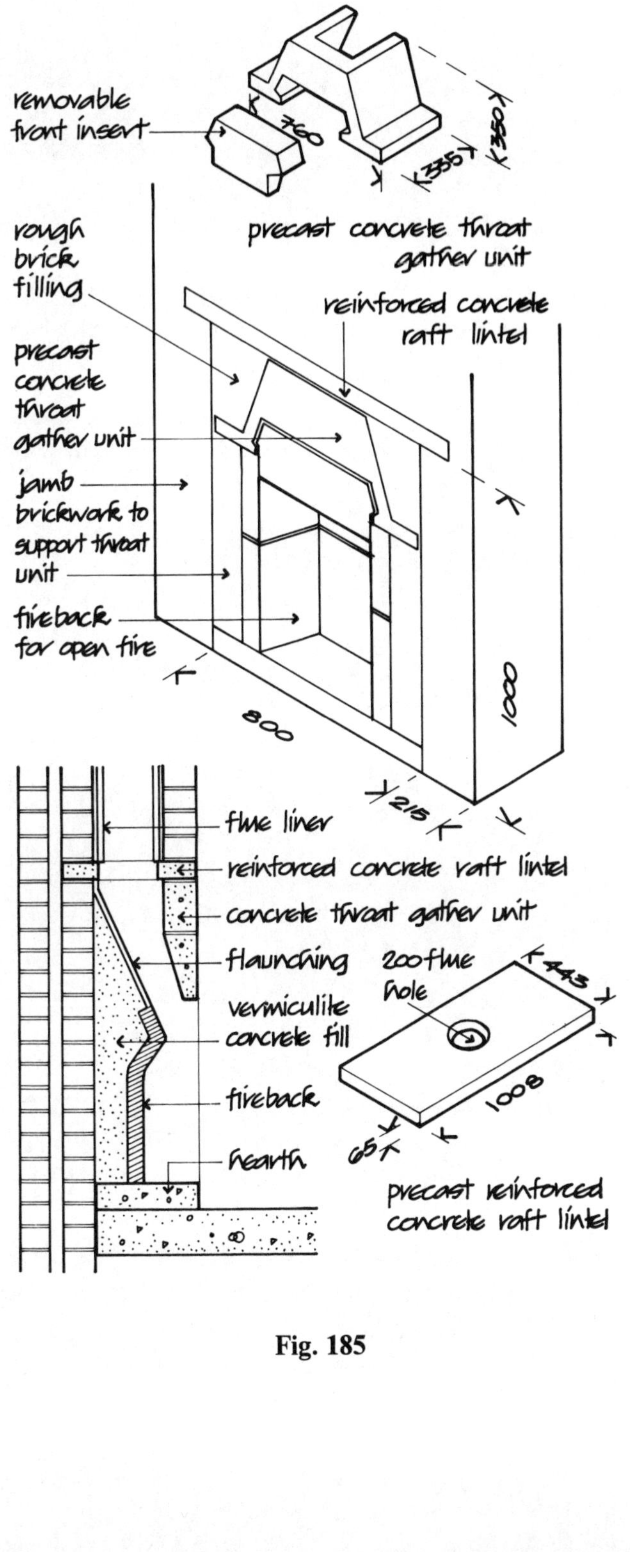

Fig. 185

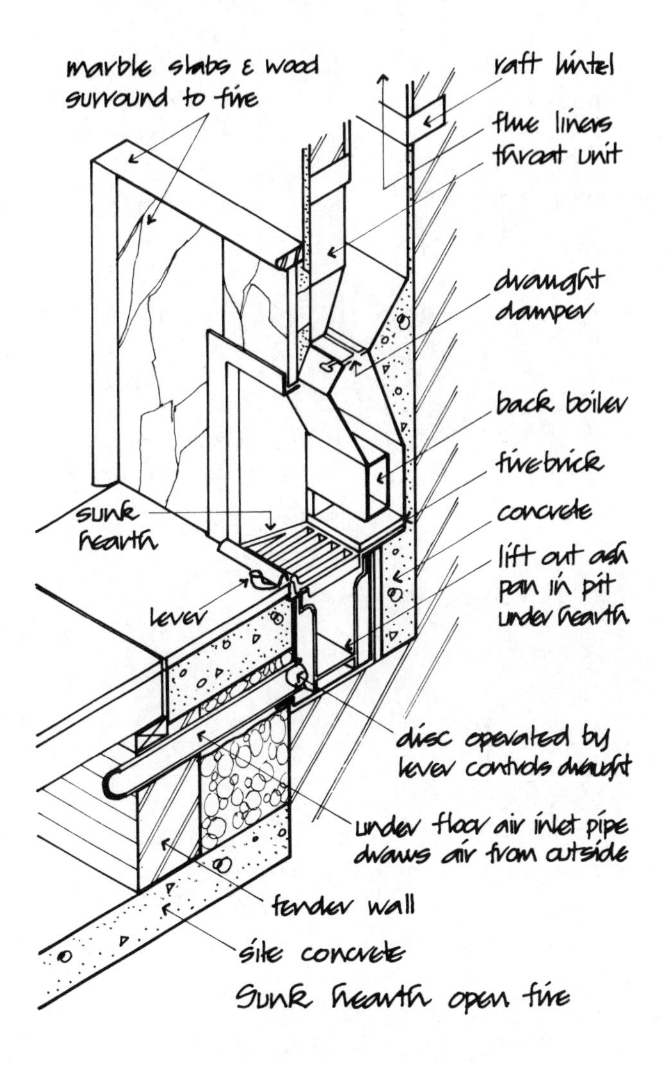

Fig. 186

Freestanding heater Fig. 187 is an illustration of an enclosed room heater freestanding in a room. The heater is set on a tiled hearth superimposed on the solid ground floor up to the fire surround. A steel register plate is sealed to the opening in the surround with angle or tee section framing behind the heater and the iron flue outlet from the heater is sealed to the register plate to discharge the products of combustion to the chimney. The appliance recess shown is built for use with either an inset fire or heater or a partly or fully freestanding heater.

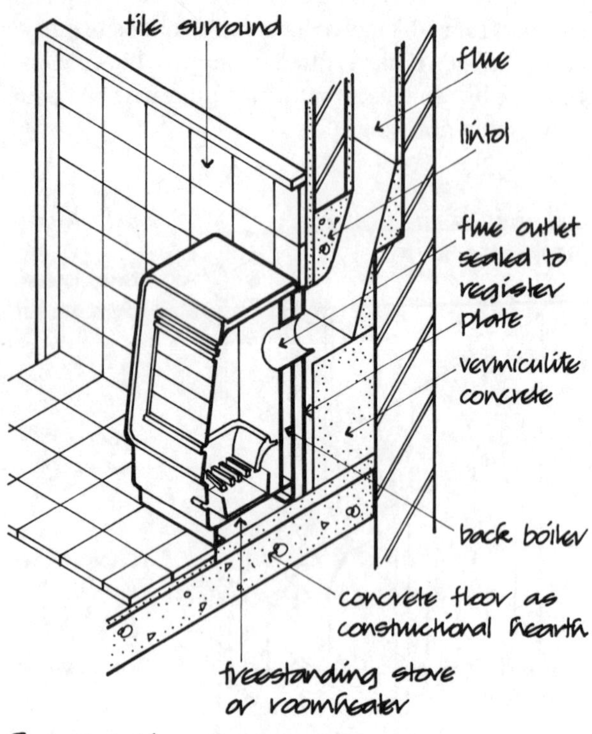

Fig. 187

CHAPTER FOUR

STAIRS

A stair is the conventional means of access between floors in buildings. It should be constructed to provide ready, easy, comfortable and safe access up and down with steps that are neither laborious nor difficult to climb within a compact area, so as not to take up excessive floor area.

The minimum dimensions for stairways are set out in Approved Document K giving practical guidance to meeting the requirements of the Building Regulations 1991.

FUNCTIONAL REQUIREMENTS

The functional requirements of a stair are:

Structure – strength and stability
Safety in use
Fire safety
Resistance to the passage of sound

Structure – strength and stability A stair and its associated landings serves much the same function as a floor in the support of the occupants of a building, with the stepped inclined plane of flights serving as support for movement between floors. The requirements for strength and stability in supporting dead and imposed loads for floors apply equally to stairs.

Safety in use The practical guidance in Approved Document K to the Building Regulations 1991 is concerned with the safety of users in determining the rise, going and headroom of stairs and the dimensions of handrails and guarding.

Flight The word flight describes an uninterrupted series of steps between floors or between floor and landing, or between landing and landing. A flight should have no fewer than three and no more than sixteen risers in it otherwise it might be dangerous, particularly for the young or elderly. The rise and going of each step in one flight and in flights and landings between floors should be equal. Variations in the rise of steps will interrupt the rhythm of going up or down, may trip people and be dangerous. The usual unobstructed width of a flight is from 800 for houses to 1000 for other buildings.

Steps The steps of a stair may be constructed as a series of horizontal open treads with a space between the treads or as enclosed steps with a vertical face between the treads, called a riser.

Tread and riser The horizontal surface of a step is described as the tread and the vertical or near vertical face as the riser. With enclosed steps the treads usually project beyond the face of the riser as a nosing to provide as wide a surface of tread as practicable. Fig. 188 illustrates the use of the terms tread, riser and nosing.

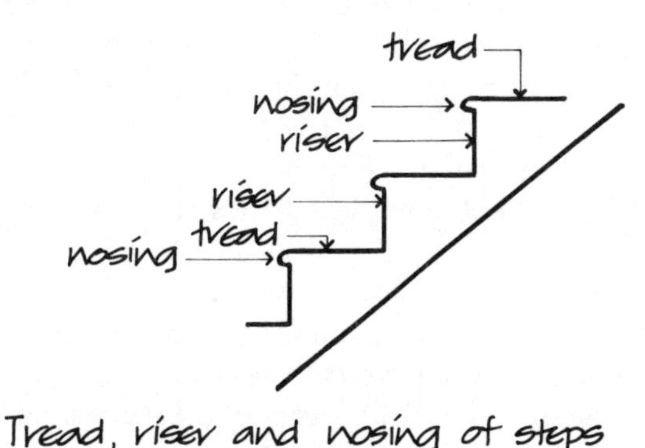

Fig. 188

Rise and going The word rise describes the distance measured vertically from the surface of one tread to the surface of the next or the distance from the bottom to the top of a flight. The word going describes the distance, measured horizontally, from the face of the nosing of one riser to the face of the nosing of the next riser, as shown in Fig. 189. The dimensions of the rise and going of steps determine whether a stair is steep or shallow. The shallow stair, illustrated in Fig. 190, would be tedious and exhausting to climb and the steep stair practically impossible to climb.

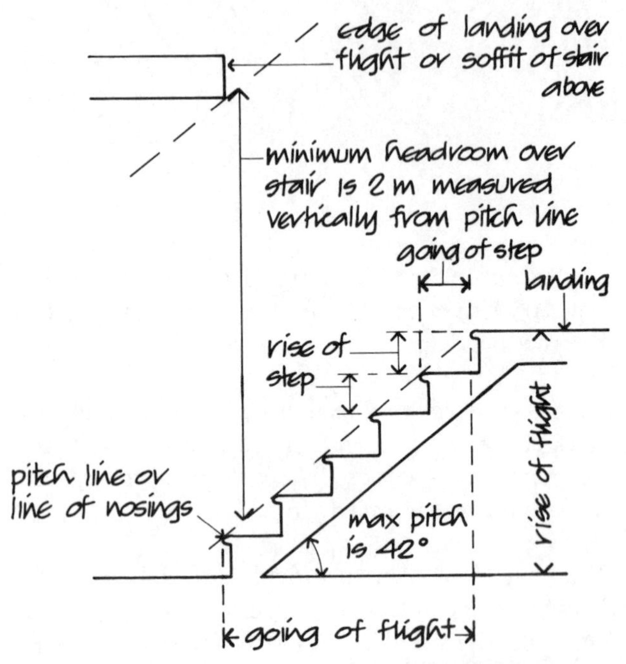

Fig. 189

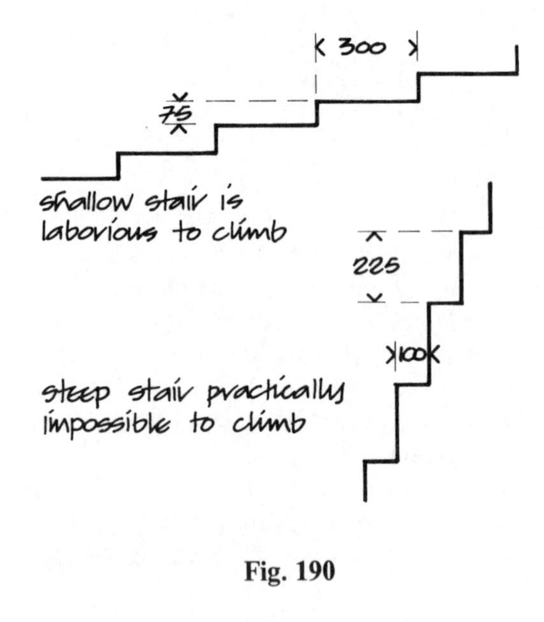

Fig. 190

Pitch The inclination of a stair can be described either by the rise and going of the steps or as the pitch of the stair, which is the angle of inclination of the stair from the horizontal, as illustrated in Fig. 189. Stairs are pitched at not more than 42° for stairs to single dwellings and not more than 38° for common stairs.

Head room and clearance For people and for moving goods and furniture a minimum head room of 2 metres, measured vertically, is recommended between the pitch line of the stair and the underside of the stairs, landings and floors above the stair.

The recommended rise and going of stairs is given in Approved Document K as set out in Table 12. It will be seen from the table that the accepted dimensions of rise and going for houses produce a steeper stair than those for public buildings. The steeper stair is accepted for houses because the occupants are familiar with the stair and a shallow stair would occupy more area on plan and thus reduce the available living area. The shallower a stair the more risers required for a given height and therefore the more treads and greater plan area occupied by the stair. The shallow stair recommended for public buildings is designed to minimise danger to the public escaping via the stairs during emergency. It was practice before standards of rise, going and pitch were recommended, to determine the pitch of a domestic stair from the simple formula: twice rise plus going equals some notional figure, between 550 and 700, where an assumption of some convenient rise was made.

To set out a stair it is necessary to select a suitable rise and adjust it, if necessary, to the height from floor to floor so that the rise of each step is the same, floor to floor, and then either select a suitable going from Table 12 or use the formula $2R + G =$ between 550 and 700 to determine the going.

Table 12. Rise and going of stairs

	Maximum Rise (mm)	Minimum Going (mm)
(1) Private stair	220†	220†
(2) Institutional and assembly stair	180**	280*
(3) Other stair	190**	250

Note:
† The maximum pitch for a private stair is 42°
* If the area of a floor of the building is less than 100m², the going may be reduced to 250mm
** For maximum rise for stairs providing the means of access for disabled people reference should be made to Approved Document M: Access and facilities for disabled people.

Taken from Approved Document K
The Building Regulations 1991, HMSO

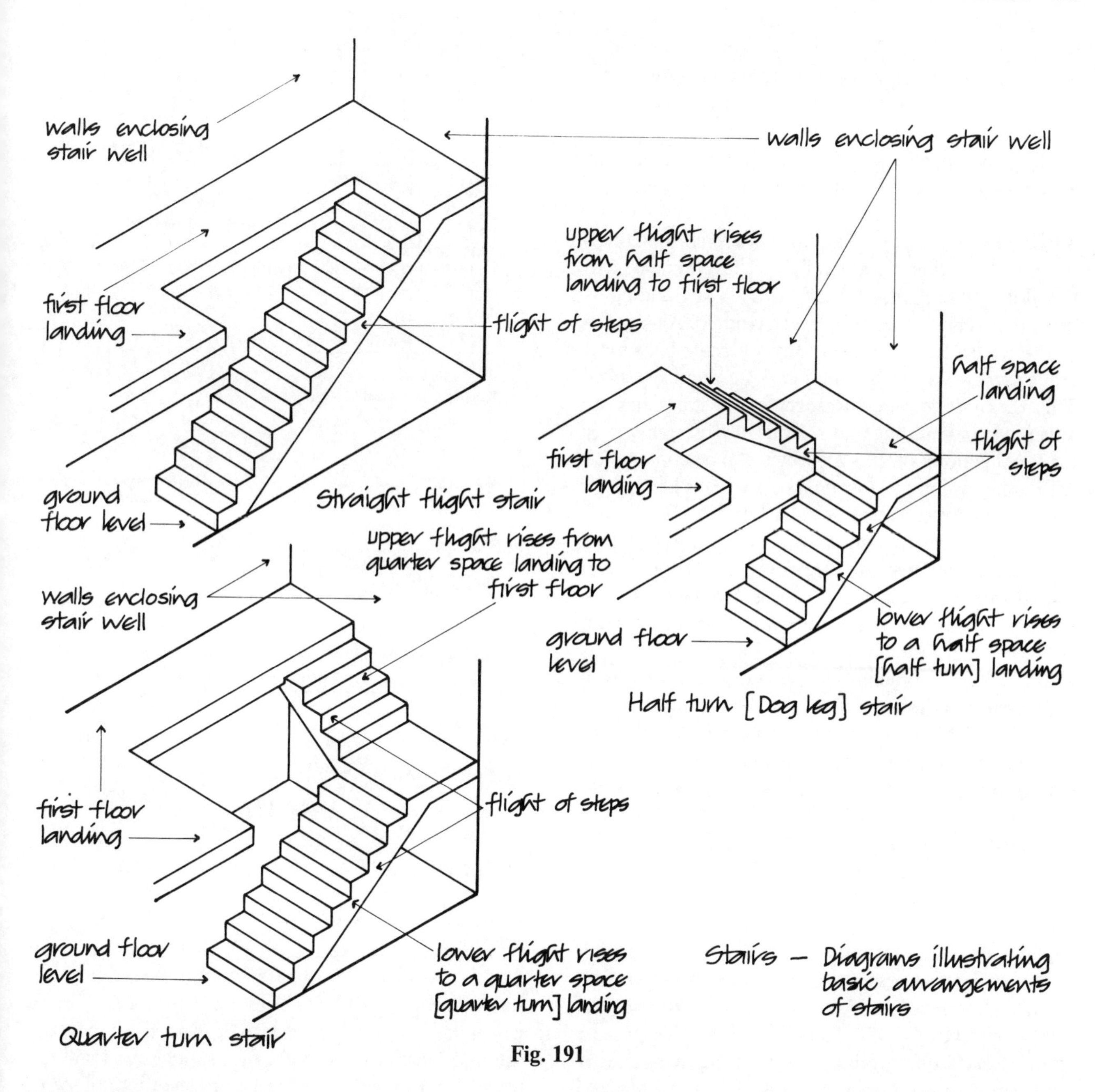

Fig. 191

TYPES OF STAIR

The three basic ways in which stairs with parallel treads are planned are illustrated in Fig. 191. These are:

A straight flight stair
A quarter turn stair
A half turn stair.

A straight flight stair rises from floor to floor in one direction with or without an intermediate landing, hence the name straight flight. A straight flight stair, sometimes called a cottage stair, was commonly used in the traditional 'two-up two-down' cottage with the stair in the centre of the plan running from front to back giving access to the two upper rooms each side of the head of the stair, with access to the two ground floor rooms from the small hall at the front of the stair. This, the most economical use of the straight flight, does rigidly determine access to the rooms top and bottom if wasteful landings are to be avoided.

A quarter turn stair rises to a landing between two floors, turns through 90°, then rises to the floor above, hence 'quarter turn'. This type of staircase was much used in the two floor semi-detached houses built in the first half of the twentieth century, for its great economy in compact planning. The quarter space or quarter turn landing shown in Fig. 191 was often replaced with winders for further economy in the use of space.

A half turn stair rises to a landing between floors, turns through 180°, then rises parallel to the lower flight to the floor above, hence 'half turn'. The landing is described as half space or half turn landing. A half turn stair is often described as a 'dog leg' stair because it looks somewhat like the hind leg of a dog in section. This, the most common arrangement of stairs, has the advantage in planning that it lands at or roughly over the starting point of the stair which can be constructed within the confines of a vertical stair well, as a means of access to and escape from similar floors.

Stairs are sometimes described as 'open well stairs'. The description refers to a space or well between flights. A half turn or dog leg stair can be arranged with no space between the flights or with a space or well between them as illustrated in Fig. 198, and this arrangement is sometimes described as an open well stair. A quarter turn stair can also be arranged with a space or well between the flights when it is also an open well stair. As the term 'open well' does not describe the arrangement of the flights of steps in a stair, it should only be used in conjunction with the more precise description straight flight, quarter or half turn stair (e.g. half turn stair with open well).

Geometrical stairs are constructed with treads that are tapered on plan, with the tapered treads around a centre support as a spiral (helical) stair, an open well circular stair or as an ellipse or part of an ellipse on plan as illustrated in Fig. 192. The spiral stair or helical stair is the most economical way of planning a stair as it takes up little floor area, but it is difficult to use and impossible for moving furniture. The effective width of taper treads is usually measured on the centre of the width of the tread. With a centre open well a circular stair is more comfortable to use.

The elliptical stair, which is extravagant in the use of space, can provide an elegant feature for grand buildings.

Stairs may be constructed of timber, stone, reinforced concrete or metal, with timber and stone being the traditional materials used before the advent of steel and reinforced concrete following the industrial revolution.

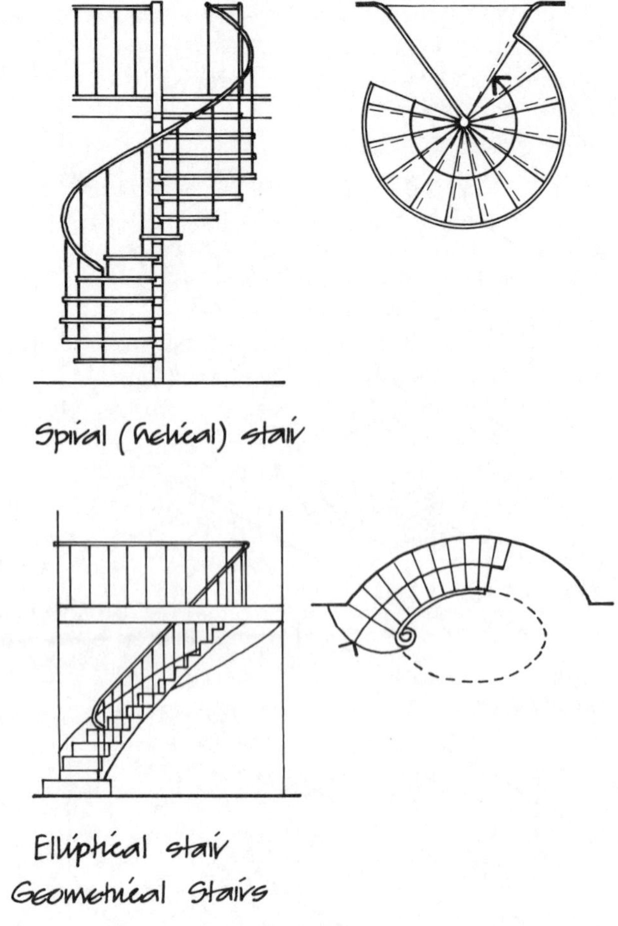

Fig. 192

TIMBER STAIRS

Staircase A staircase, which is a stair with treads and risers constructed from timber boards put together in the same way as a box or case, hence the term staircase, is the traditional stair for houses of two floors or more where the need for resistance to fire does not dictate the use of concrete. Each flight of a staircase is made up (cased) in a joiner's shop as a complete flight of steps. Landings are constructed on site and the flight or flights are fixed in position between landings and floors. The members of the staircase flight are strings (or stringers), treads and risers. The treads and risers are joined to form the steps of the flight and are housed in, or fixed to strings whose purpose is to support them. Because the members of the flight are put together like a box, thin boards can be used and yet be strong enough to carry the loads normal to stairs. The members of the flight are usually cut from timbers of the following sizes: treads 32 or 38, risers 19 or 25, and strings 38 or 44.

Fig. 193 is a view of a flight of a staircase with some of the treads and risers taken away to show the housings in the string into which they fit.

Joining risers to treads The usual method of joining risers to treads is to cut tongues on the edges of the risers and fit them to grooves cut in the treads, as illustrated in Fig. 194. Another method is to butt the top of the riser under the tread with the joint between the two, which would otherwise be visible, masked by a moulded bead housed in the tread, as illustrated in Fig. 194. The tread of the stair tends to bend under the weight of people using it. When a tread bends, the tongue on the bottom of the riser comes out of the groove in the tread and the staircase 'creaks'. To prevent this it is common practice to secure the treads to the risers with screws (Fig. 194).

Nosing on treads The nosing on treads usually projects 32, or the thickness of the tread, from the face of the riser below. A greater projection than this would increase the likelihood of the nosing splitting away from the tread and a smaller projection would reduce the width of the tread. The nosing is rounded for appearance. Fig. 194 illustrates the more usual finishes to nosings.

Strings Strings (stringers) are cut from boards 38 or 44 thick and of sufficient width to contain and support the treads and risers of a flight of steps. Staircases are usually enclosed in a stair well. The stair well is formed by an external wall or walls and partitions, to which the flights and landings are fixed. The string of a flight of steps which is fixed against a wall or partition is the wall string and the other string the outer string, unless it is also fixed to a wall or partition when it is also a wall string (this will occur with a straight flight between walls).

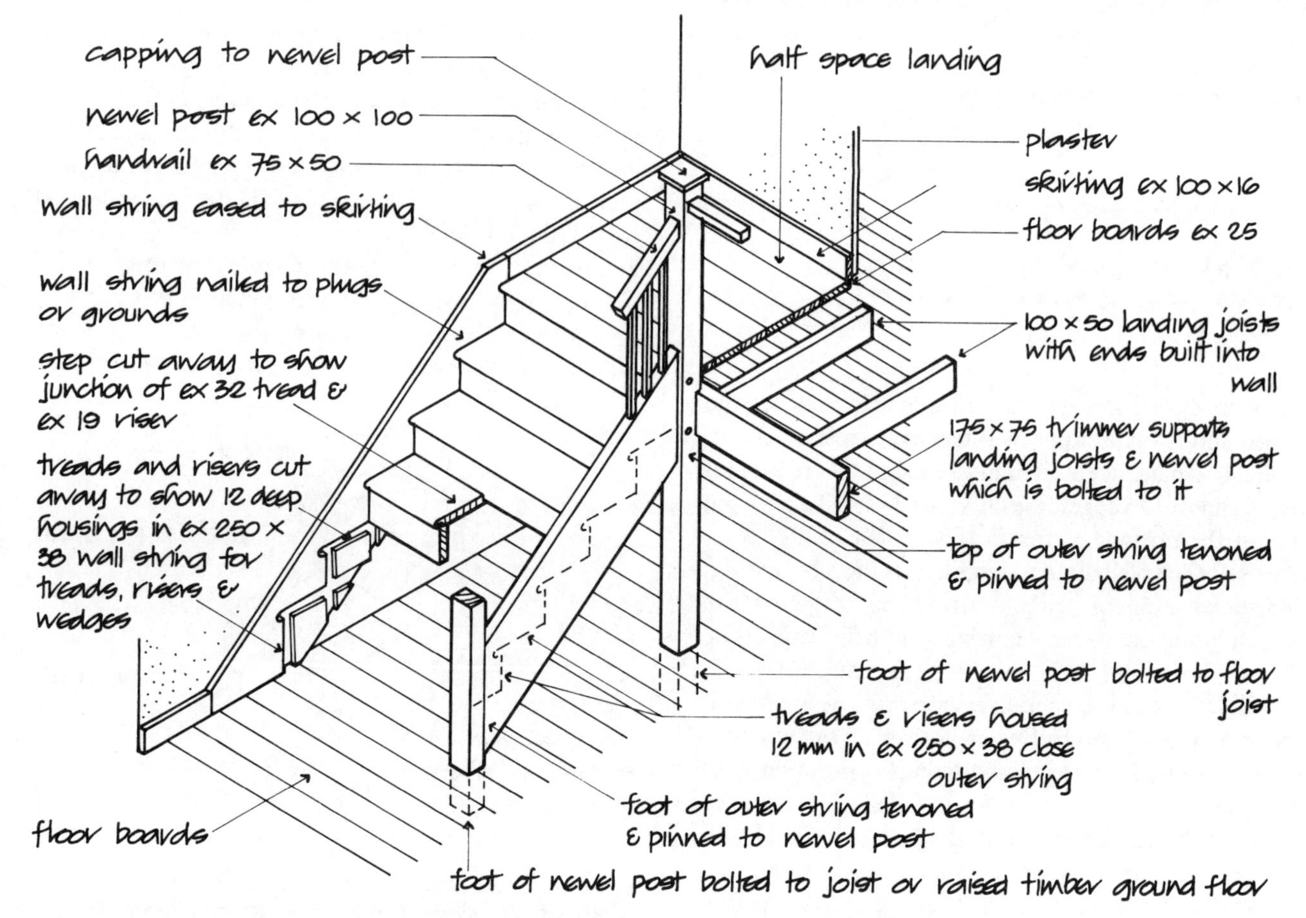

Fig. 193

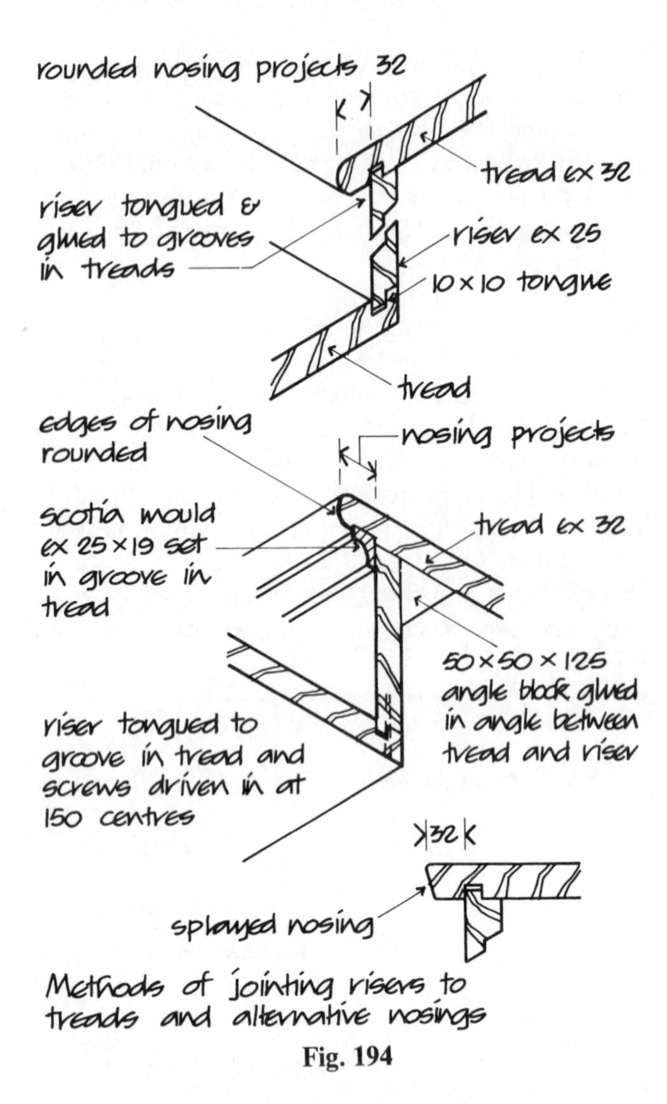

Methods of jointing risers to treads and alternative nosings

Fig. 194

Close or closed strings A string which encloses the treads and risers it supports is termed a close or closed string. It is made wide enough to enclose the treads and risers and its top edge projects some 50 or 63 above the line of the nosings of treads. The width of the string above the line of nosings is described as the margin. Fig. 195 shows a closed string. A string 250 or 280 wide is generally sufficient to contain steps with any one of the dimensions of rise and going and a 50 margin.

Wall strings are generally made as close strings so that wall plaster can be finished down on them. Outer strings can be made as closed strings or as open (cut) strings.

The ends of the treads and risers are glued and wedged into shallow grooves cut in closed strings. The grooves are cut 12 deep into strings and tapering slightly in width to accommodate treads, risers and the wedges which are driven below them, as illustrated in Fig. 195.

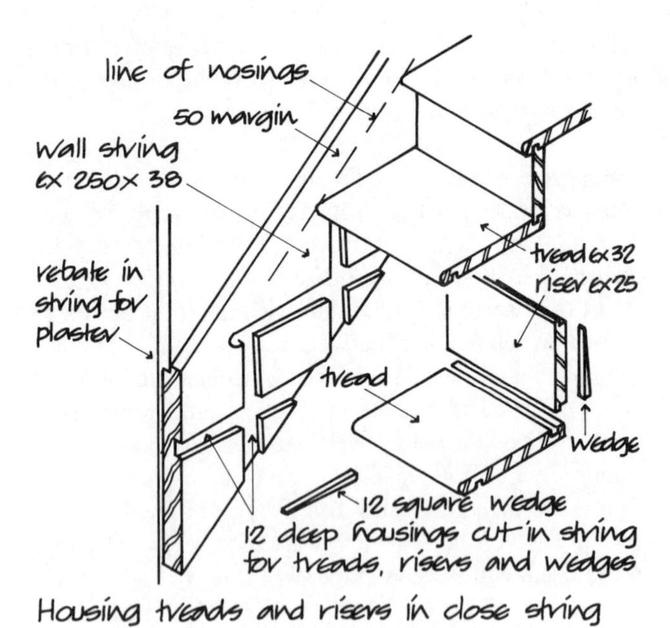

Housing treads and risers in close string

Fig. 195

Angle blocks After the treads and risers have been put together and glued and wedged into their housings in the string, angle blocks are glued in the internal angles between the underside of treads and risers and treads and risers and string. Angle glue blocks are triangular sections of softwood cut from say 50 square timber and each 120 long. Their purpose is to strengthen the right-angled joints between treads, risers and strings. Three or four blocks are used at each junction of tread and riser and one at junctions of treads, risers and string. Angle blocks are shown in Fig. 196.

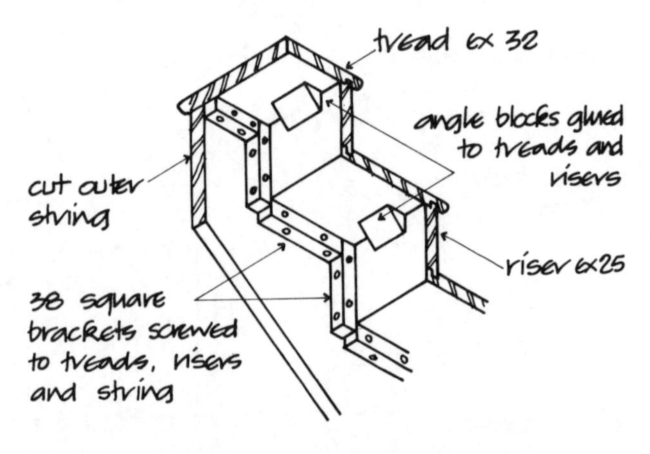

View of underside of flight to show method of fixing treads, and risers to cut outer string

Fig. 196

Open or cut string A closed outer string looks somewhat lumpy and does not show the profile of the treads and risers it encloses. The appearance of a staircase is considerably improved if the outer string is cut to the profile of the treads and risers. This type of string is termed a cut or open string. Because more labour is involved, a flight with a cut string is more expensive than one with closed strings.

As the string is cut to the outline of the treads and risers they cannot be supported in housing in the string and are secured to bearers screwed to both treads and risers and string, as illustrated in Fig. 197. It is not possible to cut a neat nosing on the end grain of treads to overhang the cut string and planted nosings are fitted as shown in Fig. 197. The planted nosings are often secured to the ends of treads by slot screwing. This is a form of secret fixing used to avoid having the heads of screws exposed. Countersunk head wood screws are driven into the ends of treads so that their heads protrude some 12. The heads of these screws fit into holes cut in the nosing. The nosing is then knocked into position so that the screw head bites into slots cut next to the holes in the nosing. It will be seen from Fig. 197 that the planted nosings are mitred to the nosing of the tread. The ends of the risers are cut at 45° to the face of the riser to fit to a matching cut on the edge of the string. Because the string is thicker than the riser it partly butts and is partly mitred to it as a mitre and butt joint.

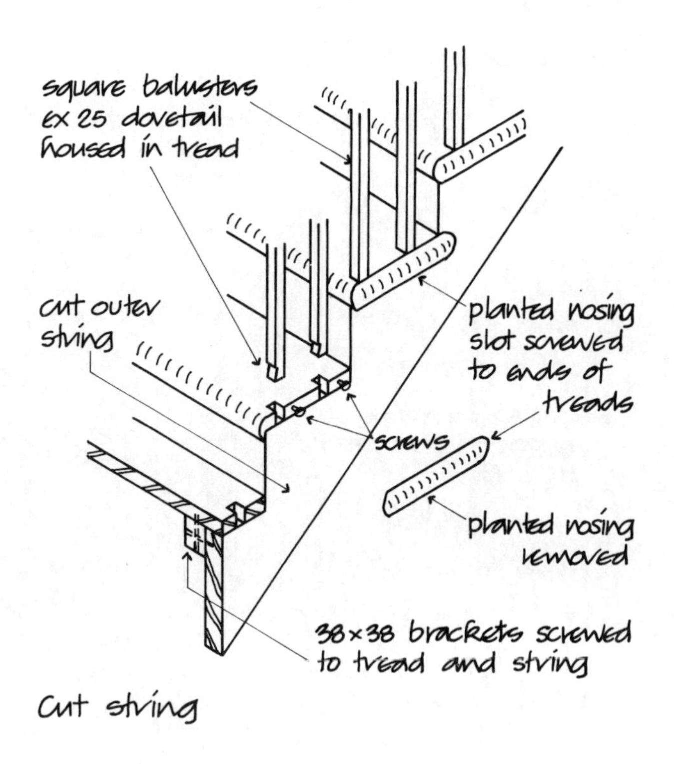

Fig. 197

Landings

Half space (turn) landing A half space landing is constructed with a sawn softwood trimmer which supports sawn softwood landing joists or bearers and floor boards, as illustrated in Figs. 193 and 198. As well as giving support to the joists of the half turn landing the trimmer also supports a newel or newel posts. Newel posts serve to support handrails and provide a means of fixing the ends of outer strings.

Newel posts The newel posts are cut from 100 × 100 timbers and are notched and bolted to the trimmer. The outer string fits to mortices cut in the newel, as illustrated in Fig. 198. For appearance the lower end of the newel post is usually finished about 100 below the flights and moulded. As it projects below the stair it is called a drop newel (Fig. 198).

Balustrade

Open balustrade The traditional balustrade consists of newel posts, handrail and timber balusters, as illustrated in Fig. 198. The newel posts at half turn landings and at landing at first floor level are housed and bolted to trimmers. These newels are fixed in position so that the faces of the risers at the foot and head of flights are in line with the centre line of the newel (Fig. 198).

Handrail The top of the handrail is usually fixed a minimum height of 840 vertically above the line of nosings and 900 above landings for domestic stairs in a single house and 900 above the line of nosings and 1000 or 1100 above landings for other stairs. The handrail is cut from 75 × 50 timber which is shaped and moulded. The ends of the handrail are tenoned to mortices in the newels.

Balusters may be 25 or 19 square or moulded. They are either tenoned or housed in the underside of the handrail and tenoned into the top of closed strings or set into housings in the treads of flights with cut strings, as shown in Fig. 198.

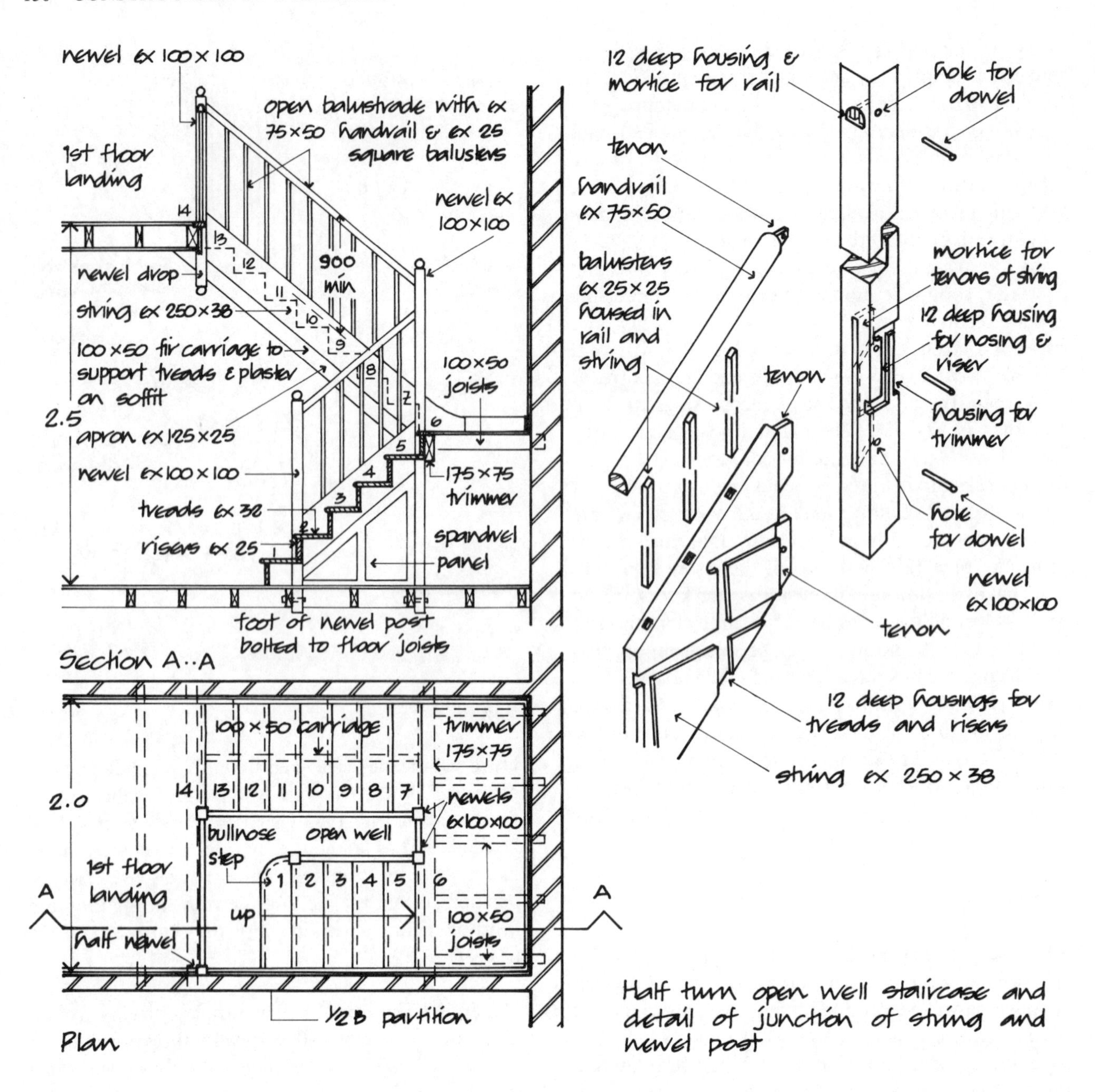

Fig. 198

Closed (enclosed) balustrade The space between the handrail and a close string can be enclosed with timber panels, plywood, hardboard, glass or any sheet material fixed to a light framework. A plywood panel, closed balustrade is illustrated in Fig. 199.

Shaped bottom steps For appearance the bottom step of a flight of wood stairs is shaped as either a quarter or a half circle, as shown in Fig. 200. The rounded end step is formed around shaped wood blocks to which the riser, which is reduced to veneer thickness, is fixed, as illustrated in Fig. 201.

Spandrel The triangular space between the underside of the lower flight of a stair and the floor is the spandrel. It may be left often but is more usually filled with timber framing as a spandrel panel (see Fig. 198).

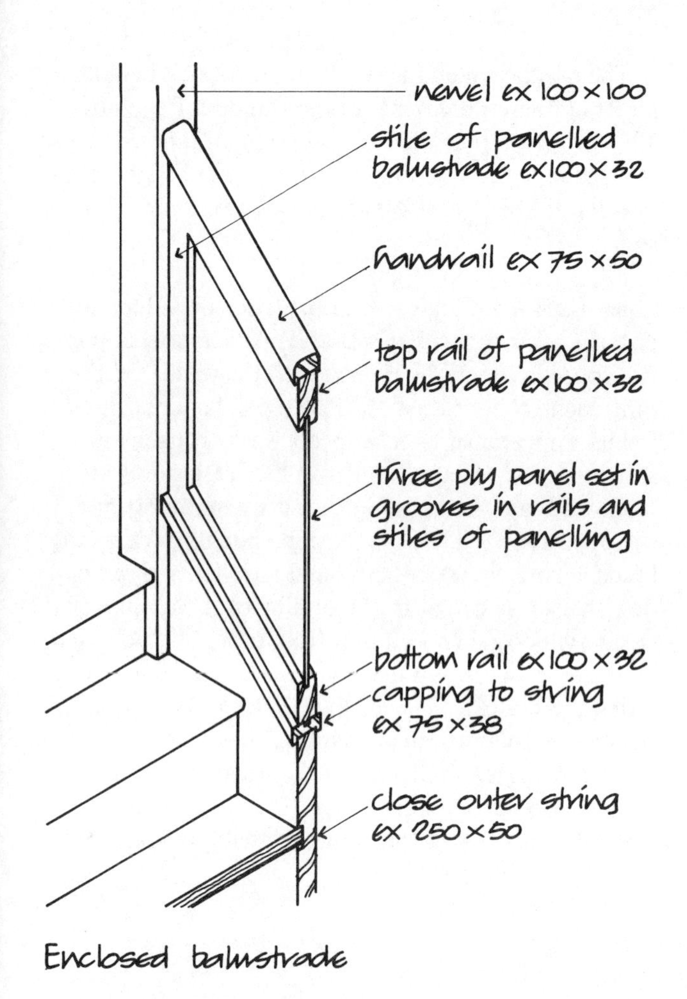

Fig. 199

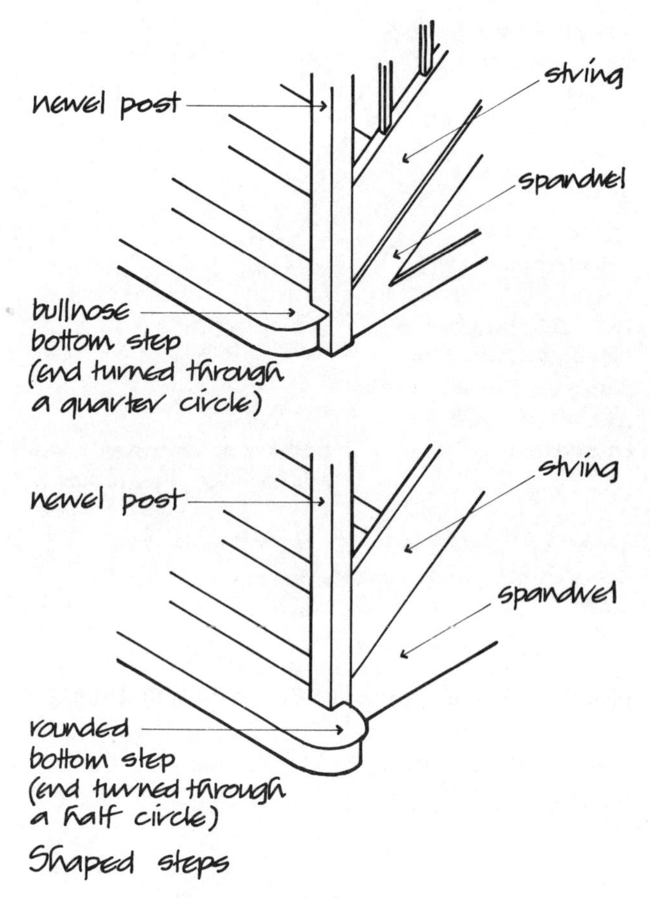

Fig. 200

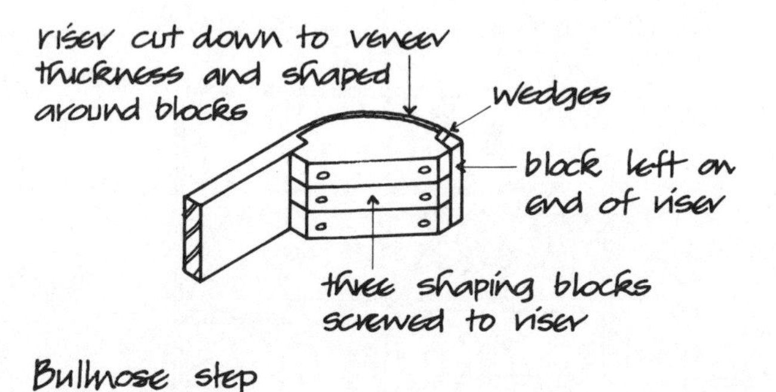

Fig. 201

Carriage A sawn softwood carriage is fixed below flights of a staircase to give support under the centre of treads and also as a fixing for plaster on the soffit or underside of flights. The fir (softwood) carriage illustrated in Fig. 202 is fixed under the centre of a staircase with brackets nailed each side of it and under the treads to reduce creaking as the stair is used. Where the soffit is to be plastered two additional carriages are fixed against the wall and next to the outer string for fixing plasterboard or lath.

Quarter space (turn) landing A quarter space landing is supported by a newel post carried down to the floor below, as illustrated in Fig. 203. The construction of the flight, balustrade and bearers is similar to that of a half space landing.

Winders Winders is the name given to tapered treads that wind round quarter or half turn stairs in place of landings to reduce the number of steps required in the rest of the stair and to economise in space. These winders may be used in domestic stairs. They present some hazard to the young and the elderly and are not recommended for use in means of escape stairs or stairs in public buildings. The winders illustrated in Fig. 203 are constructed as three taper treads at the quarter turn of a half turn stair with a quarter turn landing leading down to the lower flight. The winders are supported on

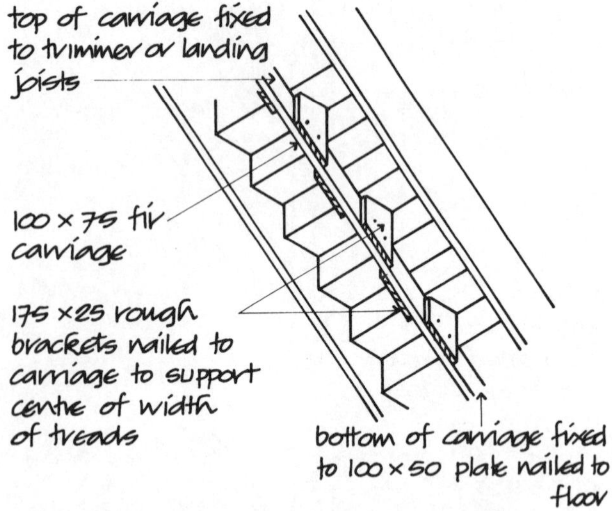

Fig. 202

bearers housed in the newel post and the wall string. To provide a housing for the treads and risers of the winders the wall string has to be built up from two boards (Fig. 203).

The practical guidance in Approved Document K to meeting the requirements of the Building Regulations 1991 proposes that the rise and going of tapered steps should be within the limits set for straight flight steps and should be at least 50 at the narrowest end of each step.

Open riser wood stair An open riser or ladder stair consists of strings with treads and no risers so that there is a space between the treads, with the treads overlapping each other at least 16. This type of stair is chosen for its appearance as it is in no way stronger than a staircase, and where a stair is to be freestanding away from walls as a straight flight stair where the strings of the open rise stair are made deep enough to carry the loads normal to a stair between floors. The strings may be either close or cut to the outline of the treads. The treads, which gain no support from risers, should be cut from 38 or 44 thick timbers which are housed in closed strings, as illustrated in Fig. 204, and secured in position with glued wood dowels.

To strengthen the fixing of the treads to the strings against shrinkage and twisting, 10 or 13 diameter steel tie rods, one to every fourth tread, are bolted under the

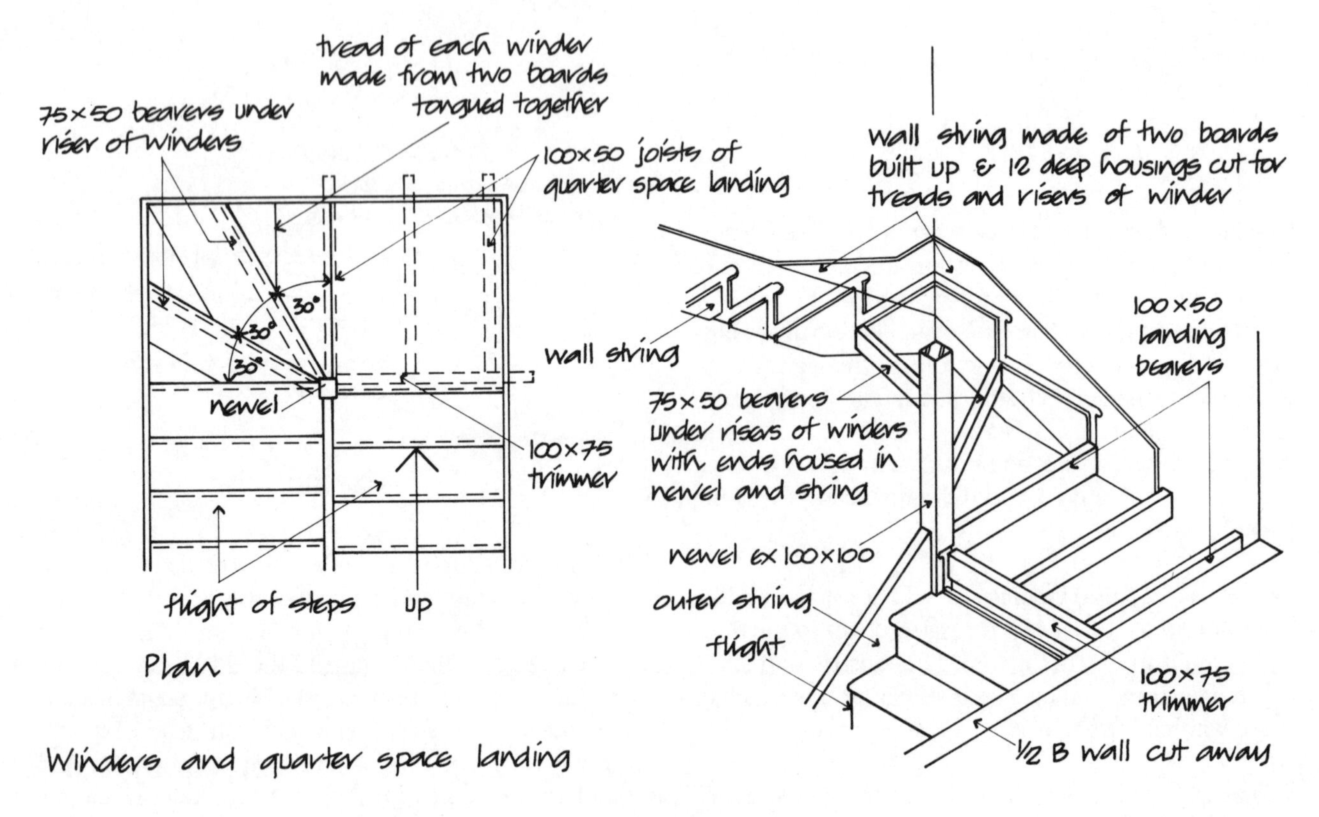

Fig. 203

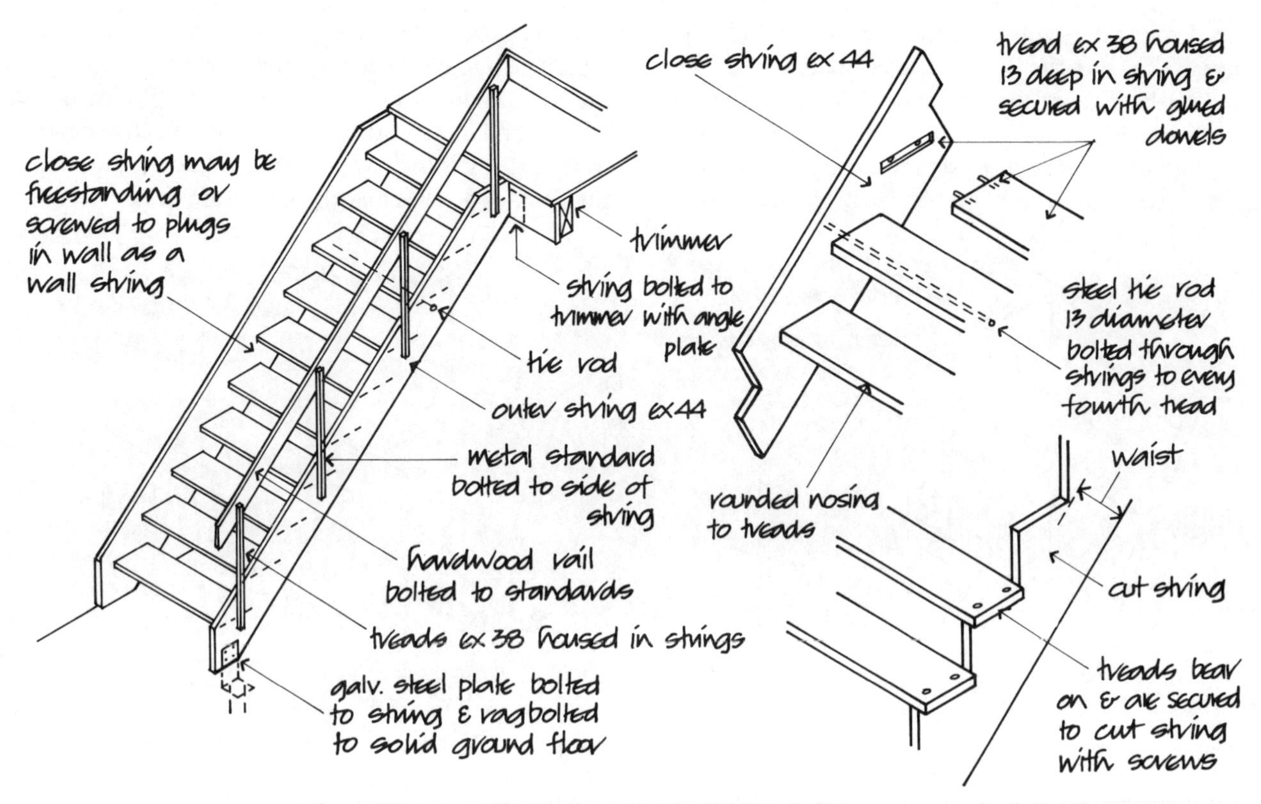

Fig. 204

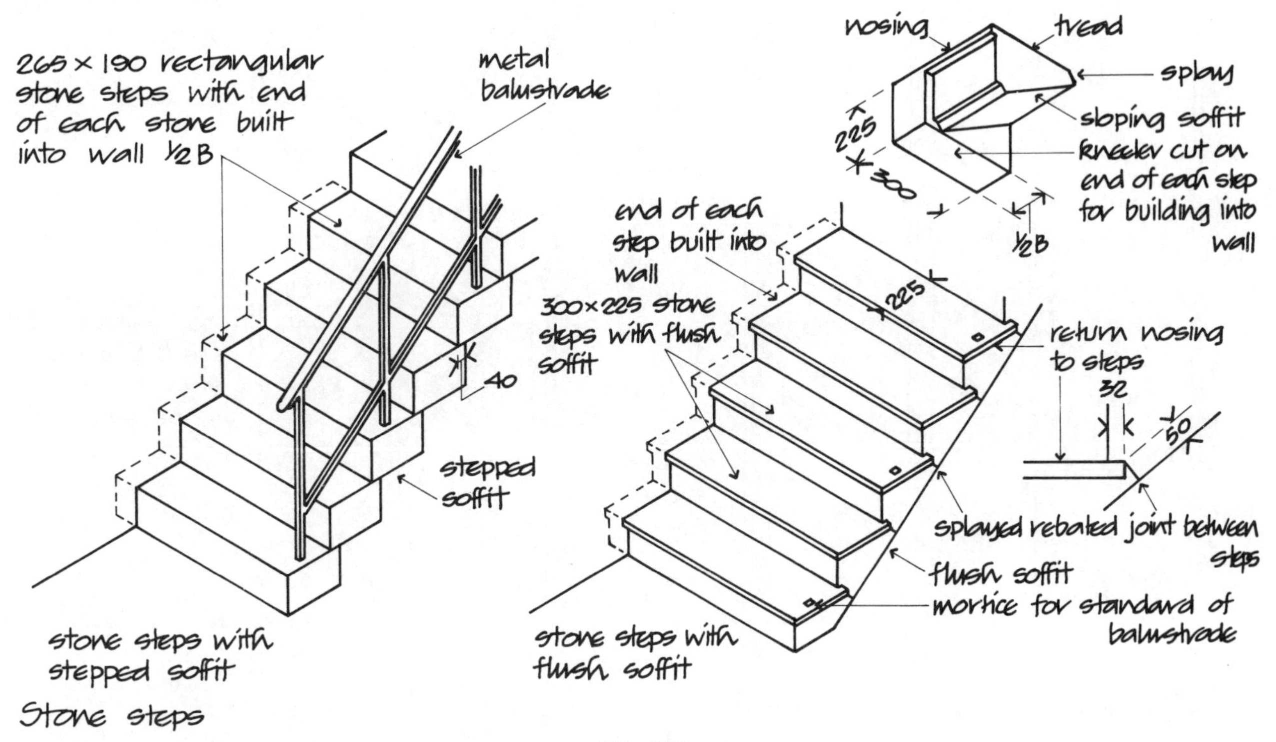

Fig. 205

treads through the strings, as illustrated in Fig. 204. The strings are fixed to the floor with steel plates which are bolted to the sides of the timber strings and bolted to timber trimmers or cast into concrete floors (Fig. 204). Where the strings are cut to provide a seating and fixing for the treads, as shown in Fig. 204, the treads are screwed to the cut top edge of the strings and this fixing is sufficient to tie the strings together without the use of the rods. A cut string will generally need to be deeper than a similar close string because the effective depth of the cut string is the narrow waist below the junction of the back of a tread and the underside of the string.

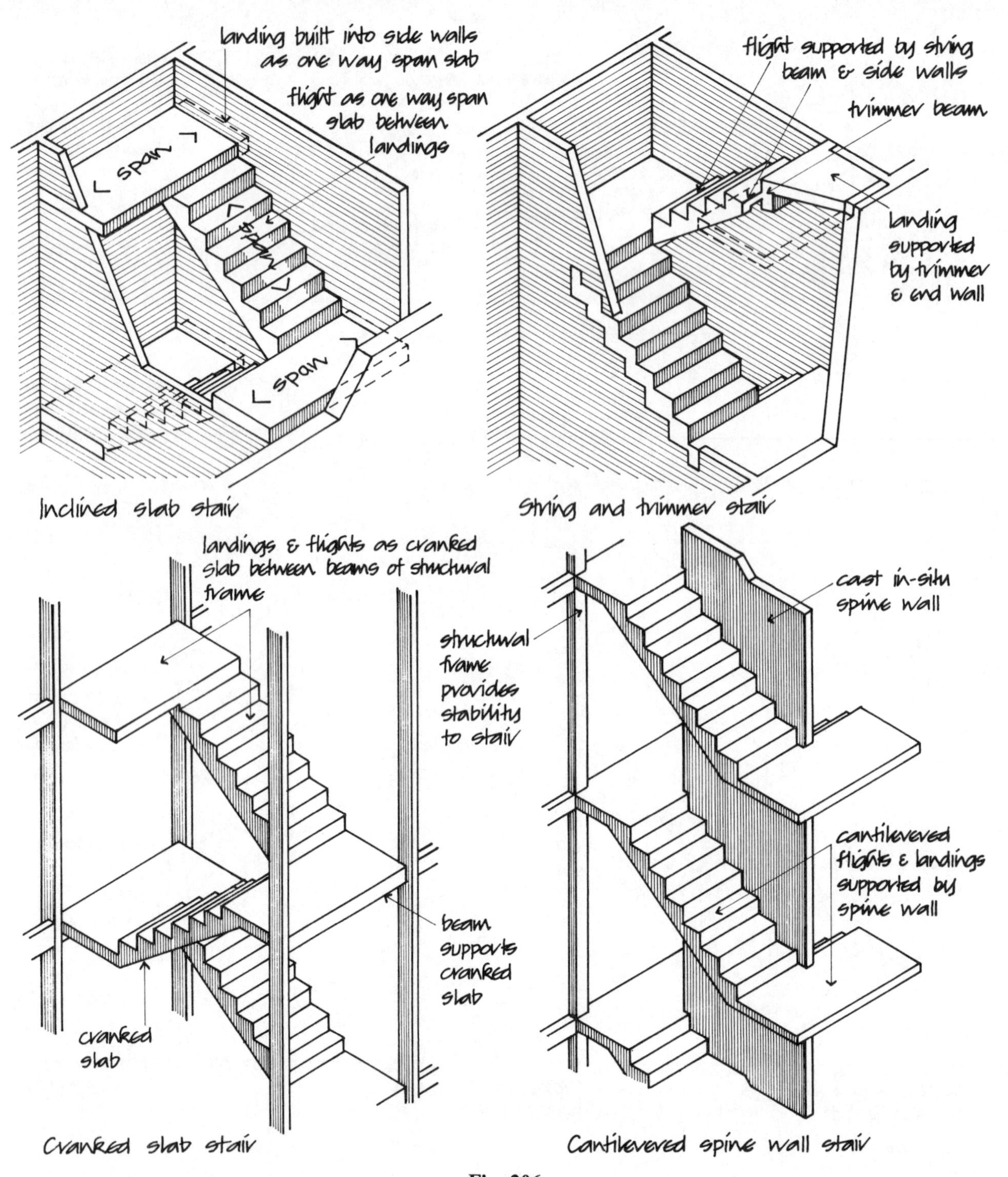

Fig. 206

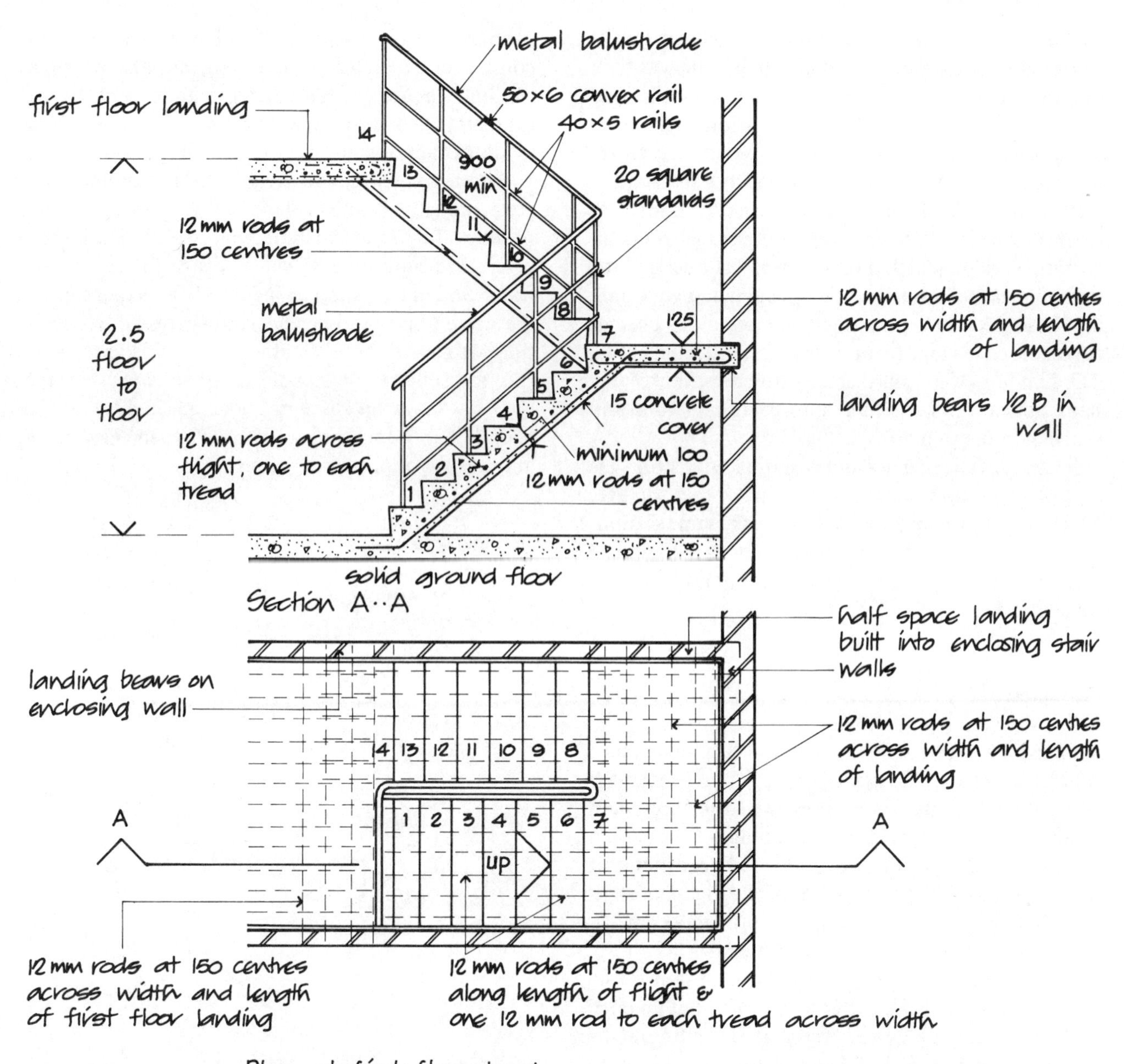

Fig. 207

Open riser wood stairs are constructed as straight flight stairs between floors and there is no newel post to provide a fixing for the handrail. The handrail and balustrade are fixed to the sides of the strings, as illustrated in Fig. 204.

STONE, CAST STONE AND CAST CONCRETE STAIRS

Stone stair Stairs were constructed of steps of natural stones of rectangular or triangular section built into an enclosing wall so that each stone was bedded on the stone below in the form of a stair. Each stone was built into the wall of the stair well from which it cantilevered and took some bearing on the stone below in the form of a prop cantilever. The steps were either of uniform rectangular section with a stepped soffit or rectangular section cut to triangular section to form a flush soffit, as illustrated in Fig. 205. The ends of the crude rectangular section steps were built into a wall. The ends of the triangular section steps had their rectangular ends built in. These steps had splayed rebated joints and nosings cut on the edge of the tread surface, as illustrated in Fig. 205. Landings were constructed with one or more large slabs of natural stone built into enclosing walls and bearing on the step below.

Because of the scarcity and cost of natural stone this type of step is now made of cast stone or cast concrete which is usually reinforced and cast in the same sections as those illustrated for natural stone, or as a combined tread and riser with a rectangular end for building into walls and a stepped soffit.

Fire safety

Reinforced concrete stairs A reinforced concrete stair which has better resistance to damage by fire than the conventional timber staircase is used for access and means of escape stairs in most buildings of more than two storeys. The width, rise, going and headroom for these stairs and the arrangement of the flights of steps as straight flight, quarter turn, half turn and geometrical stairs is the same as for timber stairs.

The usual form of a reinforced concrete stair is as a half turn (dog leg) stair either with or without an open well. The construction of the stair depends on the structural form of the building and the convenience in casting the stair in-situ or the use of reinforced concrete supports and precast steps. Where there are load-bearing walls around the stair it is generally economic to build the landings into the side walls as one-way spanning slabs and construct the flights as inclined slabs between the landings, as illustrated in Fig. 206. This form of stair is of advantage where the enclosing walls are of brick or block as it would involve a great deal of wasteful cutting of bricks or blocks were the flights to be built into the walls and the bricks or blocks cut to fit to the steps.

As an alternative the stair may be designed and constructed as a cranked (bent) slab spanning through landing, flight and landing as one slab with no side support as shown in Fig. 206. This is a more costly construction than using the landings as slabs to support the flights as the span and therefore the cost of the stair is greater. This form of construction is used where the landings cannot gain support each side of the stair.

Another construction is to form a reinforced concrete frame of beams to landings supporting inclined beams to flights, as illustrated in Fig. 206. The landing beams are supported by side walls or the beams of a frame and in turn support inclined beams that support the steps. This is a somewhat clumsy form of construction with a very untidy soffit or underside to the stair. It is best suited to the use of precast concrete steps that bear on the inclined beam under the flight with step ends built into enclosing walls or on two inclined beams and the use of precast landings.

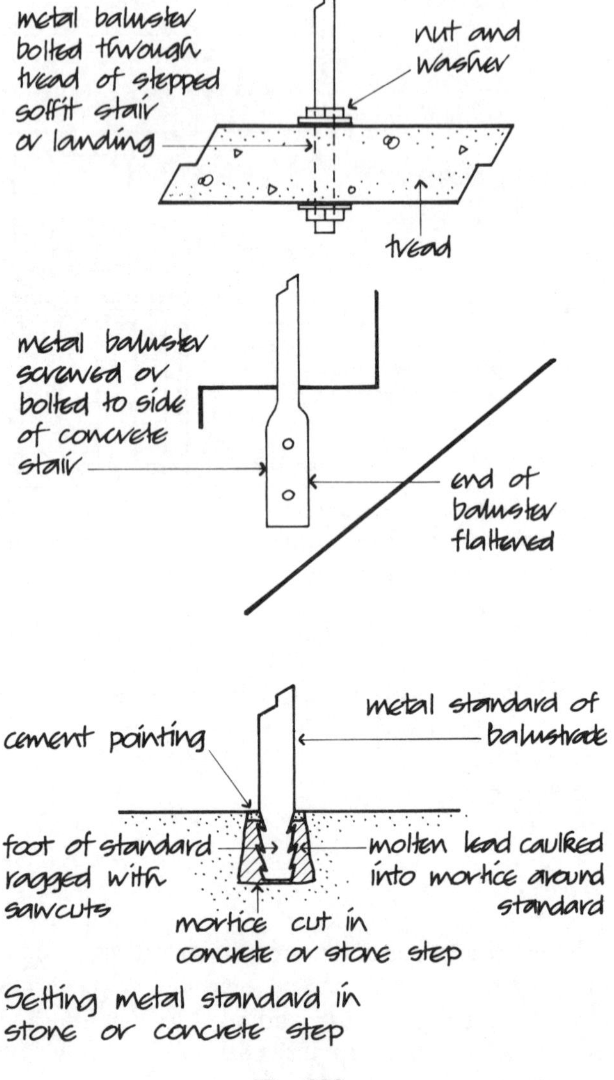

Fig. 208

Where a reinforced concrete half turn stair is constructed around a centre spine wall between the flights, the stair may be constructed to cantilever from this spine wall, as illustrated in Fig. 206 or partly cantilever from the spine wall and be supported by the enclosing frame or walls.

The reinforcement of a concrete stair depends on the system of construction adopted. The stair illustrated in Fig. 207 is designed and built with the landings built into the enclosing walls as a two-way slab, and an inclined slab as flights spanning between landings independent of the side walls. The main reinforcement of the landings is both ways across the bottom of the slab, and the main reinforcement of the flights is one way down the flights. The effective depth of the inclined slab that forms the flights is at the narrow waist formed on section by the junction of tread and riser and the soffit of the flight. It is this thickness of the slab that has constructional strength and the steps play no part in supporting loads. The reinforcement has to have cover of concrete around it to inhibit rust and protect steel rods against damage by fire.

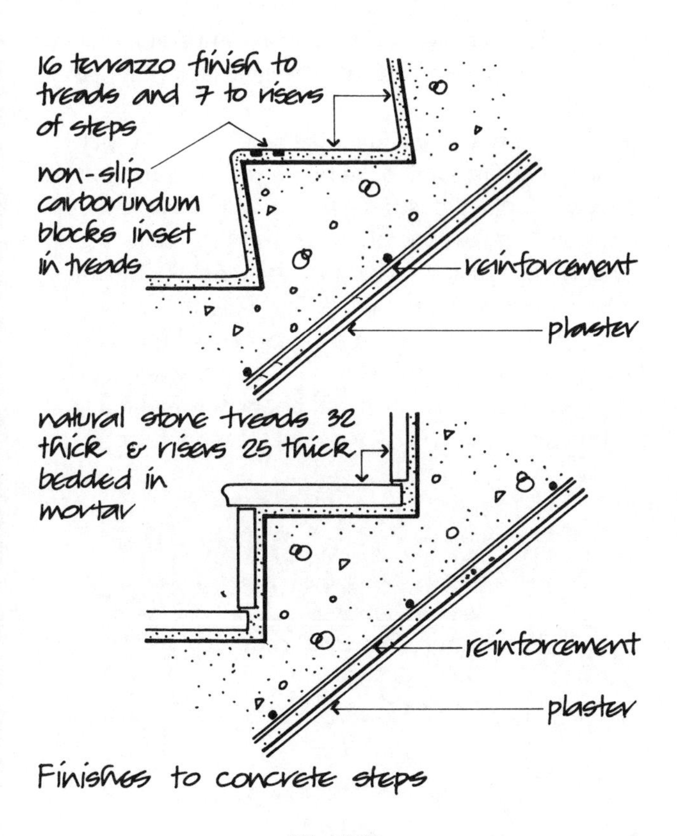

Fig. 209

Balustrade The balustrade to a stone, cast stone or reinforced concrete stair is usually of metal, the uprights of which are either bolted to the sides of the flights to studs cast or grouted into the material or bolted through the material or set in mortices either cast or cut in the material. These vertical metal supports or standards in turn support rails as a balustrade for security and a handrail. Fig. 208 illustrates ways of fixing standards.

Finishes to stone, cast stone, cast concrete or reinforced concrete steps The steps may be left as a natural finish of stone or concrete but more usually are given an applied finish to create a non-slip surface for easy cleaning and for appearance. Any one of the floor finishes used for solid floors, as described in Volume 1, may be used for stairs. In-situ or precast terrazzo is often used for its appearance and ease of cleaning, with carborundum inserts as a non-slip surface, as illustrated in Fig. 209. Wood treads of hardwood screwed to plugs in each step provide an attractive, durable and quiet-in-use surface. Stone treads and risers can be bedded as a surface finish, as illustrated in Fig. 209.

CHAPTER FIVE

INTERNAL FINISHES AND EXTERNAL RENDERING

PLASTER

The finished surface of walls of brick, concrete, stone and concrete or clay blocks is generally so coarse textured that it is an unsuitable finish for the internal walls of most buildings. These surfaces are usually rendered smooth by the application of two or three coats of plaster. Similarly the soffit (ceiling) of concrete floors and roofs is usually rendered smooth with plaster.

It is not fashionable today to leave the joists of timber floors and roofs exposed in the rooms below and they are covered with plaster spread on timber or metal lath or with plasterboard, to provide a smooth, level ceiling and to give the timbers some protection against damage by fire.

The purpose of the plaster is to provide a smooth hard level finish to walls and ceilings. But fashions change, and it is not uncommon today for one or all of the walls of rooms in modern houses to be finished with brickwork or stonework exposed or with paint applied directly to its surface.

The finished surface of plaster should at once be flat and fine textured (smooth). It would seem logical, therefore, to spread some fine grained material such as lime, mixed with water, over the surface and trowel it smooth and level. As the lime dried out it would harden into a dry smooth surface. But this is not practical because so fine-grained a material as lime, when mixed with water, cannot be spread and trowelled smooth to a thickness of much more than say 3 and the wall would absorb so much water that the lime would crack as it dried out. A thicker application of the material would sag and run down as it was being spread. Instead, some coarser-grained material is first spread on walls in one or two coats to render the surface level and when this has dried a thin coat of fine grained material is spread over it to provide a level and smooth surface.

Plaster undercoats The first coat of plaster is described as the first undercoat or render coat and it consists of sand, mixed with lime or cement, or both, or with gypsum, and water. The material is spread and struck off level with a straight edge to a thickness of about 11. If the surface on which the render coat is applied is uneven it is good practice to spread a second coat, called the second undercoat or float coat, of the same materials and mix as the render coat and finished to a thickness of about 6. Finally a coat of some fine-grained material, such as lime or gypsum plaster, mixed with water, is spread and trowelled smooth to a thickness of about 2. The final thin coat, of fine material, is described as the finishing or setting coat.

To reduce the cost of plastering it has become common practice of recent years to apply only two coats of plaster, the render and set coats (one undercoat and a finishing coat), to walls. But the finished surface of two-coat plaster applied on an irregular surface is rarely as flat as that produced by three-coat plaster properly applied. Because of variations in the sizes of bricks and blocks the surface of walls built with them is irregular, and it is to hide these irregularities that plaster is used.

To explain the disadvantage of two-coat plaster consider the surface of a wall built with very irregularly shaped bricks, as illustrated in Fig. 210. If a render coat is applied and struck off level it will tend to sag as indicated by the line at A, due to the weight of the thick, still plastic plaster at that point. This of course will occur at several places on an irregular surface, where the render coat is thickest, and in consequence the setting coat spread over it will not be finished flat. But if a thin float coat is used on top of the render coat it can

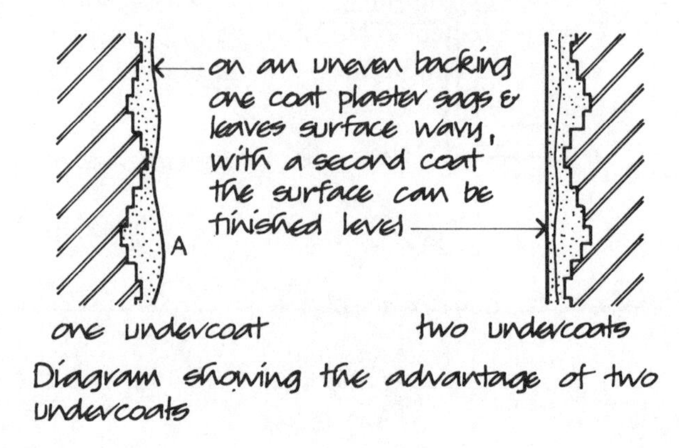

Fig. 210

be spread and finished flat as shown in Fig. 210, and then the setting coat will also be finished flat.

On very irregular backgrounds or as an alternative to the use of two undercoats it is sometimes practice to 'dub out' the deeper hollows in backgrounds by spreading a wet mix of the undercoat material in the hollows. When dry this provides a more regular background for either one or two undercoats of plaster. This operation is termed 'dubbing out'.

Materials used in plaster The undercoats in plaster, i.e. the render and float coats, consist of some coarse-grained material such as natural sand which is hard, insoluble and inert, bound with a matrix of lime, cement or gypsum.

Lime plaster Before the nineteenth century Portland cement was not manufactured and the matrix (binding agent) then used was lime. This was mixed with sand in the proportion 1 part of lime to 3 parts of sand by volume, and water, and was termed coarse stuff by plasterers.

When a plaster of lime and sand dries out it shrinks and fine hair cracks appear on the surface. To restrain this shrinkage and to reinforce lime plaster, long animal hair was mixed in with the lime and sand, 5 kg of hair being used for every square metre of coarse stuff. The resulting 'haired coarse stuff' was plastic and dried out and hardened without appreciable shrinkage. The disadvantages of lime plaster are that it is somewhat soft and easily damaged by knocks, and in time becomes dry and powdery, and it is soluble in water.

Up to the end of the nineteenth century most internal plastering was executed with render and float coats of lime and sand set with a thin coat of lime and water trowelled smooth. It has been replaced by Portland cement, for undercoats, and by gypsum plaster, for both undercoats and finishing coat. Lime may be added to the Portland cement or to some types of gypsum plaster. (Note: Portland cement should not be mixed with gypsum plaster.)

Cement plaster The properties of Portland cement were described in Volume 1. It is sold as a grey powder which, when mixed with water, hardens into a solid inert mass. It is mixed with sand and water for use as the render and float coat in plaster applied to brick, concrete and clay block walls and partitions. The mix used is 1 part of cement to 3 or 4 parts of clean washed sand.

A mixture of cement and clean sand forms a very hard surface as it sets, but it is not plastic and requires a deal of labour to spread. It is usual therefore to add either lime or a plasticiser to the cement and sand to produce a mix that is at once plastic yet dries out to form a hard surface. Usual mixes are 1 cement, ¼ lime and 3 of sand; 1 cement, 1 lime and 6 sand; 1 cement, 2 lime and 9 sand; or a mix of 1 cement to 4 of sand with a mortar plasticiser added, the proportions being by volume. Cement and sand, or cement, lime and sand undercoats (render and float coats) are the cheapest undercoats in use today for brick and block walls and in consequence are more used than other types of undercoat.

As an undercoat of cement and sand dries out it shrinks slightly and cracks may appear in the surface. In general the more cement used the greater the shrinkage and therefore cracking. The extent of the cracking that may appear depends to some extent on the strength of the surface on which the plaster is applied and the extent to which the plaster binds to the surface. For example, the surface of keyed fletton brickwork is sufficiently strong and affords sufficient key to restrain any appreciable shrinking, and therefore cracking, of plaster, but the surface of some lightweight concrete blocks is not sufficiently strong to prevent cracking of this type of plaster. A cement sand undercoat should therefore be used only on a backing of hard brick. On other bricks and clay and concrete blocks a cement, lime, sand mix (1:¼:3) and on lightweight concrete blocks a cement, lime, sand mix (1:1:6) should be used.

Gypsum plaster

During the last seventy years the use of gypsum plasters has increased greatly so that they have superseded lime as a finishing (setting) plaster and are also used as a matrix with sand for undercoats to a considerable extent. The advantage of the gypsum plasters is that they expand very slightly on setting and are not therefore likely to cause cracking as are cement and lime.

Gypsum is a chalk-like mineral mined in several parts of England. It is a crystaline combination of calcium sulphate and water ($CaSO_4.2H_2O$). If powdered gypsum is heated to about 170°C it loses about three quarters of its combined water and the result is described as hemihydrate gypsum plaster ($CaSO_4.\frac{1}{2}H_2O$). This material is better known as plaster of Paris.

If gypsum is heated at a considerably higher temperature than 170°C it loses practically all its combined water and the result is anhydrous gypsum plaster. In British Standard 1191 Part 1: 1973 four classes of gypsum plaster are noted, as described here. Classes A and B are based on plaster of Paris.

Class A, Hemi-hydrate gypsum plaster (plaster of Paris) This material is supplied in powder form as 'plaster of Paris' or 'gauging plaster'. When plaster of Paris is mixed with water it sets so quickly (10 minutes) that it is unsuitable for use as a wall or ceiling plaster. It is used for fibrous plaster work, in which the wet mix of plaster of Paris is brushed into moulds which are used for the reproduction of cornices and other decorative plaster work.

Class B, Retarded hemi-hydrate gypsum plaster To make plaster of Paris suitable for use as a wall or ceiling plaster the speed of its set is slowed down or retarded and this is done by adding 'keratin', an animal protein, in small amounts during manufacture. The amount of retarder added depends on the use of the plaster. Plaster for use in undercoats is more heavily retarded than that for use in finishing coats which require less time to spread and trowel.

This class of plaster is often described as 'board plaster' as it is the only one of the four classes of gypsum plaster which will adhere strongly to the surface of gypsum plasterboard and for that reason it is used as a finishing coat on gypsum plasterboard and gypsum lath and also in sanded undercoats.

Class B undercoat plasters are grouped in BS 1191 Part 1: 1973, as type a, and finishing or final coat plasters as type b, as follows:

Type a (1) Browning plaster
(2) Metal lathing plaster

Type b (1) Finish plaster
(2) Board finish plaster
(3) Thincoat finish plaster

Browning plaster is used with sand as an 11 thick undercoat on brick, block, concrete and stone surfaces as a background for finish plaster. Mixes are 1:3, plaster to sand by volume, on brick and clay block, 1:2 on concrete block and stone, and 1:1½ on engineering bricks.

Metal lathing plaster is used with sand as an undercoat on expanded metal lathing.

Finish plaster is used neat with water on browning plaster and sand undercoats to a trowelled finished thickness of 2.

Board finish plaster is used neat with water on plaster-board, baseboard, plank and lath to a trowelled finish of 5.

Thincoat finish plaster (thin-wall plaster) incorporates organic binders. This plaster is used for thin, single coat finishes applied to comparatively level surfaces by hand or spray application.

Class C and D gypsum plasters (BS 1191) are based on anhydrous calcium sulphate.

Class C, Anhydrous gypsum plaster (Sirapite) After gypsum has been burned at a high temperature to form anhydrous gypsum plaster it is ground to a fine powder and ½% to 1% by weight of alum or zinc sulphate is added to it. These accelerate its hardening which otherwise would be so slow as to make it unsuitable for use as a wall plaster.

This class of gypsum plaster is much used as a finishing plaster on undercoats of cement, cement and lime or gypsum and sand mixes, because it is easy to work and can be trowelled to give a smooth finish. One of its useful characteristics is that it can be 'brought back' (retempered). Once this plaster has been mixed with water and spread it becomes stiff but can then, or even several hours later, be made sufficiently plastic by sprinkling its surface with water for it to be trowelled to a smooth finish. The words 'brought back' describe the operation of making the plaster plastic by sprinkling it with water. This plaster, commonly known as Sirapite, is used only as a finishing plaster.

Class D (Keenes plaster or Keenes cement) Anhydrous gypsum plaster This class of gypsum plaster is made by heating gypsum beyond the point at which it becomes anhydrous moderately burned. The burned gypsum is ground to a fine powder and an accelerator is added as described for Class C plaster. The best known plaster in this class is Keenes cement. It is easy to work, can be trowelled to give a particularly smooth hard finish and is used as a finishing coat to a thickness of 3. It is more expensive than other types of gypsum plaster and is used in high-class work. As it dries out and hardens this plaster expands and should only be applied over strong undercoat plaster such as cement and sand (1:3) or undercoats of Keenes and sand (1:2).

Because it is particularly hard Keenes is often used as a finishing plaster to plaster angles.

Sand for plastering Sand for plastering should be clean and contain not more than 5% of clay or other soluble adherent matter. In Volume 1 an explanation was given of why sand for mortar should be clean, and this applies equally to sand for plastering. The sand should not contain any grains larger than 5.

Lightweight aggregate for plaster Recently lightweight aggregate has been used instead of natural sand in plaster. The aggregates commonly used are perlite and exfoliated vermiculite. Perlite and vermiculite are minerals which expand into multicellular lightweight materials when heated. The expanded minerals, though lightweight and cellular, have sufficient mechanical strength for use as an aggregate for plaster. The description exfoliated vermiculite describes the action of the mica-like mineral vermiculite, when heated. The thin layers of the mineral open up (exfoliate) when heated, to form a cellular mass. For use as an aggregate in plaster the expanded mineral is crushed. These lightweight aggregates are supplied ready mixed with gypsum plaster class B.

The advantages of these plasters are that they are light in weight, being less than half the weight of plasters made with sand. Because of their cellular nature these aggregates are better thermal insulators than natural sand and the insulation of a wall or ceiling can be improved by their use in plaster. The condensation of moisture, from warm moisture-laden air, on cold wall surfaces in rooms such as bathrooms and kitchens can be reduced by the use of a lightweight aggregate plaster. This is due to the insulating property of the plaster which prevents the inside face of walls being as cold as it would be if dense aggregate (sand) were used.

Lightweight aggregate plasters have good resistance to damage by fire and can be used to protect such structural parts as timber floor joists. Because of the cellular nature of the aggregate, lightweight plasters are not as resistant to damage by knocks or abrasions as dense aggregate plasters. A lightweight aggregate gypsum plaster is about twice the cost, per metre of plastered surface, of a cement, lime, sand and setting coat plaster.

These premixed lightweight plasters are supplied as type a, and b, in accordance with BS 1191 part 2:

Type a Undercoat plaster
(1) Browning plaster as undercoat for solid backgrounds
(2) Metal lathing plaster
(3) Bonding plaster as undercoat for low suction backgrounds such as concrete and plasterboards.

Type b Final coat plaster
(1) A finishing coat for *type a* undercoats.

Plaster for machine application Two types of plaster are made for application by machine to wall and ceiling backgrounds: a gypsum-based plaster for application by machine spreading in one coat to a thickness of from 5 to 13, and a plaster made of fine marble aggregate with PVA binders for spraying on to wall and ceiling finishes principally as a thin veneer of plaster in two coats 1 to 3 thick on level dense surfaces such as concrete.

The application of wet plaster by hand is a laborious, disagreeable process. The application of plaster by machine which takes out a good deal of the 'donkey work' is becoming more common.

The gypsum-based plaster is made of a blend of gypsum plasters with additives to improve water retention, plasticity and setting time. The dry prepared material is mixed with water and spread from the nozzle of a mechanical plastering machine as a ribbon over the background surface to the required thickness. It is then spread and levelled with a feather edge. When the plaster is sufficiently stiff it is finished level and smooth with the feather edge, sprayed with water and trowelled smooth with a hand or power operated float. This plaster can be used on clay bricks and blocks, dense and lightweight concrete blocks, metal lath and plasterboard backgrounds. The finished thickness depends on the irregularities in the background.

The marble aggregate and PVA binder plaster is sprayed on to the background with mechanical spray equipment in two coats as a thin veneer finish to level backgrounds such as concrete and the thicker finishes for irregular backgrounds. The specific use of this plaster is as a thin veneer finish to level backgrounds as it provides a strong bond to smooth, level, dense finishes such as in-situ cast concrete. The spray applied finish is levelled and smoothed with a spatula. For application to uneven surfaces a filler coat material is supplied that can be finished to a thickness of 10 ready for the thin finish coat.

Plaster finishes to timber joists and studs

The usual method of providing a level finished surface to the ceiling (soffit) of timber floors and roofs and on timber stud partitions is to spread plaster over timber or metal lath or to fix preformed boards to the timber ceiling or wall.

Fir lath Before the twentieth century the usual method of preparing timber ceilings and timber stud walls and partitions for plaster was to cover them with fir lath spaced about 7 to 10 apart to provide a key for the plaster. The usual size of lath is 25 wide by 5 to 7 thick, in lengths of 900. The lath is either split or sawn from Baltic fir (softwood). Split lath is usually described as riven lath and is prepared by splitting along the grain of the wood. Because the grain of wood is never absolutely straight neither is riven lath, so that when it is fixed the spaces left between the laths as a key for the plaster are not uniform. This may prevent the plaster being forced between the laths and it will not therefore bind firmly to them. Sawn lath on the other hand is uniformly straight and can be fixed with uniform spaces to give a good key for plaster. Fir lath must be adequately seasoned and free from fungal decay. The fir lath is nailed across the joists or timber studs. Obviously the ends of the laths must be fixed to a joist or stud, as illustrated in Fig. 211, and the butt end joints of laths staggered to minimise the possibility of cracks in the plaster along the joints.

Fir lath is covered with three coats of plaster. The first coat is spread and forced between the lath so that it binds to it. This coat is described as pricking up. A second undercoat, termed the float coat, is spread and finished level and then covered with the finish or setting coat.

Before the twentieth century the undercoats consisted of haired coarse stuff (1 part lime to 3 parts sand, with hair) gauged with plaster of Paris, and the finishing coat of lime and water gauged with plaster of Paris. The purpose of the gauge (addition of a small amount) of plaster of Paris is to cause the material to harden more quickly so that vibration due to the applications of the next coat, or vibrations of the floor above, will not cause the plaster to come away from the lath before it is hard.

Plastering on fir lath is generally executed with a mix of cement, lime and sand (1:2:9) or gypsum plaster, lime and sand (1:2:9) for undercoats, and gypsum plaster class C as a finishing coat. The cost of fir lath and three coats of plaster today is about three times that of a plasterboard finish and in consequence fir lath is less used now than it was.

Metal lath (EML) This lath is made by cutting thin sheets of steel so that they can be stretched into a diamond mesh of steel, as shown in Fig. 212. This lath is described as EML (expanded metal lath). The thickness of the steel sheet which is cut and expanded for plasterwork is usually 0.675 mm and the lath is described by its shortway mesh. A mesh of 6 shortway is generally used for plaster. To prevent expanded steel lath rusting it is either coated with paint or galvanised. As a background for plaster on timber joists and studs, the lath, which is supplied in sheets 2438 × 686, is fixed by nailing with galvanised clout nails or galvanised staples at intervals of about 100 along each joist or stud. During fixing, the sheet of lath should be stretched tightly across the joists. Edges of adjacent sheets of the lath should be lapped at least 25.

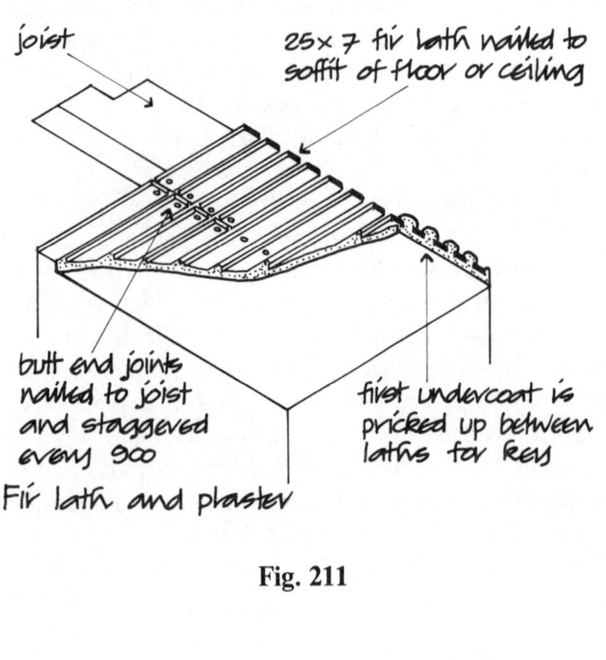

Fig. 211

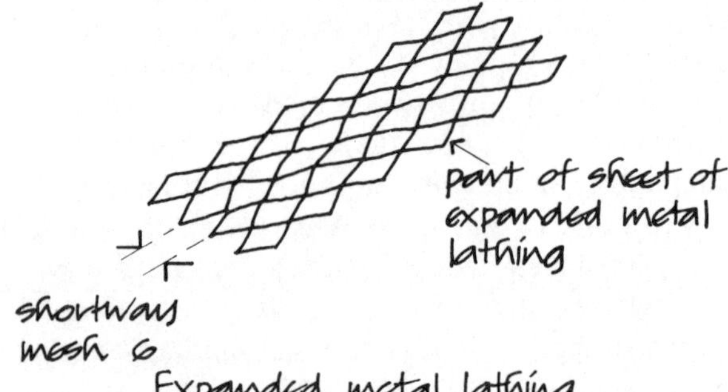

Fig. 212

The undercoat plaster generally used on this type of lath is lime and sand gauged with Portland cement or gypsum undercoat and finish. Three-coat plasterwork should be used. The following are some mixes of plaster commonly used:

Undercoats: cement, lime and sand (1:2:9)
Gypsum class B and sand.
Finishing coat: Gypsum class B or C.

The cost of three-coat plaster on metal lath is rather more than twice the cost of plasterboard finished with a skim coat.

Metal lath is principally used as a background for plaster in ceilings suspended below concrete floors and steel roofs. The lath is supported on light steel runners hung on steel hangers fixed to the floor or roof above, either in the form of a flat soffit or shaped for decorative purposes.

GYPSUM PLASTERBOARD

Gypsum plasterboard consists of a core of set (hard) gypsum plaster enclosed in and bonded to two sheets of heavy paper. The heavy paper protects and reinforces the gypsum plaster core which otherwise would be too brittle to handle and fix without damage. Plasterboard is made in thicknesses of 9.5 mm, 12.7 mm and 19 for use either as a dry lining or as a background for plaster in boards of various sizes.

Plasterboard is extensively used as a finish on the soffit (ceiling) of timber floors and roofs and on timber stud partitions. The advantage of this material as a finish are that it provides a cheaper finish, it can be fixed and plastered more speedily and provides better fire protection than lath and plaster. Its disadvantages are that, because it is a fairly rigid material, plaster finishes applied to it may crack due to vibration or movement in the joists to which it is fixed, and it is a poor sound insulator. Four types of gypsum plasterboard are manufactured; gypsum wallboard, gypsum baseboard, gypsum plank and gypsum lath.

Gypsum wallboard is a board made specifically as a dry lining for internal surfaces. The boards have one ivory coloured surface for direct decoration and one grey surface. These boards are 9.5, 12.5 and 15 thick, from 600 wide and up to 3.6 long, as illustrated in Fig. 213. The boards are made with two different edges, as illustrated in Fig. 213, tapered for smooth seamless jointing and square for cover strip jointing.

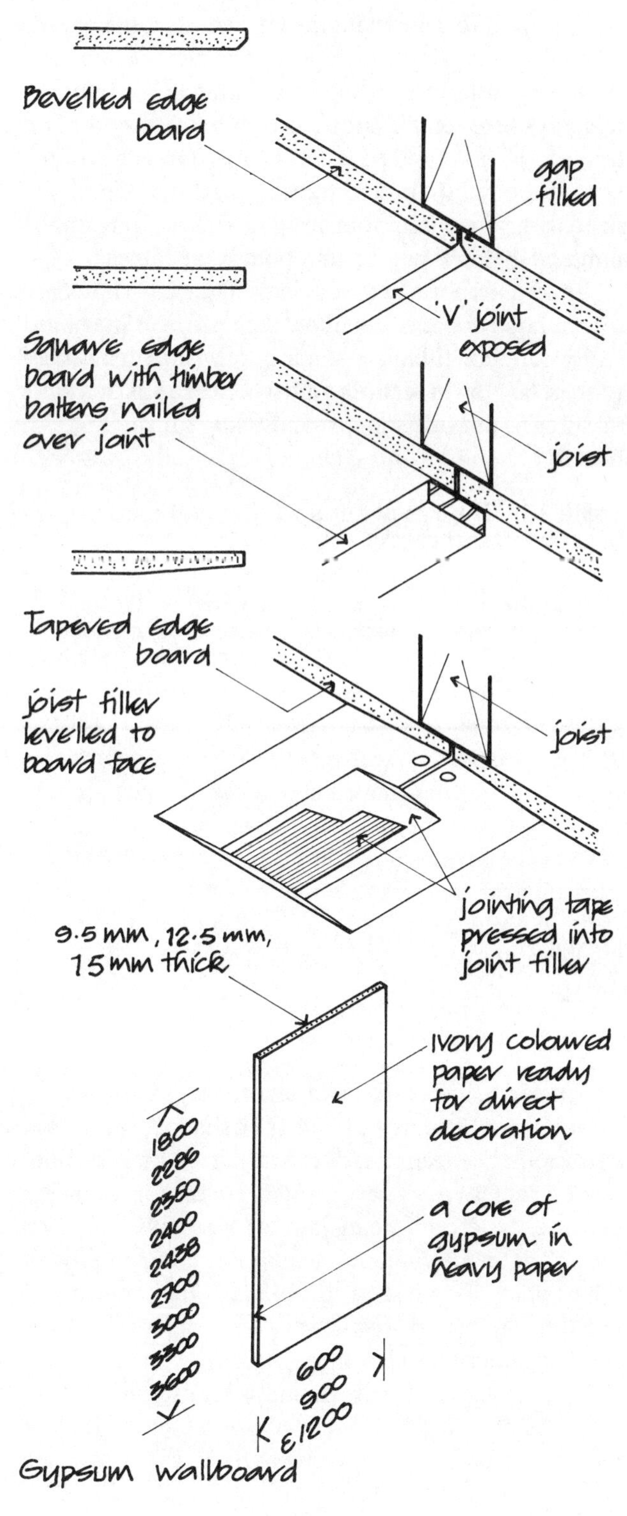

Fig. 213

Fixing and jointing Wallboard is fixed to timber supports such as floor and ceiling joists, studs or battens fixed to walls, with the long edges of the boards centred over a timber support. The timber supports should be at the centres shown in Table 13, with timber noggings between the supports to provide support and fixing to all ends and edges of boards. The edges of the boards are lightly butted together and the boards are fixed with galvanised clout-headed nails, 30 long for 9.5 mm and 40 long for 12.5 mm boards, at 150 centres to all timber supports and noggings. The nails should be driven home to leave a shallow depression in the board ready for spot filling. (Noggings are short lengths of timber, 50 × 50 in section, nailed between joists, rafters or studs to provide support and fixing for the edges of boards.)

Table 13. Centres of support for wallboard, baseboard and plank

Timber supports	Board thickness (mm)	Board width (mm)	Recommended centres (mm)
Studs or furring (walls)	9.5	900	450
		1200	400
	12.5	900	450
		1200	600
	19	600	600 Vertical
Joists or furring (ceilings)	9.5	All widths	400
	12.5	All widths	450
	19	600	750

Taken from the *British Gypsum White Book*

Tapered edge boards are designed for jointing with a smooth seamless surface finish. The shallow depression at the joint between boards is first filled with joint filler made from gypsum and water mixed to a creamy consistency. A reinforcing jointing tape 45 wide is then pressed into the filling and the joint completed with filler which is finished flush with the boards, as illustrated in Fig. 213. Nail heads are covered with filler, finished smooth as spot filling.

Square edge boards are designed for use with a cover strip of cloth, wood, metal or plastic which is glued or nailed over the joints between boards to give a panelled effect to the finished surface, as illustrated in Fig. 213.

Cracking of gypsum plasterboard finishes The two principal causes of cracking in these finishes are twisting and other moisture movements of joists or studs to which they are fixed, and deflection of timber joists under load. New timber is often not as well seasoned as it should be and as the timbers dry they tend to shrink and lose shape. Joists may wind (twist) and cause the rigid boards fixed to them to move and joints open up, as illustrated in Fig. 214.

Under the load of furniture and persons timber floor joists bend slightly. The degree to which they bend is described as their deflection under load. Even with very small deflection under load a large rigid plasterboard will bend and cracks appear at joints, as illustrated in Fig. 214. One way of minimising this cracking with large boards is to use joists some 50 deeper than they need be to carry the expected loads. This additional depth of joist reduces deflection under load and the possibility of cracking. Baseboard, plank and lath, made specifically for a plaster finish, are smaller than wallboard and therefore less liable to shrinkage and movement cracking.

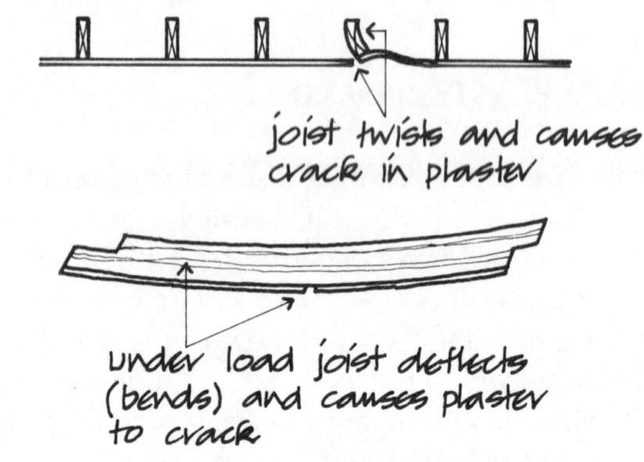

Fig. 214

Gypsum baseboard is designed for use as a base for gypsum plaster. The boards are 9.5 mm thick, 914 wide and 1200, 1219, 1350 or 1372 long, with square edges as illustrated in Fig. 215. The smaller length of these boards is chosen to minimise cracking of finishes due to movement in timber supports. The boards are fixed with a gap of about 3 at joints, with 30 nails at 150 centres. The joints are filled with gypsum and covered with jute scrim, and the surface is covered with board finish plaster to a trowelled finish thickness of 5 as illustrated in Fig. 216.

Gypsum plank is 19 thick, 600 wide and 2350, 2400, 2700, 3000 and 3200 long with either a tapered edge for seamless jointing or a square edge for plastering. The thicker plank can be fixed at 600 centres vertical and 800 horizontal with 60 nails. For direct painting the tapered edge boards are jointed as described for

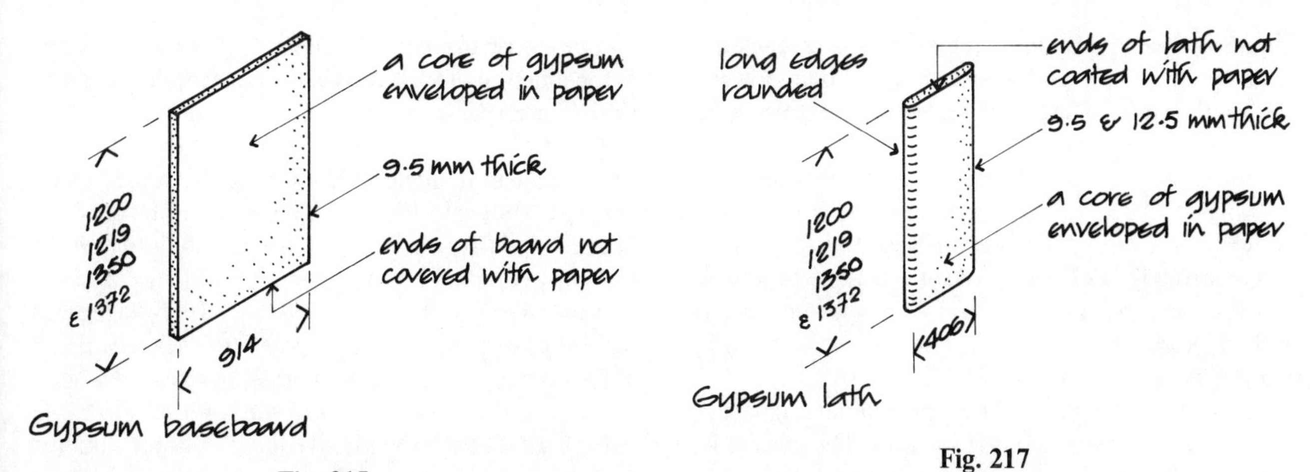

Fig. 215

Fig. 217

noggings at joints
baseboard nailed across joists with 30 galvanised nails at 150 centres with end joints staggered
one or two coat gypsum plaster
gaps, 3 wide, between boards filled and reinforced with 90 wide jute scrim
joists
timber noggings fixed between joists at perimeter of ceiling
angle reinforced with jute scrim
Fixing baseboard for plastering

Fig. 216

wallboard. For plastering the square edge boards are fixed with a gap of about 3 and the joint filled with plaster and reinforced with jute scrim, 90 wide. The surface of the boards is then covered with gypsum class B and water to a trowelled finish thickness of 5.

Gypsum lath These comparatively small boards are made as a base for plaster. The lath is 9.5 mm or 12.5 mm thick, 406 wide and 1200, 1219, 1350 or 1372 long. The long edges of the lath are rounded, as illustrated in Fig. 217. The lath is fixed to timber supports and with a gap of not more than 3 between boards. The boards are nailed at 150 centres with 30 or 40 nails for 9.5 mm and 12.5 mm thicknesses respectively. The joints do not have to be reinforced with jute scrim. The lath is covered with a finishing coat 5 thick or an undercoat of gypsum 8 thick and finish 2 thick.

Plastering on smooth dense surfaces

The walls, floors and roofs of many buildings today are constructed of reinforced concrete cast in-situ. The concrete is cast inside timber or plywood formwork. When the formwork is removed and the concrete has dried, its surface is often so smooth and dense that plaster will not adhere strongly to it. One way of preparing concrete surfaces for plastering is by hacking them with a chisel and hammer or with a power-operated hammer. The operation of hacking concrete is laborious and expensive.

Bonding liquids, which consist of an emulsion of polyvinyl acetate, are now made. These liquids are painted on to smooth dense surfaces, to which they adhere strongly, and they provide a surface to which gypsum plaster will adhere. One coat of gypsum plaster class B is generally used on surfaces treated with bonding liquids.

As an alternative to bonding liquids bonding plaster may be used on smooth dense surfaces. This plaster is a retarded hemi-hydrate gypsum plaster with low setting expansion, mixed with some selected wood fibres. The plaster is used neat and applied as one-coat finishing plaster. It will adhere to all but very smooth surfaces such as glazed tiles.

Metal beads and stops for plaster A range of galvanised steel beads and stops is made for use with plaster and plasterboard as reinforcement to angles and stops at the junction of wall and ceiling plaster and plaster to other materials.

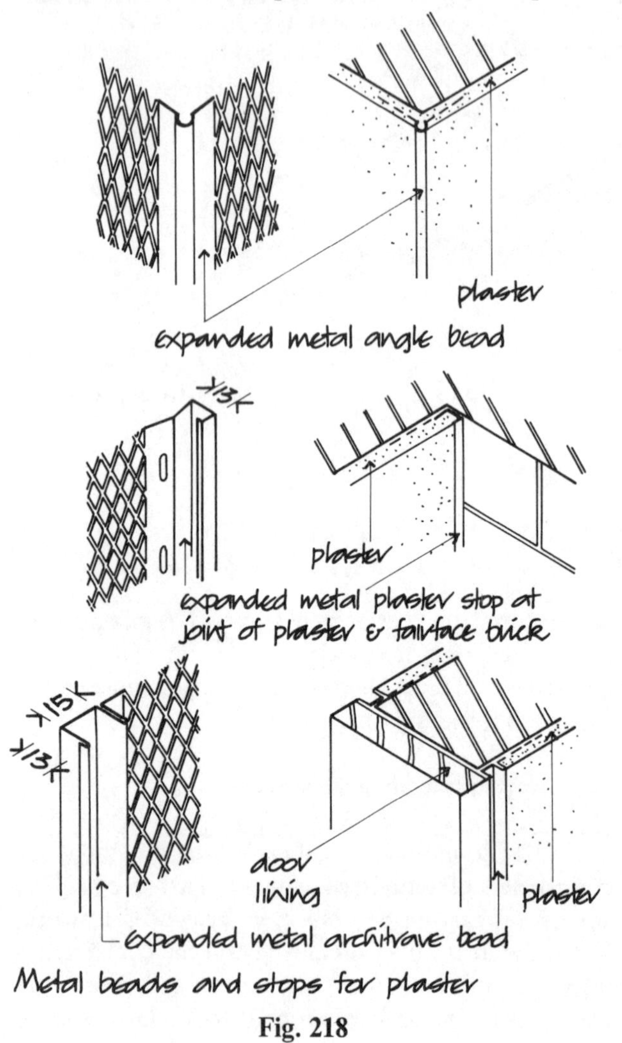

Fig. 218

An angle bead is pressed from steel strip to form reinforcement to angles. The bead has expanded metal wings, as shown in Fig. 218. The wings of the bead are bedded in plaster dabs each side of the angle. The bead is then squared and plaster run up to it (Fig. 218).

A metal stop with an expanded metal wing is pressed from steel strip and used as a stop to make a neat finish at the junction of plaster with other materials at angles, skirtings and around doors and windows, as illustrated in Fig. 218. The stop is either bedded in plaster or nailed to timber and the plaster is run up to the stop. These stops make a neat, positive break at junctions that would otherwise tend to crack or require some form of cover mould or bead to mask the joint. They are particularly useful at the junction of ceiling plaster or board on timber joists and wall plaster or fairface finishes where a crack might open or the junction of plaster to another material would be untidy.

Another bead or stop is designed for use at the junction of plaster and door and window frames to provide a definite break in surface between different materials, as illustrated in Fig. 218. The stop is either bedded in plaster dabs or nailed to wood.

SKIRTING AND ARCHITRAVES

Skirting

The skirting is a narrow band, usually projecting, formed around the base (skirt) of walls at the intersection of wall and floor. It serves to emphasise the junction of vertical and horizontal surfaces and is made from some material sufficiently hard to withstand knocks.

The types of skirting commonly used today are described here.

Timber skirting board A timber skirting is generally used at the junction of timber floors and plastered walls, to mask the junction of floor finish and plaster which would, if exposed, look ugly and collect dirt.

Softwood boards, 19 or 25 thick, from 50 to 150 wide and rounded or moulded on one edge are generally used. The skirting boards are nailed to plugs, grounds or concrete fixing blocks at the base of walls after plastering is completed. Fig. 219 illustrates some typical section of skirting board and the fixing of the board.

Plugs are wedge-shaped pieces of timber which are

driven into brick or block joints from which the mortar has been cut out. Grounds are small section lengths of sawn softwood timber, 38 or 50 wide and as thick as the plaster on the wall. These timber grounds are nailed horizontally to the wall as a background to which the skirting can be nailed. Grounds are generally fixed before plastering is commenced so that the plaster can be finished down on to and level with them. Concrete fixing blocks are either purpose-made or cut from lightweight concrete building blocks and built into brick walls as a fixing for skirtings.

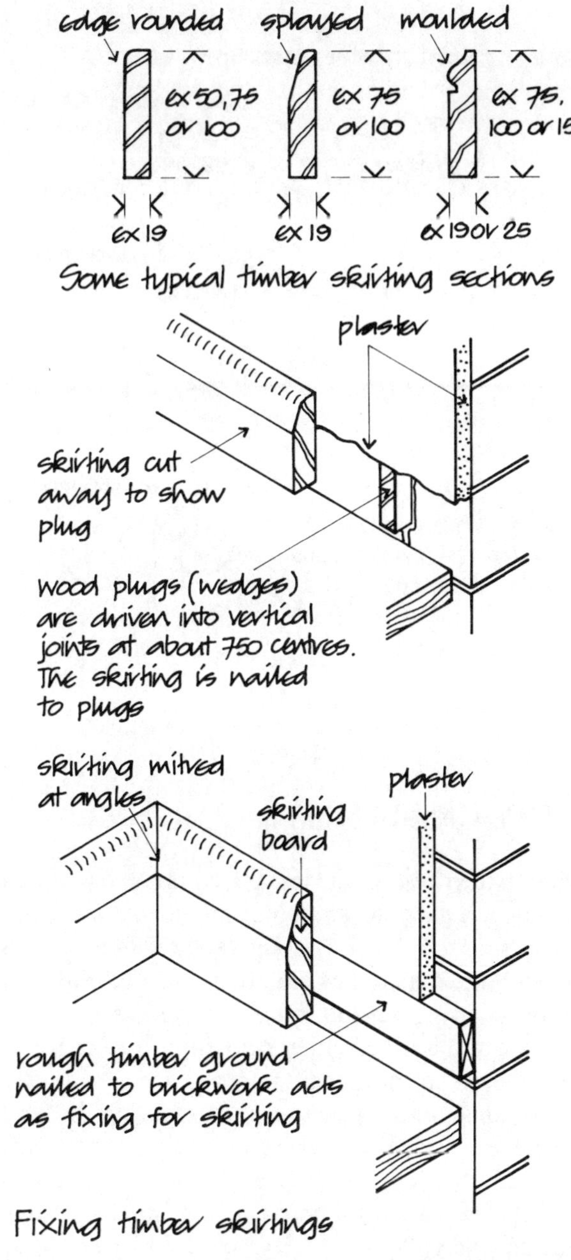

edge rounded splayed moulded

ex 50, 75 or 100

ex 75 or 100

ex 75, 100 or 150

ex 19 ex 19 ex 19 or 25

Some typical timber skirting sections

plaster

skirting cut away to show plug

wood plugs (wedges) are driven into vertical joints at about 750 centres. The skirting is nailed to plugs

skirting mitred at angles

skirting board

plaster

rough timber ground nailed to brickwork acts as fixing for skirting

Fixing timber skirtings

Fig. 219

Metal skirting A range of pressed steel skirtings is manufactured for fixing either before or after plastering. The skirting is pressed from mild steel strip and is supplied painted with one coat of red oxide priming paint. Fig. 220 illustrates these sections and their use.

It will be seen that the skirting is fixed by nailing it directly to lightweight blocks or to plugs in brick and block joints or to a timber ground. Special corner pieces to finish these skirtings at internal and external angles are supplied. The section of metal skirting manufactured does not make a particularly attractive finish at skirting level; the metal may rust due to the protective coating being damaged and the angle pieces are ugly, which is why metal skirtings are not much used.

Metal stops with expanded metal wings are also used as skirting.

Tile skirting The manufacturers of floor quarries and clay floor tiles make skirtings to match the colour and size of their products. The skirting tiles have rounded top edges and butt on to the floor finish, or they have rounded top edges and a cove base to provide an easily cleaned rounded internal angle between skirting and floor.

The skirting tiles are first thoroughly soaked in water

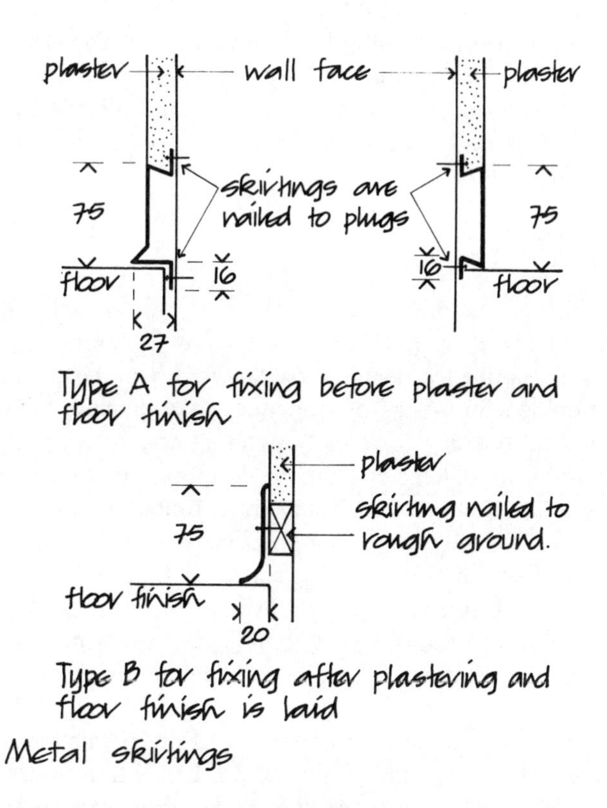

Fig. 220

and then bedded in sand and cement against walls and partitions. Special internal and external angle fittings are made. Fig. 221 illustrates the various types and use of these skirting tiles.

Skirting tiles make a particularly hardwearing, easily cleaned finish at the junction of floor and walls and are commonly used with quarry and clay tile finishes to solid floors.

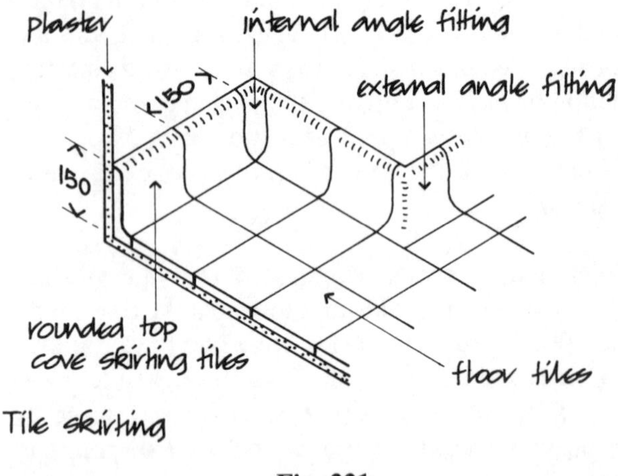

Fig. 221

Magnesite and anhydrite skirtings When solid floors are finished with one of the jointless floor finishes such as magnesium oxychloride or anhydrite (Volume 1) it is quite usual for the material to be used as a skirting with a cove formed at the junction of floor and skirting. As with cove tile skirtings this makes a neat, easily cleaned finish.

Architrave

The word architrave describes a decorative moulding fixed or cut around doors and windows to emphasise and decorate the opening. An architrave can be cut or moulded on blocks of stone, concrete or clay, built around openings externally. Internal architraves usually consist of lengths of moulded timber nailed around doors and windows. An internal timber architrave serves two purposes: to emphasise the opening and to mask the junction of wall plaster and timber door or window frame. If an architrave is not used an ugly crack tends to open up between the back of frames and wall plaster. It is to hide this crack that the architrave is fixed.

A timber architrave is usually 19 or 25 thick and from 50 to 100 wide. It may be finished with rounded edges, or splayed into the door or decorated with some moulding. Usual practice is to fix architraves so that they diminish in section towards the door or window. Narrow architraves can be fixed by nailing them to the frame or lining of the door or window. Wide architraves are usually fixed to sawn softwood grounds nailed to the wall around the frame or lining as a background to which the architrave can be securely nailed. Architraves are mitre cut (45° cut) at angles. Fig. 222 illustrates some typical sections and fixing of architraves.

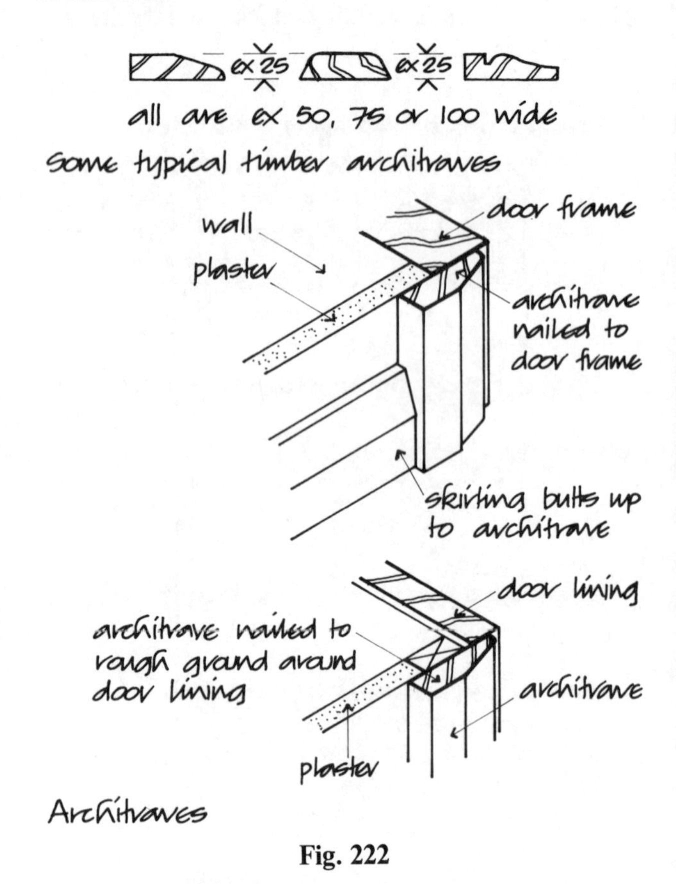

Fig. 222

EXTERNAL RENDERING

Owing to their colour and texture, common bricks and concrete and clay blocks do not provide an attractive external finish for buildings. The external faces of walls built with these materials are often rendered with two or three coats of cement and lime mixed with natural aggregate and finished either smooth or textured.

In exposed positions walls may become so saturated by rain that water penetrates to their inside face. Because an external rendering generally improves the resistance of a wall to rain penetration, the walls of buildings on the coast and on high ground are often rendered externally.

As an applied external finish to walls as additional weather protection, renderings depend on a strong bond to the background wall, on the mix used in the rendering material and on the surface finish. The rendering should have a strong bond or key to the background wall as a mechanical bond between the rendering and the wall and so that the bond resists the drying shrinkage inevitable in any wet applied mix of rendering. The surface of the background wall should provide a strong mechanical key for the rendering by the use of keyed flettons, raking out the mortar joints, hacking or scoring otherwise dense concrete surfaces and hacking smooth stone surfaces. If there is not a strong bond of rendering to background wall the rendering may shrink, crack and come away from the background and water will enter the cracks and saturate the background from which it will not readily evaporate. As a general rule the richer the mix of cement in the rendering material the stronger should be the background material and key. It is idle to cover poor quality bricks with a cement-rich rendering, which will crack, let in water and make and maintain the wall in a more saturated condition than it would be if not rendered.

The mixes for renderings depend on the background wall, lean mixes of cement and lime being used for soft porous materials and the richer cement and lime mixes for the more dense backgrounds so that the density and porosity of the rendering corresponds roughly to that of the background.

The types of external rendering used are smooth (wood float finish), scraped finish, textured finish, pebbledash (drydash), roughcast (wetdash) and machine-applied finish.

The mixes of rendering recommended for use on various backgrounds are set out in Table 14.

Smooth (wood float finish) rendering is usually applied in two coats. The first coat is spread by trowel and struck off level to a thickness of about 11. The surface of the first coat is scratched before it dries to provide key for the next coat. The first coat should be allowed to dry out. The next coat is spread by trowel and finished smooth and level to a thickness of about 8. The surface of smooth renderings should be finished with a wood float rather than a steel trowel. A steel trowel brings water and the finer particles of cement and lime to the surface which, on drying out, shrink and cause surface cracks. A wood float (trowel) leaves the surface coarse-textured and less liable to surface cracks. Three-coat rendering is used mostly in exposed positions to provide a thick protective coating to walls. The two undercoats are spread, scratched for key and allowed to dry out to a thickness of about 10 for each coat; the third or finishing coat is spread and finished smooth to a thickness of from 6 to 10.

Spatterdash Smooth, dense wall surface such as dense brick and situ-cast concrete afford a poor key and little suction for renderings. Such surfaces can be prepared for rendering by the application of a spatter-dash of wet cement and sand. A wet mix of cement and clean sand (mix 1:2, by volume) is thrown on to the surface and left to harden without being trowelled smooth. When dry it provides a surface suitable for the rendering, which is applied in the normal way.

Scraped-finish rendering An undercoat and finish coat are spread as for a smooth finish and the finished level surface, when it has set, is scraped with a steel straightedge or saw blade to remove some 2 from the surface to produce a coarse-textured finish.

Textured-finish rendering The colour and texture of smooth rendering appears dull and unattractive to some people and they prefer a broken or textured surface.

Textured rendering is usually applied in two coats. The first coat is spread and allowed to dry as previously described. The second coat is then spread by trowel and finished level. When this second coat is sufficiently hard, but still wet, its surface is textured with wood combs, brushes, sacking, wire mesh or old saw blades. A variety of effects can be obtained by varying the way in which the surface is textured.

An advantage of textured rendering is that the surface scraping removes any scum of water, cement and lime that may have been brought to the surface by trowelling and which might otherwise have caused surface cracking.

Pebbledash (drydash) finish This finish is produced by throwing dry pebbles, shingle or crushed stone on to and lightly pressed into the freshly applied finish coat of rendering so that the pebbles adhere to the rendering but are mostly left exposed as a surface of pebbles. Pebbles of from 6 to 13 gauge are used. The undercoat and finish coat are of a mix suited to the background and are trowelled and finished level. The advantage of this finish is that any hair cracks that may open due to the drying shrinkage of the rendering are masked by the pebble finish.

Roughcast (wetdash) finish A wet mix of rendering is thrown on to the matured undercoat by hand to a thickness of from 6 to 13 to produce a rough irregular

Table 14. Mixes for renderings

Mixes suitable for rendering

Mix type	Cement: lime:sand	Cement: ready-mixed lime:sand		Cement: sand (using plasticizer)	Masonry cement:sand
		Ready-mixed lime:sand	Cement: ready-mixed material		
I	1:¼:3	1:12	1:3	M	M
II	1:½:4 to 4½	1:8 to 9	1:4 to 4½	1:3 to 4	1:2½ to 3½
III	1:1:5 to 6	1:6	1:5 to 6	1:5 to 6	1:4 to 5
IV	1:2:8 to 9	1:4½	1:8 to 9	1:7 to 8	1:5½ to 6½

Recommended mixes for external renderings in relation to background materials, exposure conditions and finish required.

Background material	Type of finish	First and subsequent undercoats Exposure			Final coat Exposure		
		Severe	Moderate	Sheltered	Severe	Moderate	Sheltered
(1) Dense, strong, smooth	Wood float	II or III	II or III	II or III	III	III or IV	III or IV
	Scraped or textured	II or III	II or III	II or III	III	III or IV	III or IV
	Roughcast	I or II	I or II	I or II	II	II	II
	Drydash	I or II	I or II	I or II	II	II	II
(2) Moderately strong, porous	Wood float	II or III	III or IV	III or IV	III	III or IV	III or IV
	Scraped or textured	III	III or IV	III or IV	III	III or IV	III or IV
	Roughcast	II	II	II	as undercoats		
	Drydash	II	II	II			
(3) Moderately weak, porous*	Wood float	III	III or IV	III or IV			
	Scraped or textured	III	III or IV	III or IV			
	Drydash	III	III	III	as undercoats		
(4) No fines concrete concrete†	Wood float	II or III	II, III or IV	II, III or IV	II or III	III or IV	III or IV
	Scraped or textured	II or III	II, III or IV	II, III or IV	III	III or IV	III or IV
	Roughcast	I or II	I or II	I or II	II	II	II
	Drydash	I or II	I or II	I or II	II	II	II
(5) Woodwool slabs *‡	Wood flat	III or IV	III or IV	III or IV	IV	IV	IV
	Scraped or textured	III or IV	III or IV	III or IV	IV	IV	IV
(6) Metal lathing	Wood float	I, II or III	I, II or III	I, II or III	II or III	II or III	II or III
	Scraped or textured	I, II or III	I, II or III	I, II or III	III	III	III
	Roughcast	I or II	I or II	I or II	II	II	II
	Drydash	I or II	I or II	I or II	II	II	II

* Finishes such as roughcast and drydash require strong mixes and hence are not advisable on weak backgrounds

† If proprietary lightweight aggregates are used, it may be desirable to use the mix weaker than the recommended type.

‡ Three-coat work is recommended, the first undercoat being thrown on like a spatterdash coat

Taken from *Building Research Establishment Digest*, **196** (1976)

textured finish. The finish is determined by the gauge of the aggregate used in the wet mix.

Machine-applied finish A wet mix of rendering is thrown on to a matured undercoat by machine to produce a regular coarse-textured finish. The texture of the finish is determined by the gauge of the aggregate used in the final wetdash finish which may have the natural colour of the materials or be coloured to produce what are called Tyrolean finishes.

INDEX